GEOMETRY WITH TRIGONOMETRY

"All things stand by proportion."
George Puttenham (1529-1590)

"Mathematics possesses not only truth, but supreme beauty - a beauty cold and austere like that of sculpture, and capable of stern perfection, such as only great art can show."
Betrand Russell in *The Principles of Mathematics* (1872-1970)

DEDICATION

To my wife Fran, and Conor, Una and Brian

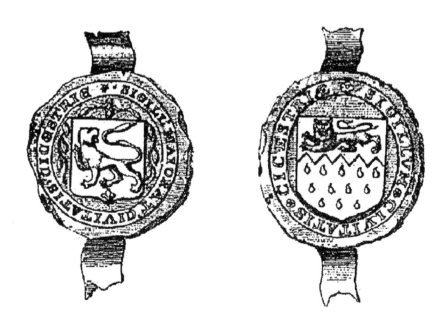

SICILLLUM MAIORAT CIVITATIS CICESTRIE
Mediaeval Seals of Mayors of Chichester, 1502 & 1530,
the design *motif* for the Horwood Publishing colophon

ABOUT DR. PADDY BARRY

Paddy Barry was born in Co. Westmeath in 1934 and his family moved to Co. Cork five years later. After his secondary school education at Patrician Academy, Mallow he studied at University College, Cork from 1952 to 1957, obtaining a degree of BSc in Mathematics and Mathematical Physics in 1955 and the degree of MSc in 1957. He then did research in complex analysis under the supervision of Professor W.K. Hayman, FRS, at Imperial College of Science and Technology, London (1957-1959) for which he was awarded the degree of PhD in 1960. He took first place in the Entrance Scholarships Examination to University College, Cork in 1952 and was awarded a Travelling Studentship in Mathematical Science at the National University of Ireland in 1957.

He was appointed Instructor in Mathematics at Stanford University, California in 1959-1960. Returning to University College, Cork in 1960 he made his career there, becoming Professor of Mathematics and Head of Department in 1964. He has participated in the general administration of the College, being a member of the Governing Body on a number of occasions, and was the first modern Vice-President of the College from 1975 to 1977. He was also a member of the Senate of the National University of Ireland from 1977 to 1982. University College, Cork was upgraded to National University of Ireland, Cork in 1977.

His mathematical interests expanded in line with his extensive teaching experience. As examiner for matriculation for many years he had to keep in contact with the detail of secondary school mathematics and the present book arose from that context, as it seeks to give a thorough account of the geometry and trigonometry that is done, necessarily incompletely, at school.

Geometry with Trigonometry

Patrick D. Barry
Professor of Mathematics
National University of Ireland
Cork

Horwood Publishing
Chichester

First published in 2001 by
HORWOOD PUBLISHING LIMITED
International Publishers in Science and Technology
Coll House, Westergate, Chichester, West Sussex, PO20 6QL England

British Library Cataloguing in Publication Data
A catalogue record of this book is available from the British Library

ISBN 1-898563-69-1

Printed in Great Britain by Martins Printing Group, Bodmin, Cornwall

Preface

I have for a long time held the view that whereas university courses in algebra, number systems and analysis admirably consolidate the corresponding school material, this is not the case for geometry and trigonometry. These latter topics form an important core component of mathematics, as they underpin analysis in its manifold aspects and applications in classical applied mathematics and sundry types of science and engineering, and motivate other types of geometry, and topology. Yet they are not well treated as university topics, being either neglected or spread over a number of courses, so that typically a student picks up a knowledge of these incidentally and relies mainly on the earlier intuitive treatment at school.

Clearly the treatment of geometry has seriously declined over the last fifty years, in terms of both quantity and quality. Lecturers and authors are faced with the question of what, if anything, should be done to try to restore it to a position of some substance. Bemoaning its fate is not enough, and surely authors especially should ponder what kinds of approach are likely to prove productive.

Pure or synthetic geometry was the first mathematical topic in the field and for a very long time the best established. It was natural for authors to cover as much ground as was feasible, and ultimately there was a large bulk of basic and further geometry. That was understandable in its time but perhaps a different overall strategy is now needed.

Synthetic geometry seems very difficult. In it we do not have the great benefit of symbolic manipulations. It is very taxing to justify diagrams and to make sure of covering all cases. From the very richness of its results, it is difficult to plan a productive approach to a new problem. In the proofs that have come down to us, extra points and segments frequently need to be added to the configuration. It is true that, as in any approach, there are some results which are handled very effectively and elegantly by synthetic methods, but that is certainly not the whole story. On the other hand, what is undeniable is that synthetic geometry really deals with geometry, and it forces attention to, and clarity in, geometrical concepts. It encourages the careful layout of sequential proof. Above all, it has a great advantage in its intuitive visualisation and concreteness.

The plan of this book is to have a basic layer of synthetic geometry, essentially five chapters in all, because of its advantages, and thereafter to diversify as much as possible to other techniques and approaches because of its difficulties. More than that, we assume strong axioms (on distance and angle-measure) so as to have an efficient approach from the start. The other approaches that we have in mind are the use of coordinates, trigonometry, position-vectors and complex numbers. Our emphasis is on clarity of concepts, proof and systematic and complete development of material. The synthetic geometry that we need is what is sufficient to start coordinate geometry and trigonometry, and that takes us as far as the ratio results for triangles and Pythagoras' theorem. In all, a considerable portion of traditional ground involving straight-lines and circles is covered. The overall approach is innovative as is the detail on trigonometry in Chapter 9 and on what are termed 'mobile coordinates' in Chapter 11. Some new concepts and substantial new notation have been introduced. There is enough for a two-semester course; a one-semester one could be made from Chapters 2-9, with Chapter 7 trimmed back.

My object has been to give an account at once accessible and unobtrusively rigorous. Preparation has been in the nature of unfinished business, stemming from my great difficulties when young in understanding the then textbooks in geometry. I hold that the reasoning in geometry should be as convincing as that in other parts of mathematics. It is too much to hope that there are no errors, mathematical or typographical. I should be grateful to be told of any at the email address *pdb@ucc.ie*.

Acknowledgements I am grateful to students in a succession of classes who responded to this material in its nascent stages, to departmental colleagues, especially Finbarr Holland and Des MacHale, who attended presentations of it, and to American colleagues David Rosen of Swarthmore and John Elliott of Fort Kent, Maine, who read an earlier approach of mine to geometry. I am especially grateful to Dr. P.A.J. Cronin, a colleague in our Department of Ancient Classics, for preparing the translations from Greek and Latin in the Glossary on pp. xiv-xv.

Paddy Barry

National University of Ireland, Cork,
November, 2000.

Contents

Glossary
of
Greek and Latin roots of mathematical words

acute < L *acutus*, sharp-pointed (perf. partic. of *acuere*, to sharpen).

addition < L *additio*, an adding to (*addere*, to add).

angle < L *angulus*, corner < Gk *agkylos*, bent.

area < L *area*, a vacant space.

arithmetic < Gk *arithmetike* (sc. *tekhne*), the art of counting (*arithmein*, to count; *arithmos*, number).

axiom < Gk *axioma*, self-evident principle (*axioun*, to consider worthy; *axios*, worthy).

calculate < L *calculatus*, reckoned (perf. partic. of *calculare* < *calculus*, pebble).

centre < Gk *kentron*, sharp point (*kentein*, to spike).

chord < L *khorde*, string of gut.

circle < L *circulus*, ring-shaped figure (related to Gk *kyklos*, ring; *kirkos* or *krikos*, ring).

congruent < L *congruens* (gen. *congruentis*), agreeing with (pres. partic. of *congruere*, to agree with).

curve < L *curvus* (*curvare*, to bend).

decimal < L(Med) *decimalis*, of tenths (*decima* (sc. *pars*), tenth part; *decem*, ten).

degree < OF *degre* < L *degredi*, descend (*de*, down; *gradi*, to step).

diagonal < L *diagonalis*, diagonal < Gk *diagonios*, from angle to angle (*dia*, through ; *gonia*, angle).

diagram < Gk *diagramma*, plan, figure indicated by lines (*dia*, through ; *gramma*, a thing which is drawn; *graphein*, to draw).

diameter < Gk *diametros*, diametrical (*dia*, through ; *metron*, measure).

distance < L *distantia*, remoteness (*distare*, to stand apart).

divide < L *dividere*, to separate.

equal < L *equalis*, equal (*aequare*, to make equal ; *aequus*, equal).

example < L *exemplum*, sample < *eximere*, to take out.

exponent < L *exponens* (gen. *exponentis*), setting forth (pres. partic. of *exponere*, to set forth ; *ex*, out ; *ponere* to place).

factor < L *factor*, maker, doer (*facere*, to make).

focus < L *focus*, hearth.

fraction < L *fractio*, a breaking into pieces (*frangere*, to break).

geometry < Gk *geometria*, measuring of land (*ge*, land; *metrein*, to measure).

graph < Gk *graphos*, drawing, picture (*graphein*, to draw).

hypotenuse < Gk *hypoteinousa*(sc. *gramme*), the line extending underneath (pres. partic. of *hypoteinein*, to extend under; *gramme*, line).

hypothesis < Gk *hypothesis*, supposition, assumption (*hypotithenai*, to place beneath).

inclination < L *inclinatio*, a leaning to one side (*inclinare*, to cause to lean).

induction < L *inductio*, a leading into (*inducere*, to lead in).

isosceles < Gk *isoskeles*, having equal legs (*isos*, equal; *skelos*, leg).

line < L *linea*, a linen thread (*linum* , flax < Gk. *linon*).

logic < Gk *logike* (sc. *techne*), the art of reasoning (*logos*, reason; *logikos*, endowed with reason; *techne*, art).

magnitude < L *magnitudo*, size, greatness (*magnus*, great).

mathematics < Gk *mathematika*, things that require mathematical or scientific reasoning (*mathema*, lesson; *mathematikos*, mathematical or scientific; *manthanein*, to learn).

measure < F *mesure* < L *mensura*, measure (*metiri*, to measure).

minus < L *minus*, less.

multiply < L *multiplicare*, multiply (*multus*, much; *plicare*, to lay together).

negative < L *negativus*, denying (*negare*, to deny).

number < F *nombre* < L *numerus*, number.

oblong < L *oblongus*, longish.

obtuse < L *obtusus*, blunted (perf. partic. of *obtundere*, to blunt).

orthogonal < Gk *orthogonios*, rectangular (*orthos*, right; *gonia*, angle).

parallel < Gk *parallelos*, beside one another (*para*, beside; *allelous*, one another).

perimeter < Gk *perimetron*, circumference (*peri* around; *metron*, measure).

perpendicular < L *perpendiculum*, plumbline (*perpendere*, to weigh precisely).

plane < L *planum*, level ground (*planus*, level).

point < L *punctum*, small hole (*pungere*, to pierce).

polygon < Gk *polygonon*, thing with many angles (*polys*, many; *gonia*, angle).

positive < L *positivus*, settled (*ponere*, to place).

postulate < L *postulare*, to ask for.

power < OF *poeir* < L *posse*, to be able.

product < L *productus*, brought forth (perf. partic. of *producere*, to bring forth).

proportion < L *portio*, comparative relationship (*pro*, according to; *portio*, part).

quadrangle < L *quadrangulum*, thing with four angles (*quattuor*, four; *angulus*, angle).

quotient < L *quotiens*, how often.

radius < L *radius*, rod, spoke of wheel.

rectangle < L *rectiangulum*, right-angled (*rectus*, right; *angulus*, angle).

rhombus < Gk *rhombos*, a device whirled round (*rhembein*, to whirl round).

science < L *scientia*, knowledge (*scire*, to know).

secant < L *secans* (gen. *secantis*), cutting (*secare*, to cut).

square < OF *esquarre* < L *quattuor*, four.

subtract < L *subtractus*, withdrawn (perf. partic. of *subtrahere*, to withdraw).

sum < L *summa*, top.

tangent < L *tangens*(sc. *linea*) (gen. *tangentis*), touching line (*tangere*, to touch; *linea*, line).

technical < Gk *tekhnikos*, artistic, skilful (*tekhne*, art).

theorem < Gk *theorema*, thing observed, deduced principle (*theorein*, to observe).

total < L *totus*, whole, all.

trapezium < Gk *trapezion*, small table (*trapeza*, table).

triangle < L *triangulum*, triangle (*tris*, three; *angulus*, angle).

trigonometry < Gk *trigonometria*, measurement of triangles (*trigonon*, triangle; *metrein*, to measure).

vector < L *vector*, bearer (*vehere*, to bear).

vertex < L *vertex*, summit.

volume < OF *volum* < L *volumen*, roll, book (*volvere*, to roll).

1

Preliminaries

1.1 HISTORICAL NOTE

This is one in a long line of textbooks on geometry. While all civilisations seem to have had some mathematical concepts, the most significant very old ones historically were the linked ones of Sumer, Akkad and Babylon, largely in the same region in what is now southern Iraq, and the separate one of Egypt. These are the ones which have left substantial traces of their mathematics, which was largely arithmetic, and geometrical shapes and measurement.

The outstanding contribution to mathematics was in Greece about 600B.C.–200B.C. The earlier mathematics conveyed techniques by means of examples, but the Greeks stated general properties of the mathematics they were doing, and organised proof of later properties from ones taken as basic. There was astonishing progress in three centuries and the fruit of that was written up in Euclid's *The Elements*, c.300B.C. He worked in Alexandria in Egypt, which country had come into the Greek sphere of influence in the previous century.

Euclid's *The Elements* is one of the most famous books in the world, certainly the most famous on mathematics. But it was influential widely outside mathematics too, as it was greatly admired for its logical development. It is the oldest writing on geometry of which we have copies by descent, and it lasted as a textbook until after 1890, although it must be admitted that in lots of places and for long periods not very many people were studying mathematics. It should probably be in the Guinness Book of Records as the longest lasting textbook in history.

The Elements shaped the treatment of geometry for 2,000 years. Its style would be unfamiliar to us today, as apart from using letters to identify points and hence line-segments, angles and other figures in diagrams, it consisted totally of words. Thus it did not use symbols as we do. It had algebra different from ours in that it said things in words written out in full. Full symbolic algebra as we know it was not perfected until about 1600A.D. in France, by Vieta and later Descartes. Another very significant feature of *The Elements* was that it did not have numbers ready-made, and used distance or length, angle-measure and area as separate quantities, although links between them were worked out.

Among prominent countries, *The Elements* lasted longest in its original style in the U.K., until about 1890. They had started chipping away at it in France in

1

the 16th century, beginning with one Petrus Ramus (1515-1572). There is a very
readable account of the changes which were made in France in Cajori [3, pages 275
– 289]. These changes mainly involved dis-improvements logically; authors brought
in concepts which are visually obvious, but they did not provide an account of the
properties of these concepts. Authors in France, and subsequently elsewhere, started
using our algebra to handle the quantities and this was a major source of advance.
One very prominent textbook of this type was *Elements of Geometry* by Legendre,
(first edition 1794), which was very influential on the continent of Europe and in the
U.S.A. All in all, these developments in France shook things up considerably, and
that was probably necessary before a big change could be made.

Although *The Elements* was admired widely and for a long time for its logic,
there were in fact logical gaps in it. This was known to the leading mathematicians
for quite a while, but it was not until the period 1880-1900 that this geometry was
put on what is now accepted as an adequate logical foundation. Another famous
book *Foundations of Geometry* by Hilbert (1899) was the most prominent in doing
this. The logical completion made the material very long and difficult, and this
type of treatment has not filtered down to school-level at all, or even to university
undergraduate level except for advanced specialised options.

Another sea-change was started in 1932 by G.D. Birkhoff; instead of building up
the real number system via geometrical quantities, he assumed a knowledge of numbers and used that from the start in geometry; this appeared in his 'ruler postulate'
and 'protractor postulate'. His approach allowed for a much shorter, easier and more
efficient treatment of geometry.

In the 1960's there was the world-wide shake-up of the 'New Mathematics', and
since then there are several quite different approaches to geometry available. In this
Chapter 1 we do our best to provide a helpful introduction and context, and suggest
a re-familiarisation with the geometrical knowledge already acquired.

Logically organised geometry dates from c.600-300B.C. in Greece; by c.350B.C.
there was already a history of geometry by Eudemus. From the same period and
earlier, date positive integers and the treatment of positive fractions via ratios. The
major mathematical topics date from different periods: geometry as just indicated;
full algebra from c.1600A.D.; full coordinate geometry from c.1630A.D.; full numbers
(negative, rational, decimals) from c.1600A.D.; complex numbers from c.1800A.D.;
calculus from c.1675A.D.; trigonometry from c.200B.C., although circles of fixed
length of radius were used until c.1700A.D. when ratios were introduced.

There is an account of the history of geometry of moderate length by H. Eves in
[2, pages 165-192]

It should be clear from this history that the Greek contribution to geometry greatly
influenced all later mathematics. It was transmitted to us via the Latin language,
and we have included a Glossary on pp. xiv-xv showing the Greek or Latin roots of
mathematical words.

1.2 NOTE ON DEDUCTIVE REASONING

The basic idea of a logical system is that we list up-front the terms and properties
that we start with, and thereafter proceed by way of definitions and proofs. There
are two main aspects to this.

1.2.1 Definitions

The first aspect concerns specifying what we are dealing with. A *definition* identifies a new concept in terms of accepted or known concepts. In practice a definition of a word, symbol or phrase E is a statement that E is to be used as a substitute for F, the latter being a phrase consisting of words and possibly symbols or a compound symbol. We accept ordinary words of the English language in definitions and what is at issue is the meaning of technical mathematical words or phrases. In attempting a definition, there is no progress if the technical words or symbols in F are not all understood at the time of the definition.

The disconcerting feature of this situation is that in any one presentation of a topic there must be a first definition and of its nature that must be in terms of accepted concepts. Thus we must have terms which are accepted without definition, that is there must be *undefined* or *primitive* terms. This might seem to leave us in a hopeless position but it does not, as we are able to assume properties of the primitive terms and work with those.

There is nothing absolute about this process, as a term which is taken as primitive in one presentation of a topic can very well be a defined term in another presentation of that topic, and vice versa. We need *some* primitive terms to get an approach under way.

1.2.2 Proof

The second aspect concerns the properties of the concepts that we are dealing with. A *proof* is a finite sequence of statements the first of which is called the *hypothesis*, and the last of which is called the *conclusion*. In this sequence, each statement after the hypothesis must follow logically from one or more statements that have been previously accepted. Logically there would be a vicious circle if the conclusion were used to help establish any statement in the proof.

There is also a disconcerting feature of this, as in any presentation of a topic there must be a first proof. That first proof must be based on some statements which are not proved (at least the hypothesis), which are in fact properties that are accepted without proof. Thus any presentation of a topic must contain unproved statements; these are called *axioms* or *postulates* and these names are used interchangeably.

Again there is nothing absolute about this, as properties which are taken as axiomatic in one presentation of a topic may be proved in another presentation, and vice versa. But we must have *some* axioms to get an approach under way.

1.3 EUCLID'S *The Elements*

1.3.1

The Elements involved the earliest surviving deductive system of reasoning, having axioms or postulates and common notions, and proceeding by way of careful statements of results and proofs. Up to c.1800, geometry was regarded as the part of mathematics which was best-founded logically. But its position was overstated, and its foundations not completed until c.1880-1900. Meanwhile the foundations of algebra and calculus were properly laid in the 19th century. From c.1800 on, some editions used algebraic notation in places to help understanding.

1.3.2   Definitions

The Greeks did not appreciate the need for primitive terms, and *The Elements* started
with an attempt to define a list of basic terms.

DEFINITIONS

1. A **POINT** is that which has no parts, or which has no magnitude.

2. A **LINE** is length without breadth.

3. The **EXTREMITIES** of a line are points.

4. A **STRAIGHT LINE** is that which lies evenly between its extreme points.

5. A **SUPERFICIES** is that which has only length and breadth.

6. The **EXTREMITIES** of a superficies are lines.

7. A **PLANE SUPERFICIES** is that in which any two points being taken, the
   straight line between them lies wholly in that superficies.

8. A **PLANE ANGLE** is the inclination of two lines to one another in a plane,
   which meet together, but are not in the same direction.

9. A **PLANE RECTILINEAL ANGLE** is the inclination of two straight lines
   to one another, which meet together, but are not in the same straight line.

10. When a straight line standing on another straight line, makes the adjacent angles
    equal to one another, each of the angles is called a **RIGHT ANGLE**; and the
    straight line which stands on the other is called a **PERPENDICULAR** to it.

11. An **OBTUSE ANGLE** is that which is greater than a right angle.

12. An **ACUTE ANGLE** is that which is less than a right angle.

13. A **TERM** or **BOUNDARY** is the extremity of anything.

14. A **FIGURE** is that which is enclosed by one or more boundaries.

15. A **CIRCLE** is a plane figure contained by one line, which is called the **CIR-
    CUMFERENCE**, and is such that all straight lines drawn from a certain
    point within the figure to the circumference are equal to one another.

16. And this point is called the **CENTRE** of the circle.

17. A **DIAMETER** of a circle is a straight line drawn through the centre, and
    terminated both ways by the circumference.

18. A **SEMICIRCLE** is the figure contained by a diameter and the part of the
    circumference cut off by the diameter.

19. A **SEGMENT** of a circle is the figure contained by a straight line and the circumference which cuts it off.

20. **RECTILINEAL FIGURES** are those which are contained by straight lines.

21. **TRILATERAL FIGURES**, or **TRIANGLES**, by three straight lines.

22. **QUADRILATERAL FIGURES** by four straight lines.

23. **MULTILATERAL FIGURES**, or **POLYGONS**, by more than four straight lines.

24. Of three-sided figures, an **EQUILATERAL TRIANGLE** is that which has three equal sides.

25. An **ISOSCELES TRIANGLE** is that which has two sides equal.

26. A **SCALENE TRIANGLE** is that which has three unequal sides.

27. A **RIGHT-ANGLED TRIANGLE** is that which has a right angle.

28. An **OBTUSE-ANGLED TRIANGLE** is that which has an obtuse angle.

29. An **ACUTE-ANGLED TRIANGLE** is that which has three acute angles.

30. Of four-sided figures, a **SQUARE** is that which has all its sides equal, and all its angles right angles.

31. An **OBLONG** is that which has all its angles right angles, but not all its sides equal.

32. A **RHOMBUS** is that which has all its sides equal, but its angles are not right angles.

33. A **RHOMBOID** is that which has its opposite sides equal to one another, but all its sides are not equal, nor its angles right angles.

34. All other four-sided figures besides these are called **TRAPEZIUMS**.

35. **PARALLEL STRAIGHT LINES** are such as are in the same plane, and which being produced ever so far both ways do not meet.

Although in *The Elements* these definitions were initially given, some of these were treated just like motivations (for instance, for a point where no use was made of the fact that 'it has no parts') whereas some were genuine definitions (like that of a circle, where the defining property was used). Our definitions will differ in some respects from these.

1.3.3 Postulates and Common Notions

The Greeks understood the need for axioms, and these were laid out carefully in *The Elements*. *The Elements* has two lists, a first one of POSTULATES and a second one of COMMON NOTIONS. It is supposed by some writers that Euclid intended his list of POSTULATES to deal with concepts which are mathematical or geometrical, and the second list to deal with concepts which applied to science generally.

POSTULATES

Let it be granted,

1. That a straight line may be drawn from any one point to any other point.

2. That a terminated straight line may be produced to any length in a straight line.

3. And that a circle may be described from any centre, at any distance from that centre.

4. All right angles are equal to one another.

5. If a straight line meet two straight lines, so as to make the two interior angles on the same side of it taken together less than two right angles, these straight lines, being continually produced, shall at length meet on that side on which are the angles which are less than two right angles.

COMMON NOTIONS

1. Things which are equal to the same thing are equal to one another.

2. If equals be added to equals the wholes are equals.

3. If equals be taken from equals the remainders are equal.

4. Magnitudes which coincide with one another, [that is, which exactly fill the same space] are equal to one another.

5. The whole is greater than its part.

1.3.4

The Elements attempted to be a logically complete deductive system. There were earlier Elements but these have not survived, presumably because they were outclassed by Euclid's.

The Elements are charming to read, proceed very carefully by moderate steps and within their own terms impart a great sense of conviction. The first proposition is to describe an equilateral triangle on $[A, B]$. With centre A a circle is described passing through B, and with centre B a second circle is described passing through A. If C is a point at which these two circles cut one another, then we take the triangle with vertices A, B, C.

It is ironical that, with *The Elements* being so admired for their logical proceeding, there should be a gap in the very first proposition. The postulates and common notions did not make any provisions which would ensure that the two circles in the proof would have a point in common. This may seem a curious choice as a first proposition, dealing with a very special figure. But in fact it is used immediately in Proposition 2, from a given point to lay off a given distance.

The main logical lack in *The Elements* was that not enough assumed properties were listed, and this fact was concealed through the use of diagrams.

1.3.5 Congruence

Two types of procedure in *The Elements* call for special comment. The first is the method of *superposition* by which one figure was envisaged as being moved and placed exactly on a second figure. The second is the process of *construction* by which figures were not dealt with until it was shown by construction that there actually was such a type of figure. In the physical constructions, what were allowed to be used were straight edges and compasses.

The notion of *superposition* is basic to Euclid's treatment of figures. It is visualised that one figure is moved physically and placed on another, fitting perfectly. We use the term *congruent* figures when this happens. Common Notion 4 is to be used in this connection. This physical idea is clearly extraneous to the logical set-up of primitive and defined terms, assumed and proved properties, and is not a formal part of modern treatments of geometry. However it can be used in motivation, and properties motivated by it can be assumed in axioms.

1.3.6 Quantities or magnitudes

The Elements spoke of one segment (then called a line) being equal to or greater than another, one region being equal to or greater than another, and one angle being equal to or greater than another, and this indicates that they associated a magnitude with each segment (which we call its length), a magnitude with each region (which we call its area), and a magnitude with each angle (which we call its measure). They did not define these magnitudes or give a way of calculating them, but they gave sufficient properties for them to be handled as the theory was developed. In the case of each of them the five common notions were supposed to apply.

Thus in *The Elements*, the *quantities* for which the Common Notions are intended are *distance* or equivalently *length* of a segment, *measure* of an angle and *area* of a region. These are not taken to be known, either by assumption or definition, but congruent segments are taken to have equal lengths, congruent angles are taken to have equal measures, and congruent triangles are taken to have equal areas. Thus equality of lengths of segments, equality of measures of angles, and equality of areas of triangles are what is started with. Treatment of area is more complicated than the other two, and triangles equal in area are not confined to congruent triangles. Addition and subtraction of lengths are to be handled using Common Notions 1, 2, 3 and 4; so are addition and subtraction of angle measures; so are addition and subtraction of area.

Taking a unit length, there is a long build-up to the length of any segment. They reached the stage where if some segment were to be chosen to have length 1 the length of any segment which they encountered could be found, but this was not actually done in *The Elements*.

Taking a right-angle as a basic unit, there was a long build-up to handling any angle. They reached the stage where if a right-angle was taken to have measure 90°, the measure of any angle which they encountered could be found, but this was not actually done.

There is a long build up to the area of figures generally. The regions which they considered were those which could be built up from triangles, and they reached the stage where if some included region were chosen to have area 1 the area of any included region could be found. This is not actually completed in *The Elements* but the materials are there to do it with.

All this shows that *The Elements* although very painstaking, thorough and exact were also rather abstract. It should be remembered that the Greeks did not have algebra as we have, and used geometry to do a lot of what we do by algebra. In particular, considering the area of a rectangle was their way of handling multiplication of quantities. Traditionally in arithmetic the area of a rectangle was dealt with as the product of the length and the breadth, that is by multiplication of two numbers. However, reconciling the geometrical treatment of area with the arithmetical does not seem to have been handled very explicitly in books, not even when from 1600A.D. onwards real numbers were being detached from the 'quantities' of Euclid.

1.4   OUR APPROACH

1.4.1   Type of course

Very scholarly courses in geometry assume as little as possible, and as a result are long and difficult. Shorter and easier courses have more or stronger assumptions, and correspondingly less to prove. What is difficult in a thorough course of geometry is not the detail of proof usually included, but rather is, first of all, locational viz. to prove that points are where diagrams suggest they are, that is to verify the diagrams, and secondly to be sure of covering all cases.

In particular, the type of approach which assumes that distance, angle-measure and area are different 'quantities' leads to a very long and difficult treatment of geometry. To make things much easier and shorter, we shall suppose that we know what numbers are, and deal with distance/length and angle-measure as basic concepts given in terms of numbers, and develop their properties. Moreover, we shall *define* area in terms of lengths.

What we provide, in fact, is a combination of Euclid's original course and a modification of an alternative treatment due to the American mathematician G.D. Birkhoff in 1932.

1.4.2   Need for preparation

What this course aims to do is to revise and extend the geometry and trigonometry that has been done at school. It gives a careful, thorough and logical account of familiar geometry and trigonometry. At school, a complete, logically adequate treatment of geometry is out of the question. It would be too difficult and too long, unattractive and not conducive to learning geometry; it would tend rather to put pupils off.

Thus this is not a first course in geometry. It is aimed at third level students, who should have encountered the basic concepts at secondary, or even primary, school. It starts geometry and trigonometry from scratch, and thus is self-contained to that extent.

But it is demanding because of a sustained commitment to deductive reasoning. In preparation the reader is strongly urged to start by revising the geometry and trigonometry which was done at school, at least browsing through the material. It would also be a good idea to read in some other books some descriptive material on geometry, such as the small amount in Ledermann and Vajda [10, pages 1 – 26], or the large amount in Wheeler and Wheeler [13, Chapters 11 – 15]. Similarly trigonometry and vectors can be revised from McGregor, Nimmo and Stothers [11, pages 99 –123, 279 – 331].

It would moreover be helpful to practise geometry by computer, e.g. by using software systems such as *The Geometer's Sketchpad* or *Cabri-Géomètre*. Material which can be found in elementary books should be gone over, and also a look forwards could be had to the results in this book.

1.5   REVISION OF GEOMETRICAL CONCEPTS

1.5.1

As part of the preliminary programme, we now include a review of the basic concepts of geometry. Geometry should be thought of as arising from an initial experimental and observational stage, where the figures are looked at and there is a great emphasis on a visual approach.

1.5.2   The basic shapes

1. The plane Π is a set, the elements of which are called *points*. Certain subsets of Π are called *lines*.

By observation, given any distinct points $A, B \in \Pi$, there is a unique line to which A and B both belong. It is denoted by AB.

2. Given distinct points A and B, the set of points consisting of A and B themselves and all the points of the line AB which are between A and B is called a *segment*, and denoted by $[A, B]$.

Figure 1.1. A line AB.
The arrows indicate that the line
is to be continued unendingly.

Figure 1.2. A segment $[A, B]$.

NOTE. Note that the modern mathematical terminology differs significantly from that in *The Elements*. What was called a 'line' is now called a segment, and we have added the new concept of 'line'. This is confusing, but the practice is well established. In ordinary English and in subjects cognate to mathematics, 'line' has its old meaning.

3. The set consisting of the point A itself and all the points of the line AB which are on the same side of A as B is, is called a *half-line*, and denoted by $[A, B$. If A is between B and C, then the half-lines $[A, B$ and $[A, C$ are said to be *opposite*.

Figure 1.3. A half-line $[A, B$.　　　　Opposite half-lines.

4. If the points B, C are distinct from A, then the pair of half-lines $\{[A, B , [A, C \}$ is called an *angle-support* and denoted by $|\underline{BAC}$; if $[A, B$ and $[A, C$ are opposite half-lines, then $|\underline{BAC}$ is called a *straight* angle-support. In each case A is called its *vertex*, $[A, B$ and $[A, C$ its *arms*.

Figure 1.4. An angle-support.　　　　A straight angle-support.

5. The set of all the points on, or to one side of, a line AB is called a *closed half-plane*, with *edge AB*.

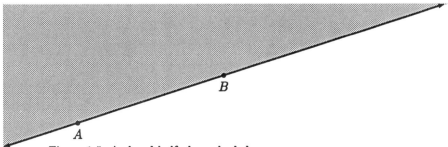

Figure 1.5. A closed half-plane shaded.

6. If the points A, B, C are not collinear, then the set of points which are in both the closed half-plane with edge AB, containing C, and the closed half-plane with edge AC, containing B, is called the *interior region* of $|\underline{BAC}$ and denoted by $\mathcal{IR}(|\underline{BAC})$; also $(\Pi \setminus \mathcal{IR}(|\underline{BAC})) \cup |\underline{BAC}$ is called the *exterior region* of $|\underline{BAC}$ and denoted by $\mathcal{ER}(|\underline{BAC})$. When $C \in [A, B$ the interior and exterior regions of $|\underline{BAC}$ are taken to be $[A, B$ and Π, respectively.

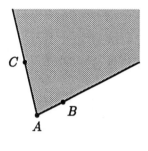
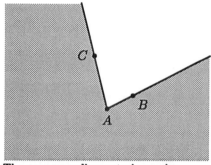

Figure 1.6. An interior region.   The corresponding exterior region.

7. If $|\underline{BAC}$ is a non-straight angle-support, then the couples $(|\underline{BAC}, \mathcal{IR}(|\underline{BAC}))$, $(|\underline{BAC}, \mathcal{ER}(|\underline{BAC}))$, are called the *wedge-angle* and *reflex-angle*, respectively, with support $|\underline{BAC}$; this wedge-angle is denoted by $\angle BAC$. Thus a wedge-angle is a pair of arms in association with an interior region, while a reflex-angle is a pair of arms combined with an exterior region.

If $|\underline{BAC}$ is a straight angle-support, and $\mathcal{H}_1, \mathcal{H}_2$ are the closed half-planes with edge the line AB, then the couples $(|\underline{BAC}, \mathcal{H}_1)$, $(|\underline{BAC}, \mathcal{H}_2)$, are called the *straight-angles* with support $|\underline{BAC}$. If $C \in [A, B$ then the wedge-angle $\angle BAC = \angle BAB$ is called a *null-angle*, and the reflex-angle with support $|\underline{BAB}$ is called a *full-angle*.

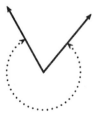

Figure 1.7. A wedge-angle.   A reflex-angle.

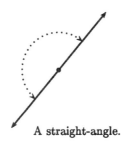

A straight-angle.

NOTE. The reason we call $|\underline{BAC}$ an angle-support and not an angle is that it sup-
ports two angles. If we were confining ourselves to pure geometry, and not concerned
to go forward to coordinate geometry and trigonometry, we could confine ourselves to
wedge and straight angles. Even more if we were to confine ourselves to the angles in
triangles, we could take $|\underline{BAC} = [A, B \, \cup \, [A, C$. However when A is between B and
C, this would result in a straight-angle being a line, and it would not have a unique
vertex. In the early part of our course, we can confine our attention to wedge and
straight angles.

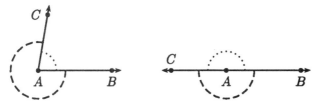

Figure 1.8. Supports bearing two angles each.

8. If A is between B and C and $D \notin BC$, the wedge-angles $\angle BAD$, $\angle CAD$ are
called *supplementary*. If A, B, C are not collinear, and A is between B and B_1, and
A is between C and C_1, then the wedge-angles $\angle BAC, \angle B_1 AC_1$ are called *opposite
angles at a vertex.*

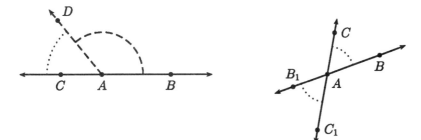

Figure 1.9. Supplementary angles.   Opposite angles at a vertex.

9. If A, B, C are non-collinear points and $[A, D$ is in the interior region of $|\underline{BAC}$,
then $[A, D$ is said to be *between* $[A, B$ and $[A, C$.

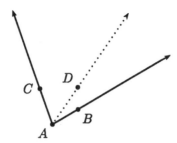

Figure 1.10. $[A, D$ between $[A, B$ and $[A, C$.

10. If A, B, C are non-collinear points, let $\mathcal{H}_1$ be the closed half-plane with edge BC, containing A, $\mathcal{H}_3$ be the closed half-plane with edge CA, containing B, $\mathcal{H}_5$ be the closed half-plane with edge AB, containing C. Then the intersection $\mathcal{H}_1 \cap \mathcal{H}_3 \cap \mathcal{H}_5$ is called a *triangle*. The points A, B, C are called its *vertices* and the segments $[B, C], [C, A], [A, B]$ its *sides*. If a vertex is not the end-point of a side (e.g. the vertex A and the side $[B, C]$), then the vertex and side are said to be *opposite* each other. We denote the triangle with vertices A, B, C by $[A, B, C]$.

If at least two sides of a triangle have equal lengths, then the triangle is called *isosceles*.

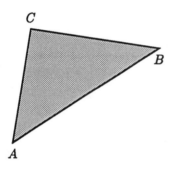

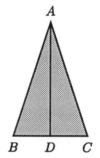

Figure 1.11. A triangle $[A, B, C]$.   An isosceles triangle.

11. Let A, B, C, D be points no three of which are collinear, and such that $[A, C] \cap [B, D] \neq \emptyset$. For this let $\mathcal{H}_1$ be the closed half-plane with edge AB, containing C, $\mathcal{H}_3$ be the closed half-plane with edge BC, containing D, $\mathcal{H}_5$ be the closed half-plane with edge CD, containing A, $\mathcal{H}_7$ be the closed half-plane with edge DA, containing B. Then the intersection $\mathcal{H}_1 \cap \mathcal{H}_3 \cap \mathcal{H}_5 \cap \mathcal{H}_7$ is called a *convex quadrilateral*.

The points A, B, C, D are called its *vertices*, the segments $[A, B]$, $[B, C]$, $[C, D]$, $[D, A]$ its *sides*, and the segments $[A, C]$, $[B, D]$ its *diagonals*. Two vertices which have a side in common are said to be *adjacent*, and two vertices which have a diagonal in common are said to be *opposite*. Thus A and B are adjacent as they both belong to $[A, B]$ which is a side; A and C are opposite as they both belong to $[A, C]$ which is a diagonal.

Two sides which have a vertex in common are said to be *adjacent*, and two sides which do not have a vertex in common are said to be *opposite*. Thus the sides $[A, B]$, $[D, A]$ are adjacent as the vertex A belongs to both, while the sides $[A, B]$, $[C, D]$ are opposite as none of the vertices belongs to both of them. The wedge angles $\angle DAB$, $\angle ABC$, $\angle BCD$, $\angle CDA$ are called the *angles* of the convex quadrilateral; two of these angles are said to be adjacent or opposite according as their two vertices are adjacent or opposite vertices of the convex quadrilateral.

We denote the convex quadrilateral with vertices A, B, C, D, with A and C opposite, by $[A, B, C, D]$.

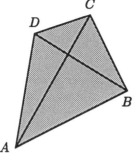

Figure 1.12. A convex quadrilateral.

1.5.3 Distance; degree-measure of an angle

1. With each pair (A, B) of points we associate a non-negative real number $|A, B|$, called the *distance* from A to B or the *length* of the segment $[A, B]$. In all cases $|B, A| = |A, B|$. By observation, given any non-negative real number k, and any half-line $[A, B$ there is a unique point $P \in [A, B$ such that $|A, P| = k$.

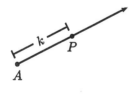

Laying off a distance k.

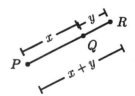

Figure 1.13. Addition of distances.

By observation, if $Q \in [P, R]$ then $|P, Q| + |Q, R| = |P, R|$. In all cases $|A, A| = 0$, while $|A, B| > 0$ if $A \neq B$.

2. Given distinct points A and B, choose the point $C \in [A, B$ so that $|A, C| = \frac{1}{2}|A, B|$. Then C is between A and B and

$$|C, B| = |A, B| - |A, C| = |A, B| - \frac{1}{2}|A, B| = \frac{1}{2}|A, B| = |A, C|.$$

The point C which is on the line AB and equidistant from A and B, is called the *mid-point* of A and B. It is also called the mid-point of the segment $[A, B]$.

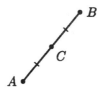

Figure 1.14. Mid-point of A and B.

3. With each wedge-angle $\angle BAC$ we associate a non-negative number, called its *degree-measure*, denoted by $|\angle BAC|^\circ$, and for each straight-angle α we take $|\alpha|^\circ = 180$.

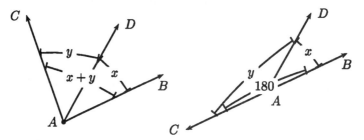

Figure 1.15. Addition of angle-measures.

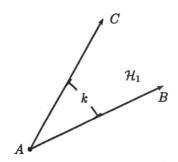

Figure 1.16. Laying off an angle.

By observation, we note that if A, B, C are non-collinear and $[A, D$ is between $[A, B$ and $[A, C$, then $|\angle BAD|^\circ + |\angle CAD|^\circ = |\angle BAC|^\circ$, while if $[A, B$ and $[A, C$ are opposite and $D \notin AB$, then $|\angle BAD|^\circ + |\angle CAD|^\circ = 180$.

By observation, given any number k with $0 \le k < 180$ and any half-line $[A, B$, on each side of the line AB there is a unique wedge-angle $\angle BAC$ with $|\angle BAC|^\circ = k$. In all cases $|\angle BAB|^\circ = 0$, so that the degree-measure of each null angle is 0, while if $\angle BAC$ is not null then $|\angle BAC|^\circ > 0$.

It follows from the foregoing, that if $\angle BAD$ is any wedge-angle then $|\angle BAD|^\circ < 180$, and that if $\angle BAD$, $\angle CAD$ are *supplementary* angles, then $|\angle CAD|^\circ = 180 - |\angle BAD|^\circ$.

4. Given points B and C distinct from A such that $C \notin [A, B$, we can choose a point D such that $|\angle BAD|^\circ$ is equal to half the degree-measure of the wedge or

straight angle with support $|\underline{BAC}$. Then for all points $P \neq A$ on the line AD we have $|\angle BAP|° = |\angle PAC|°$. We call AP the *mid-line* or *bisector* of the support $|\underline{BAC}$.

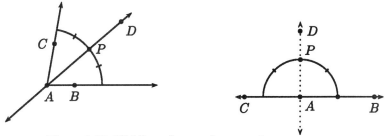

Figure 1.17. Mid-line of an angle-support.

5. Any angle $\angle BAC$ such that $0 < |\angle BAC|° < 90$ is called *acute*, such that $|\angle BAC|° = 90$ is called *right*, and such that $90 < |\angle BAC|° < 180$ is called *obtuse*.

If $\angle BAC$ is a right-angle, then the lines AB and AC are said to be *perpendicular* to each other, written $AB \perp AC$.

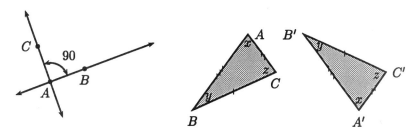

Figure 1.18. Perpendicular lines.   Figure 1.19. Congruent triangles.

1.5.4 Our treatment of congruence

If $[A, B, C]$, $[A', B', C']$ are triangles such that

$$|B, C| = |B', C'|, \ |C, A| = |C', A'|, \ |A, B| = |A', B'|,$$
$$|\angle BAC|° = |\angle B'A'C'|°, \ |\angle CBA|° = |\angle C'B'A'|°, \ |\angle ACB|° = |\angle A'C'B'|°,$$

then we say by way of definition that the triangle $[A, B, C]$ is *congruent* to the triangle $[A', B', C']$. Behind this concept is the physical idea that $[A, B, C]$ can be placed on $[A', B', C']$, fitting it exactly.

By observation if $[A, B, C]$, $[A', B', C']$ are such that

$$|C, A| = |C', A'|, \ |A, B| = |A', B'|, \ |\angle BAC|° = |\angle B'A'C'|°,$$

then $[A, B, C]$ is congruent to $[A', B', C']$. This is known as the SAS (side, angle, side) condition for congruence of triangles.

Similarly by observation if $[A, B, C]$, $[A', B', C']$ are such that

$$|B, C| = |B', C'|, \ |\angle CBA|° = |\angle C'B'A'|°, \ |\angle BCA|° = |\angle B'C'A'|°,$$

then $[A, B, C]$ is congruent to $[A', B', C']$. This is known as the ASA (angle, side, angle) condition for congruence of triangles.

It can be proved that if T and T' are triangles with vertices $\{A, B, C\}, \{A', B', C'\}$, respectively, for which

$$|B, C| = |B', C'|, \ |C, A| = |C', A'|, \ |A, B| = |A', B'|,$$

then T is congruent to T'. This is known as the SSS(side-side-side) principle of congruence.

1.5.5   Parallel lines

1. Distinct lines l, m are said to be *parallel* if $l \cap m = \emptyset$; this is written as $l \parallel m$. We also take $l \parallel l$.

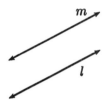

Figure 1.20. Parallel lines.

By observation, given any line l and any point P there cannot be more than one line m through P which is parallel to l.

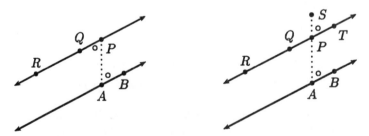

Figure 1.21. Alternate angles for a transversal.   Corresponding angles.

It can be shown that two lines are parallel if and only if *alternate* angles made by a transversal, as indicated, are equal in magnitude, or equivalently, if and only if *corresponding* angles made by a transversal are equal in magnitude.

2. A convex quadrilateral in which opposite side-lines are parallel to each other is called a *parallelogram*. A parallelogram in which adjacent side-lines are perpendicular to each other is called a *rectangle*.

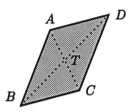

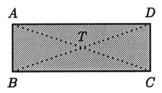

Figure 1.22. A parallelogram.   A rectangle.

1.6   PRE-REQUISITIES

Although this book re-starts geometry and trigonometry from the beginning, it does not take mathematics from a start. Consequently there is material from other parts of mathematics which is assumed known. This also should be revised at the start, or at the appropriate time when it is needed.

At the beginning, we presuppose a moderate knowledge of set theory, sufficient to deal with sets, relations and functions, in particular order and equivalence relations. From Chapter 3 on we assume a knowledge of the real number system, and the elementary algebra involved. Later requirements come in gradually.

1.6.1   Set notation

For set notation we refer to Smith [12, pages 1 – 38]. We mention that we use the word *function* where it uses *map*. We should also like to emphasise the difference between a set $\{a, b\}$ and a couple or ordered pair (a, b). In a set, the order of the elements does not matter, so that $\{a, b\} = \{b, a\}$ in all cases, and

$$\{a, b\} = \{c, d\}$$

if and only if either

$$a = c \quad \text{and} \quad b = d$$

or

$$a = d \quad \text{and} \quad b = c.$$

In a couple (a, b) it matters which is first and which is second. Thus $(a, b) \neq (b, a)$ unless $a = b$, and

$$(a, b) = (c, d)$$

if and only if

$$a = c \quad \text{and} \quad b = d.$$

1.6.2   Classical algebra

We need a knowledge of the real number system and the complex number system, and the classical algebra involving these, up to dealing with quadratic equations and two simultaneous linear equations in two unknowns. For this very elementary material there is an ample provision of textbooks entitled *College Algebra* by international publishers.

1.6.3 Other algebra

For matrices and determinants we refer to Smith [12, pages 95 – 124] and McGregor, Nimmo and Stothers [11, pages 243 – 278], and for the little that we use on group theory to Smith [12, pages 125 – 152].

1.6.4 Distinctive property of real numbers

For the properties that distinguish the field of real numbers from other ordered fields, we refer to Smith [12, pages 153 – 196].

2

Basic shapes of geometry

COMMENT. Geometry deals with our intuitions as to the physical space in which we exist, with the properties of the shapes and sizes of bodies as mathematically abstracted. It differs from set theory in that in geometry there are distinguished or special subsets, and relations involving them. To start with we presuppose a moderate knowledge of set theory, sufficient to deal with sets, relations and functions. From Chapter 3 on we assume a knowledge of the real number system, and the elementary algebra involved.

In this first chapter we introduce the plane, points, lines, natural orders on lines, and open half-planes as primitive concepts, and in terms of these develop other special types of geometrical sets.

2.1   LINES, SEGMENTS AND HALF-LINES

2.1.1   Plane, points, lines

Primitive Terms. Assuming the terminology of sets, the **plane**, denoted by Π, is a universal set the elements of which are called **points**. Certain subsets of Π are called (straight) **lines**. We denote by Λ the set of all these lines.

AXIOM A_1. *Each line is a proper non-empty subset of Π. For each set $\{A,B\}$ of two distinct points in Π, there is a unique line in Λ to which A and B both belong.* |

We denote by AB the unique line to which distinct points A and B belong, so that $A \in AB$ and $B \in AB$. It is an immediate consequence of Axiom A_1 that $AB = BA$; that if C and D are distinct points and both belong to the line AB, then $AB = CD$; and that if A, B are distinct points, both on the line l and both on the line m, then $l = m$.

Furthermore if l, m are any two lines in Λ, then either

$$l \cap m = \emptyset,$$

or

$$l \cap m \quad \text{is a singleton,}$$

or

$$l = m \quad \text{and in this last case} \quad l \cap m = l = m.$$

Moreover the plane Π is not a line, as each line is a proper subset of Π.

If three or more points lie on one line we say that these points are **collinear**. If one point lies on three or more lines we say that these lines are **concurrent**.

2.1.2   Natural order on a line

COMMENT. The two intuitive senses of motion along a line give us the original examples of linear (total) orders, and we refer to these as the two natural orders on that line. On a diagram the sense of a double arrow gives one natural order on l, while the opposite sense would yield the other natural order on l. We now take natural order as a primitive term, and go on to define segments and half-lines in terms of this and our existing terms.

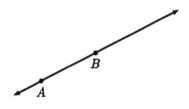

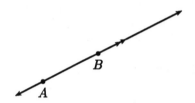

Figure 2.1. A line AB.
The arrows indicate that the line
is to be continued unendingly.

Figure 2.2: The double arrow indicates
a sense along the line AB.

Primitive Term. On each line $l \in \Lambda$ there is a binary relation $\leq_l$, which we refer to as a **natural order** on l. We read $A \leq_l B$ as 'A precedes or coincides with B on l '.

AXIOM A_2. *Each natural order $\leq_l$ has the properties:-*

(i) *$A \leq_l A$ for all points $A \in l$;*

(ii) *if $A \leq_l B$ and $B \leq_l C$ then $A \leq_l C$;*

(iii) *if $A \leq_l B$ and $B \leq_l A$, then $A = B$;*

(iv) *for any points $A, B \in l$, either $A \leq_l B$ or $B \leq_l A$.* |

COMMENT. We refer to (i), (ii), (iii) in A_2 as the reflexive, transitive and anti-symmetric properties, respectively, of a binary relation; property (iii) can be reworded as, if $A \leq_l B$ and $A \neq B$ then $B \not\leq_l A$. A binary relation with these three properties is commonly called a partial order. A binary relation with all four properties (i), (ii), (iii) and (iv) in A_2 is commonly called a linear order or a total order.

2.1.3   Reciprocal orders

If $A \leq_l B$ we also write $B \geq_l A$ and read this as 'B succeeds or coincides with A on l'. Then $\geq_l$ is also a total order on l, i.e. $\geq_l$ satisfies A_2(i), (ii), (iii) and (iv), as can readily be checked as follows.

First, on interchanging A and A in A_2(i), we have $A \geq_l A$ for all $A \in l$. Secondly, suppose that $A \geq_l B$ and $B \geq_l C$; then $C \leq_l B$ and $B \leq_l A$, so by A_2(ii) $C \leq_l A$; hence $A \geq_l C$. Thirdly, suppose that $A \geq_l B$ and $B \geq_l A$; then $B \leq_l A$ and $A \leq_l B$ so by A_2(iii) $A = B$. Finally, let A, B be any points on l; by A_2(iv), either $A \leq_l B$ or $B \leq_l A$ and so either $B \geq_l A$ or $A \geq_l B$.

We say that $\geq_l$ is **reciprocal** to $\leq_l$. If now we start with $\geq_l$ and let $\succeq_l$ be its reciprocal we have $A \succeq_l B$ if $B \geq_l A$; then we have $A \succeq_l B$ if and only if $A \leq_l B$. Thus $\succeq_l$ coincides with $\leq_l$, and so the reciprocal of $\geq_l$ is $\leq_l$.

The upshot of this is that $\leq_l$ and $\geq_l$ are a pair of total orders on l, each the reciprocal of the other. There is no natural way of singling out one of $\leq_l$, $\geq_l$ over the other, and the notation is equally interchangeable as we could have started with $\geq_l$. Having this pair is a nuisance but it is unavoidable, and we try to minimise the nuisance as follows. Given distinct points A and B, let $l = AB$. Then exactly one of $A \leq_l B$, $A \geq_l B$ holds; for by A_2(iv) either $A \leq_l B$ or $A \geq_l B$, and by A_2(iii) both cannot hold as that would imply that $A = B$. Thus we can choose the natural order on l in which A precedes B, by taking $\leq_l$ when $A \leq_l B$, and by taking $\geq_l$ when $A \geq_l B$; we will use the notation $\leq_l$ for this natural order.

Let A and B be distinct points in Π, let $l = AB$ and $A \leq_l B$. Let C be a point of l, distinct from A and B. Then exactly one of

$$\text{(a) } C \leq_l A \leq_l B, \quad \text{(b) } A \leq_l C \leq_l B, \quad \text{(c) } A \leq_l B \leq_l C,$$

holds.

Proof. If $C \leq_l A$ then clearly (a) holds. If $C \leq_l A$ is false, then by A_2(iv) $A \leq_l C$; by A_2(iv) we have moreover either $C \leq_l B$ or $B \leq_l C$, and these yield, respectively, (b) and (c). Thus at least one of (a), (b), (c) holds.

On the other hand, if (a) and (b) hold, we have $A = C$ by A_2(iii) and this contradicts our assumptions. Similarly if (b) and (c) hold we would have $B = C$. Finally if (a) and (c) hold, from (a) we have $C \leq_l B$ by A_2(ii) and then $B = C$.

2.1.4   Segments

Definition. For any points $A, B \in \Pi$, we define the **segments** $[A, B]$ and $[B, A]$ as follows. Let l be a line such that $A, B \in l$ and $\leq_l$, $\geq_l$ a pair of reciprocal natural orders on l. Then if

$$A \leq_l B \quad \text{so that} \quad B \geq_l A, \tag{2.1.1}$$

we define

$$[A, B] = \{P \in l : A \leq_l P \leq_l B\} = \{P \in l : A \leq_l P \text{ and } P \leq_l B\},$$
$$[B, A] = \{P \in l : B \geq_l P \geq_l A\},$$

while if

$$B \leq_l A \quad \text{so that} \quad A \geq_l B, \tag{2.1.2}$$

we define

$$[B, A] = \{P \in l : B \leq_l P \leq_l A\},$$
$$[A, B] = \{P \in l : A \geq_l P \geq_l B\}.$$

We should use a more complete notation such as $[A, B]_{\leq_l, \geq_l}$, $[B, A]_{\leq_l, \geq_l}$, but make do with the less precise one. Note that (2.1.2) comes from (2.1.1) on interchanging A and B, or on interchanging $\leq_l$ and $\geq_l$.

When $A \neq B$, by A_1 $l = AB$; by A_2(iv) at least one of (2.1.1) and (2.1.2) holds, and by A_2(iii) only one of (2.1.1) and (2.1.2) holds.

When $A = B$, l can be any line through A, and we find that
$$\{P \in l : A \leq_l P \leq_l A\} = \{A\},$$
$$\{P \in l : A \geq_l P \geq_l A\} = \{A\},$$
for the singleton $\{A\}$. To see this we note that $A \leq_l A \leq_l A$ by A_2(i), while if $A \leq_l P \leq_l A$ then $P = A$ by A_2(iii). The same argument holds for $\geq_l$. Thus $[A, A] = \{A\}$.

Figure 2.3. A segment $[A, B]$.

Segments have the following properties:-

(i) *If $A \neq B$, then $[A, B] \subset AB$.*

(ii) *$A, B \in [A, B]$ for all $A, B \in \Pi$.*

(iii) *$[A, B] = [B, A]$ for all $A, B \in \Pi$.*

(iv) *If $C, D \in [A, B]$ then $[C, D] \subset [A, B]$.*

(v) *If A, B, C are distinct points on a line l, then precisely one of*

$$A \in [B, C], \quad B \in [C, A], \quad C \in [A, B],$$

holds.

Proof. In each case we suppose that $A \leq_l B$ so that we have (2.1.1) above; otherwise replace $\leq_l$ by $\geq_l$ throughout to cover (2.1.2).

(i) By A_1, $l = AB$ so $[A, B]$ is a set of points on AB.

(ii) By A_2(i) $A \leq_l A \leq_l B$ and $A \leq_l B \leq_l B$.

(iii) As $A \leq_l B$, then $B \geq_l A$ so $[B, A] = \{P \in l : B \geq_l P \geq_l A\}$. Now if $P \in [A, B]$, then $A \leq_l P$ and $P \leq_l B$. It follows that $B \geq_l P$ and $P \geq_l A$. Thus $P \in [B, A]$ and so $[A, B] \subset [B, A]$. By a similar argument $[B, A] \subset [A, B]$ and so $[A, B] = [B, A]$.

(iv) Let $C, D \in [A, B]$ so that $A \leq_l C \leq_l B$ and $A \leq_l D \leq_l B$. By A_2(iv) either $C \leq_l D$ or $D \leq_l C$.

If $C \leq_l D$ and $P \in [C, D]$, then $C \leq_l P \leq_l D$. Thus $A \leq_l C$, $C \leq_l P$ so by A_2(ii), $A \leq_l P$. Also $P \leq_l D$, $D \leq_l B$ so by A_2(ii) $P \leq_l B$. Thus $P \in [A, B]$.

If $D \leq_l C$, we interchange C and D in the last paragraph.

(v) This follows immediately from 2.1.3.

2.1.5   Half-lines

Definition. Given a line $l \in \Lambda$, a point $A \in l$ and a natural order $\leq_l$ on l, then the set

$$\rho(l, A, \leq_l) = \{P \in l : A \leq_l P\},$$

is called a **half-line** or **ray** of l, with **initial point** A.

Given distinct points A, B let $\leq_l$ be the natural order on $l = AB$ for which $A \leq_l B$; then we also use the notation $[A, B$ for $\rho(l, A, \leq_l)$.

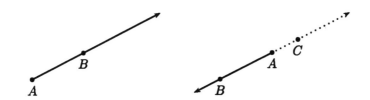

Figure 2.4. A half-line $[A, B$.   Opposite half-lines.

As $\geq_l$ is also a natural order on l,

$$\rho(l, A, \geq_l) = \{P \in l : A \geq_l P\} = \{P \in l : P \leq_l A\}$$

is also a half-line of l, with initial point A. We say that $\rho(l, A, \leq_l)$ and $\rho(l, A, \geq_l)$ are **opposite** half-lines.

Half-lines have the following properties:-

(i) *In all cases $\rho(l, A, \leq_l) \subset l$.*

(ii) *In all cases $A \in \rho(l, A, \leq_l)$.*

(iii) *If $B, C \in \rho(l, A, \leq_l)$, then $[B, C] \subset \rho(l, A, \leq_l)$.*

Proof.
(i) By the definition of $\rho(l, A, \leq_l)$, we have $P \in l$ for all $P \in \rho(l, A, \leq_l)$ and so $\rho(l, A, \leq_l) \subset l$.

(ii) By A$_2$(i) $A \leq_l A$, so $A \in \rho(l, A, \leq_l)$.

(iii) As $B, C \in \rho(l, A, \leq_l)$ we have $A \leq_l B$ and $A \leq_l C$. Since $B, C \in l$, by A$_2$(iv) either $B \leq_l C$ or $C \leq_l B$. When $B \leq_l C$, we have $B \leq_l P$ for all $P \in [B, C]$; with $A \leq_l B$ this gives $A \leq_l P$ by A$_2$(ii), and so $P \in \rho(l, A, \leq_l)$. When $C \leq_l B$, we have a similar proof.

2.2   OPEN AND CLOSED HALF-PLANES

2.2.1   Convex sets

Definition. A set $\mathcal{E}$ is said to be **convex** if for every $P, Q \in \mathcal{E}$, $[P, Q] \subset \mathcal{E}$ holds.

NOTE. By 2.1.4(iv) every segment is a convex set; by 2.1.5(iii) so is every half-line. In preparation for the next subsection, we note that by A$_1$, for each line $l \in \Lambda$ we have $\Pi \setminus l \neq \emptyset$.

2.2.2   Open half-planes

Primitive Term. Corresponding to each line $l \in \Lambda$, there is a pair $\{\mathcal{G}_1, \mathcal{G}_2\}$ of non-empty sets called **open half-planes with common edge** l.

AXIOM A_3. *Open half-planes $\mathcal{G}_1, \mathcal{G}_2$ with common edge l have the properties:-*

(i) $\Pi \setminus l = \mathcal{G}_1 \cup \mathcal{G}_2$;

(ii) *$\mathcal{G}_1$ and $\mathcal{G}_2$ are both convex sets;*

(iii) *if $P \in \mathcal{G}_1$ and $Q \in \mathcal{G}_2$, then $[P,Q] \cap l \neq \emptyset$.* |

We note the following immediately.

Open half-planes $\{\mathcal{G}_1, \mathcal{G}_2\}$ with common edge l have the properties:-

(i) $l \cap \mathcal{G}_1 = \emptyset$, $l \cap \mathcal{G}_2 = \emptyset$.

(ii) $\mathcal{G}_1 \cap \mathcal{G}_2 = \emptyset$.

(iii) *If $P \in \mathcal{G}_1$ and $[P, Q] \cap l \neq \emptyset$ where $Q \notin l$, then $Q \in \mathcal{G}_2$.*

(iv) *Each line l determines a unique pair of open half-planes.*

Proof.

(i) By A_3(i), $l \cap (\mathcal{G}_1 \cup \mathcal{G}_2) = \emptyset$ and as $\mathcal{G}_1 \subset \mathcal{G}_1 \cup \mathcal{G}_2$ it follows that $l \cap \mathcal{G}_1 = \emptyset$. The other assertion is proved similarly.

(ii) If $\mathcal{G}_1 \cap \mathcal{G}_2 \neq \emptyset$, there is some point R in both $\mathcal{G}_1$ and $\mathcal{G}_2$. By A_3(iii) with $P = R$, $Q = R$, we have that $[R, R] \cap l \neq \emptyset$. But R is the only point in $[R, R]$ so $R \in l$. This contradicts the fact that $l \cap \mathcal{G}_1 = \emptyset$.

(iii) For otherwise by A_3(i), $Q \in \mathcal{G}_1$ and then by A_3(ii) $[P, Q] \subset \mathcal{G}_1$. As $l \cap \mathcal{G}_1 = \emptyset$, it follows that $[P, Q] \cap l = \emptyset$ which contradicts the assumptions.

(iv) Suppose that

$$\Pi \setminus l = \mathcal{G}_1 \cup \mathcal{G}_2 = \mathcal{G}'_1 \cup \mathcal{G}'_2,$$

where $\{\mathcal{G}_1, \mathcal{G}_2\}$ and $\{\mathcal{G}'_1, \mathcal{G}'_2\}$ are both sets of open half-planes with common edge l. Then

$$\mathcal{G}_1 \subset \mathcal{G}_1 \cup \mathcal{G}_2 = \mathcal{G}'_1 \cup \mathcal{G}'_2$$

so either

(a) $\mathcal{G}_1 \subset \mathcal{G}'_1$ or   (b) $\mathcal{G}_1 \subset \mathcal{G}'_2$ or   (c) $\mathcal{G}_1 \cap \mathcal{G}'_1 \neq \emptyset$, $\mathcal{G}_1 \cap \mathcal{G}'_2 \neq \emptyset$.

In (c) we have $P \in \mathcal{G}_1$, $P \in \mathcal{G}'_1$ and $Q \in \mathcal{G}_1$, $Q \in \mathcal{G}'_2$ for some P and Q. But then we have $[P, Q] \subset \mathcal{G}_1$, by A_3(ii) applied to $\mathcal{G}_1$, and $[P, Q] \cap l \neq \emptyset$, by A_3(iii) applied to $\{\mathcal{G}'_1, \mathcal{G}'_2\}$. This gives a contradiction as $l \cap \mathcal{G}_1 = \emptyset$. Thus (c) cannot happen.

By similar reasoning, we must have either

(d) $\mathcal{G}'_1 \subset \mathcal{G}_1$ or   (e) $\mathcal{G}'_1 \subset \mathcal{G}_2$.

Now (a) and (d) give $\mathcal{G}_1 = \mathcal{G}'_1$ and it follows that $\mathcal{G}_2 = \mathcal{G}'_2$ as

$$(\mathcal{G}_1 \cup \mathcal{G}_2) \setminus \mathcal{G}_1 = \mathcal{G}_2, \ (\mathcal{G}'_1 \cup \mathcal{G}'_2) \setminus \mathcal{G}'_1 = \mathcal{G}'_2.$$

Similarly (b) and (e) give $\mathcal{G}_1 = \mathcal{G}'_2$ and it follows that $\mathcal{G}_2 = \mathcal{G}'_1$.
Finally, we cannot have (a) and (e) as that would imply $\mathcal{G}_1 \subset \mathcal{G}_2$. Neither can we have (b) and (d).

TERMINOLOGY. If two points are both in $\mathcal{G}_1$ or both in $\mathcal{G}_2$ they are said to be **on the one side of the line** l, while if one of the points is in $\mathcal{G}_1$ and the other is in $\mathcal{G}_2$ they are said to be **on different sides of** l.

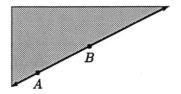

Figure 2.5. A closed half-plane shaded.

2.2.3   Closed half-planes

Definition. If $\mathcal{G}_1, \mathcal{G}_2$ are open half-planes with common edge l, we call

$$\mathcal{H}_1 = \mathcal{G}_1 \cup l, \ \mathcal{H}_2 = \mathcal{G}_2 \cup l,$$

closed half-planes with common edge l.

Closed half-planes $\mathcal{H}_1, \mathcal{H}_2$ with common edge l have the properties:-

(i) $\mathcal{H}_1 \cup \mathcal{H}_2 = \Pi$.

(ii) $\mathcal{H}_1 \cap \mathcal{H}_2 = l$.

(iii) *Each of $\mathcal{H}_1, \mathcal{H}_2$ is a convex set.*

(iv) *If $A \in l$ and $B \neq A$ is in $\mathcal{H}_1$, then $[A, B \subset \mathcal{H}_1$.*

Proof.
    (i) By A_3(i), $\Pi = \mathcal{G}_1 \cup \mathcal{G}_2 \cup l = (\mathcal{G}_1 \cup l) \cup (\mathcal{G}_2 \cup l) = \mathcal{H}_1 \cup \mathcal{H}_2$.
    (ii) For $(\mathcal{G}_1 \cup l) \cap (\mathcal{G}_2 \cup l) = (\mathcal{G}_1 \cap \mathcal{G}_2) \cup (\mathcal{G}_1 \cap l) \cup (\mathcal{G}_2 \cap l) \cup (l \cap l) = l \cap l = l$.
    (iii) We prove that $\mathcal{H}_1$ is convex; proof for $\mathcal{H}_2$ is similar.
    Let $A, B \in \mathcal{H}_1$; we wish to show that $[A, B] \subset \mathcal{H}_1$.
    CASE (a). Let $A, B \in \mathcal{G}_1$. Then the conclusion follows from A_3(iii).
    CASE (b). Let $A, B \in l$. Then $[A, B] \subset l \subset \mathcal{H}_1$, so the result follows.
    CASE (c). Let one of A, B be on l and the other in $\mathcal{G}_1$, say $A \in l, B \in \mathcal{G}_1$.
    Suppose that $[A, B]$ is not a subset of $\mathcal{H}_1$. Then there is some point $C \in [A, B]$ such that $C \in \mathcal{G}_2$. Note that $C \neq A, C \neq B$ as $A, B \notin \mathcal{G}_2, C \in \mathcal{G}_2$.
    Now $B \in \mathcal{G}_1, C \in \mathcal{G}_2$ so by A_3(iii) there is some point D of $[B, C]$ on l, so that $D \in [B, C], D \in l$. Now A, B, C are collinear and distinct, and $C \in [A, B]$ so by 2.1.4 we cannot have $A \in [B, C]$. Hence $A \neq D$.
    But $A \in l, D \in l$ so by $A_1, AD = l$. However $AB = BC$ and $D \in BC$, so $D \in AB$. Then $AB = AD = l$, so $B \in l$. This gives a contradiction. Thus the original supposition is untenable so $[A, B] \subset \mathcal{H}_1$, and this proves (iii).

(iv)

CASE (a). Let $B \in l$. Then $[A, B \subset l \subset \mathcal{H}_1$, which gives the desired conclusion.

CASE (b). Let $B \in \mathcal{G}_1$. Suppose that $[A, B$ is not a subset of $\mathcal{H}_1$. Then there is some point $C \in [A, B$ such that $C \in \mathcal{G}_2$. Clearly $C \neq A, C \neq B$. Now A, B, C are distinct collinear points, so by 2.1.4 precisely one of

$$A \in [B, C], \ B \in [C, A], \ C \in [A, B],$$

holds. We cannot have $A \in [B, C]$ as that would put B, C in different half-lines with initial-point A, whereas they are both in $[A, B$. This leaves us with two subcases.

Subcase 1. Let $C \in [A, B]$. We recall that $A, B \in \mathcal{H}_1$ so by part (iii) of the present result $[A, B] \subset \mathcal{H}_1$. As $C \in [A, B], C \in \mathcal{G}_2$, we have a contradiction.

Subcase 2. Let $B \in [A, C]$. We recall that $A \in \mathcal{H}_2, C \in \mathcal{H}_2$ so by part (iii) of the present result, $[A, C] \subset \mathcal{H}_2$. Then $B \in \mathcal{H}_2, B \in \mathcal{G}_1$ which gives a contradiction. Thus the original supposition is untenable, and this proves (iv).

NOTE. The terms 'open' and 'closed' are standard in analysis and point-set topology. What is significant is that an open half-plane contains none of the points of the edge, while a closed half-plane contains all of the points of the edge.

2.3   ANGLE-SUPPORTS, INTERIOR AND EXTERIOR REGIONS, ANGLES

2.3.1   Angle-supports, interior regions

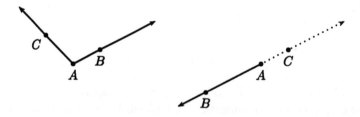

Figure 2.6. An angle-support.   A straight angle-support.

Definition. We call a pair $\{[A, B , [A, C \}$ of co-initial half-lines an **angle-support**. For this we use the notation $|\underline{BAC}$. When $A \in [B, C]$, this is called a **straight angle-support**. We call the half-lines $[A, B$ and $[A, C$ the **arms**, and the point A the **vertex**, of $|\underline{BAC}$. Note that we are assuming $B \neq A$ and $C \neq A$ from the definition of half-lines. In all cases we have $|\underline{BAC} = |\underline{CAB}$.

COMMENT. The reason that we do not call $|\underline{BAC}$ an *angle* is that there are two angles associated with this configuration.

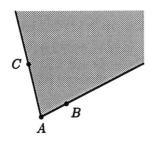
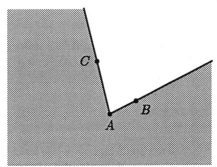

Figure 2.7. An interior region.   The corresponding exterior region.

Definition. Consider an angle-support $|\underline{BAC}$ which is not straight. When A, B, C are not collinear, let $\mathcal{H}_1$ be the closed half-plane with edge AB in which C lies, and $\mathcal{H}_3$ the closed half-plane with edge AC in which B lies. Then $\mathcal{H}_1 \cap \mathcal{H}_3$ is called **the interior region of** $|\underline{BAC}$, and we denote it by $\mathcal{IR}(|\underline{BAC})$. When A, B, C are collinear we have $[A, B = [A, C$ and we define $\mathcal{IR}(|\underline{BAC}) = [A, B$.

Interior regions have the following properties:-

(i) $[A, B$ *and* $[A, C$ *are both subsets of* $\mathcal{IR}(|\underline{BAC})$.

(ii) *If* $P, Q \in \mathcal{IR}(|\underline{BAC})$ *then* $[P, Q] \subset \mathcal{IR}(|\underline{BAC})$, *so that an interior region is a convex set.*

(iii) *If* $P \in \mathcal{IR}(|\underline{BAC})$ *and* $P \neq A$, *then* $[A, P \subset \mathcal{IR}(|\underline{BAC})$.

Proof.

(i) When A, B, C are non-collinear, by 2.1.5 $[A, B \subset AB \subset \mathcal{H}_1$ and by 2.2.3 $[A, B \subset \mathcal{H}_3$ so $[A, B \subset \mathcal{H}_1 \cap \mathcal{H}_3$. Similarly for $[A, C$. When $[A, B = [A, C$ the result is trivial.

(ii) When A, B, C are non-collinear, we have that $[P, Q]$ is a subset of $\mathcal{H}_1$ by 2.2.3. It is a subset of $\mathcal{H}_3$ similarly, and so is a subset of the intersection of these closed half-planes. When $[A, B = [A, C$, the result follows from 2.1.5.

(iii) When A, B, C are non-collinear, by 2.2.3 $[A, P$ is a subset of each of $\mathcal{H}_1$ and $\mathcal{H}_3$, and so of their intersection. When $[A, B = [A, C$ we have $\mathcal{IR}(|\underline{BAC}) = [A, B$ and $[A, P = [A, B$.

2.3.2   Exterior regions

Definition. If $|\underline{BAC}$ is an angle-support which is not straight and $\mathcal{IR}(|\underline{BAC})$ is its interior region, then

$$\{\Pi \setminus \mathcal{IR}(|\underline{BAC})\} \cup [A, B \cup [A, C$$

is called **the exterior region of** $|\underline{BAC}$, and denoted by $\mathcal{ER}(|\underline{BAC})$. Thus the interior and exterior regions have in common only the arms.

2.3.3   Angles

Figure 2.8. A wedge-angle.   A reflex-angle.   A straight-angle.

Definition. Let $|\underline{BAC}$ be an angle-support which is not straight, with interior region $\mathcal{IR}(|\underline{BAC})$ and exterior region $\mathcal{ER}(|\underline{BAC})$. Then the pair $\left(|\underline{BAC},\ \mathcal{IR}(|\underline{BAC})\right)$ is called a **wedge-angle**, and the pair $\left(|\underline{BAC},\ \mathcal{ER}(|\underline{BAC})\right)$ is called a **reflex-angle**. If $|\underline{BAC}$ is a straight-angle support and $\mathcal{H}_1, \mathcal{H}_2$ are the closed half-planes with common edge AB, then each of the pairs $(|\underline{BAC}, \mathcal{H}_1)$, $(|\underline{BAC}, \mathcal{H}_2)$ is called a **straight-angle.**
In each case the point A is called the **vertex** of the angle, the half-lines $[A, B$ and $[A, C$ are called the **arms** of the angle, and $|\underline{BAC}$ is called the **support** of the angle.

We denote a wedge-angle with support $|\underline{BAC}$ by $\angle BAC$. The wedge-angle $\angle BAB$ is said to be a **null-angle.**

2.4   TRIANGLES AND CONVEX QUADRILATERALS

2.4.1   Terminology

COMMENT. The terminology which we have used hitherto is established, apart from 'angle-support' and 'wedge-angle' which we have coined. Now we are reaching terminology which is of long standing but is used in slightly varying senses.

In Euclidean geometry it is generally accepted that the concept of triangle is associated with:

(i) a set $\{A, B, C\}$ of three points which are not collinear;

(ii) a union of segments $[B, C] \cup [C, A] \cup [A, B]$, where the points A, B, C are as in (i);

(iii) an intersection of half-planes $\mathcal{H}_1 \cap \mathcal{H}_3 \cap \mathcal{H}_5$, where A, B, C are as in (i), $\mathcal{H}_1$ is the closed half-plane with edge BC in which A lies, $\mathcal{H}_3$ is the closed half-plane with edge CA in which B lies, and $\mathcal{H}_5$ is the closed half-plane with edge AB in which C lies.

However in some courses the actual definition of a triangle is taken to be (i), in other courses it is taken to be (ii), and in other courses it is taken to be (iii), with (ii) and (iii) very common. In yet other courses a combination of (i) and (ii) is taken.

Having to make a choice for the sake of precision, we opt for (iii); then for us (i) will be the set of vertices of our triangle, and (ii) will be the perimeter of our triangle, with the individual segments being the sides. We shall then be able to refer naturally to the area of a triangle and the length of its perimeter.

Consideration similar to (i), (ii) and (iii) for a triangle surround each of the terms quadrilateral, parallelogram, rectangle and square, and we adopt our terminology consistently.

2.4.2   Triangles

NOTE. Let A, B, C be points which do not lie on one line. Then by A_1, A, B, C are distinct points, and $A \notin BC$, $B \notin CA$, $C \notin AB$. In fact these lines are not concurrent; for BC and CA cannot have a point P in common other than $P = C$, while $C \notin AB$.

Definition. For non-collinear points A, B, C let $\mathcal{H}_1$ be the closed half-plane with edge BC in which A lies, $\mathcal{H}_3$ the closed half-plane with edge CA in which B lies, and $\mathcal{H}_5$ the closed half-plane with edge AB in which C lies. Then the intersection $\mathcal{H}_1 \cap \mathcal{H}_3 \cap \mathcal{H}_5$ is called a **triangle**, and is denoted by $[A, B, C]$.

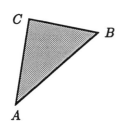
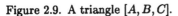
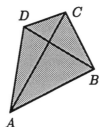

Figure 2.9. A triangle $[A, B, C]$.   Figure 2.10. A convex quadrilateral.

The points A, B, C are called its **vertices**; the segments $[B, C], [C, A], [A, B]$ are called its **sides**; the lines BC, CA, AB are called its **side-lines**. The union $[B, C] \cup [C, A] \cup [A, B]$ of its sides is called its **perimeter**. A side and a vertex not contained in it are said to be **opposite**; thus A is opposite $[B, C]$ but is not opposite $[C, A]$ or $[A, B]$.

Triangles have the following properties:-

(i) *$[A, B, C]$ is independent of the order of the points A, B, C.*

(ii) *Each of the vertices A, B, C is an element of $[A, B, C]$.*

(iii) *If $P, Q \in [A, B, C]$, then $[P, Q] \subset [A, B, C]$ so that a triangle is a convex set.*

(iv) *Each of the sides $[B, C], [C, A], [A, B]$ is a subset of $[A, B, C]$.*

Proof.
(i) As $\cap$ is commutative, $\mathcal{H}_1 \cap \mathcal{H}_3 \cap \mathcal{H}_5$ is independent of the order of $\mathcal{H}_1, \mathcal{H}_3, \mathcal{H}_5$.
(ii) The vertex A is in $\mathcal{H}_1$ by definition. It is also in the edge of each of $\mathcal{H}_3$ and $\mathcal{H}_5$, so by 2.2.3 it is in each of these closed half-planes. The vertices B and C are treated similarly.
(iii) By definition of an intersection, P and Q are in each of $\mathcal{H}_1, \mathcal{H}_3, \mathcal{H}_5$. By 2.2.3, $[P, Q]$ is a subset of each of these closed half-planes, and so it is a subset of their intersection.
(iv) This follows from parts (ii) and (iii) of the present result.

2.4.3 Pasch's property, 1882

PASCH'S PROPERTY. *If a line cuts one side of a triangle, not at a vertex, then it will either pass through the opposite vertex, or cut one of the other two sides.*

   Proof. Let $[A, B, C]$ be the triangle and l a line which cuts the side $[A, B]$ at a point which is not a vertex. If $C \in l$ we have the first conclusion. Otherwise suppose that l does not cut $[B, C]$. Then A and B are on different sides of l, but B and C are on the same side of l. It follows that A and C are on different sides if l, so by A_3(iii) a point of $[A, C]$ lies on l.

2.4.4 Convex quadrilaterals

Definition. Let A, B, C, D be four points in Π, no three of which are collinear, and such that $[A, C] \cap [B, D] \neq \emptyset$. Let $\mathcal{H}_1$ be the closed half-plane with edge AB in which D lies, $\mathcal{H}_3$ the closed half-plane with edge BC in which A lies, $\mathcal{H}_5$ the closed half-plane with edge CD in which B lies, and $\mathcal{H}_7$ the closed half-plane with edge DA in which C lies. Then the intersection $\mathcal{H}_1 \cap \mathcal{H}_3 \cap \mathcal{H}_5 \cap \mathcal{H}_7$ of these four half-planes is called a **convex quadrilateral**, and we denote it by $[A, B, C, D]$.

   Each of the four points A, B, C, D is called a **vertex** ; the segments $[A, B], [B, C]$, $[C, D], [D, A]$ are called the **sides**, and AB, BC, CD, DA are called the **side-lines** ; the union of the sides $[A, B] \cup [B, C] \cup [C, D] \cup [D, A]$ is called **the perimeter**. The segments $[A, C], [B, D]$ are called the **diagonals**, and AC, BD the **diagonal lines**. Vertices which are the end-points of a side are called **adjacent** while vertices which are the end-points of a diagonal are called **opposite**; thus A and B are adjacent as $[A, B]$ is a side, and A and C are opposite as $[A, C]$ is a diagonal. Sides which have a vertex in common are said to be **adjacent** while sides which do not have a vertex in common are said to be **opposite**; thus the sides $[A, B], [A, D]$ are adjacent as the vertex A is in both, while the sides $[A, B], [C, D]$ are opposite as neither C nor D is in AB and so neither of them could be A or B. If we write

then two vertices in $[A, B, C, D]$ will be adjacent if the letters for them in this diagram are linked.

Exercises

   2.1 Let P be a fixed point in Π. Identify the union of all lines $l \in \Lambda$ such that $P \in l$.

   2.2 Prove that segments have the following properties:-

   (i) If $C \in [A, B]$, then $[A, C] \cup [C, B] = [A, B]$ and $[A, C] \cap [C, B] = \{C\}$.

   (ii) If $C \in [A, B]$ and $B \in [A, C]$ then $B = C$.

   (iii) If $C \in [A, B]$ and $D \in [A, C]$, then $C \in [D, B]$.

(iv) If $B \neq A$, $B \in [A, C]$ and $B \in [A, D]$, then either $C \in [B, D]$ or $D \in [B, C]$.

2.3 Prove that half-lines have the following properties:-

(i) If $B \in \rho(l, A, \leq_l)$, then $\rho(l, B, \leq_l) \subset \rho(l, A, \leq_l)$.

(ii) If $B \in \rho(l, A, \leq_l)$, then $\rho(l, A, \leq_l) = [A, B] \cup \rho(l, B, \leq_l)$ and $[A, B] \cap \rho(l, B, \leq_l) = \{B\}$.

(iii) Let $B \in \rho(l, A, \leq_l)$, $A \neq B$ and $A \in [B, C]$. Then $C \in \rho(l, A, \leq_l)$ only if $C = A$.

(iv) Let $B \in \rho(l, A, \leq_l)$ and $A \neq B$. Then $C \in \rho(l, A, \leq_l)$ if and only if either $B \in [A, C]$ or $C \in [A, B]$.

(v) In all cases

$$\rho(l, A, \leq_l) \cup \rho(l, A, \geq_l) = l \text{ and } \rho(l, A, \leq_l) \cap \rho(l, A, \geq_l) = \{A\}.$$

(vi) Let $B \in \rho(l, A, \leq_l)$ and $A \neq B$. Then $C \in \rho(l, A, \geq_l)$ if and only if $A \in [B, C]$.

(vii) Let $B \in \rho(l, A, \leq_l)$ and $A \neq B$. Then

$$\rho(l, A, \leq_l) \cup \rho(l, B, \geq_l) = l, \ \rho(l, A, \leq_l) \cap \rho(l, B, \geq_l) = [A, B],$$

$$\rho(l, A, \geq_l) \cap \rho(l, B, \leq_l) = \emptyset, \ \rho(l, A, \geq_l) \cup \rho(l, B, \leq_l) \cup [A, B] = l.$$

(viii) If $A \neq B, A \neq C$ and $C \in [A, B$, then $[A, B = [A, C$.

2.4 If $[A, B], [C, D]$ are both segments of a line l such that $[A, B] \cap [C, D] \neq \emptyset$, show that $[A, B] \cap [C, D]$ and $[A, B] \cup [C, D]$ are both segments.

2.5 Show that if $A \neq B$ and C, D are both in $AB \setminus [A, B]$, then either $[A, B] \cap [C, D] = \emptyset$ or $[A, B] \subset [C, D]$.

2.6 Let $\leq_E$ be a total order on the set E and $f : E \rightarrow F$ a 1:1 onto function. If for $a, b \in F, a \leq_F b$ when $f^{-1}(a) \leq_E f^{-1}(b)$, show that $\leq_F$ is a total order on F.

2.7 Use Ex.2.6 to show that if F is an infinite set and there is a total order on F, then there are infinitely many total orders on F.

2.8 Show that interior regions have the following properties:-

(i) If $P \in \mathcal{IR}(|BAC)$ and $P \neq A$, then $AP \cap \mathcal{IR}(|BAC) = [A, P$.

(ii) If A, B, C are non-collinear and $U \in [A, B$, $V \in [A, C$ but neither U nor V is A, then $UV \cap \mathcal{IR}(|BAC) = [U, V]$.

(iii) If A, U, V are distinct collinear points, and U and V are both in $\mathcal{IR}(|BAC)$, then $V \in [A, U$.

2.9 Show that an exterior region has the following properties:-

(i) The arms $[A, B$ and $[A, C$ are both subsets of $\mathcal{ER}(|BAC)$.

(ii) If $P \in \mathcal{ER}(|BAC)$ and $P \neq A$, then $[A, P \subset \mathcal{ER}(|BAC)$.

2.10 Show that convex quadrilaterals have the following properties:-

(i) Each of

$\langle 2 \rangle [A, D, C, B], \qquad \langle 3 \rangle [C, B, A, D], \qquad \langle 4 \rangle [C, D, A, B],$

$\langle 5 \rangle [B, A, D, C], \qquad \langle 6 \rangle [B, C, D, A], \qquad \langle 7 \rangle [D, A, B, C],$

$\langle 8 \rangle [D, C, B, A],$

is equal to $\langle 1 \rangle [A, B, C, D]$.

(ii) Each of the vertices A, B, C, D is an element of $[A, B, C, D]$.

(iii) If $P, Q \in [A, B, C, D]$, then $[P, Q] \subset [A, B, C, D]$ so that $[A, B, C, D]$ is a convex set.

(iv) Each side and each diagonal is a subset of $[A, B, C, D]$.

(v) Any pair of opposite sides are disjoint.

3

Distance; degree-measure of an angle

COMMENT. In this chapter we introduce distance as a primitive concept, relate it to the properties of segments, and define the notion of the mid-point of two points. We also introduce as a primitive concept the notion of the degree-measure of a wedge-angle and of a straight-angle, relate it to the properties of interior-regions and half-planes, and define the notion of the mid-line of an angle-support.

3.1 DISTANCE

3.1.1 Axiom for distance

Notation. We denote by $\mathbf{R}$ the set of real numbers.

Primitive Term. There is a function $|\ | : \Pi \times \Pi \to \mathbf{R}$ called **distance**. We read $|A, B|$ as the distance from A to B. We also refer to $|A, B|$ as the **length of the segment** $[A, B]$.

AXIOM A_4. *Distance has the following properties:-*

(i) $|A, B| \geq 0$ *for all* $A, B \in \Pi$;

(ii) $|A, B| = |B, A|$ *for all* $A, B \in \Pi$;

(iii) *if* $Q \in [P, R]$, *then* $|P, Q| + |Q, R| = |P, R|$;

(iv) *given any* $k \geq 0$ *in* $\mathbf{R}$, *any line* $l \in \Lambda$, *any point* $A \in l$ *and either natural order* $\leq_l$ *on* l, *there is a unique point* $B \in l$ *such that* $A \leq_l B$ *and* $|A, B| = k$, *and a unique point* $C \in l$ *such that* $C \leq_l A$ *and* $|A, C| = k$. |

COMMENT. Note that A_4(iv) states that we can lay off a distance k, uniquely, on l on either side of A. The fact that different letters A, B, C are used is not to be taken as a claim that A, B, C are distinct in all cases. Axiom A_4(iv) implies that each line l contains infinitely many points and this supersedes the specification in A_1 that $l \neq \emptyset$;

nevertheless it was convenient to stipulate the latter to avoid a trivial situation. In A_4(iii) addition $+$ of real numbers is involved.

Figure 3.1. Addition of distances.   Laying off a distance k.

3.1.2   Derived properties of distance

Distance has the following properties:-

(i) *For all $A \in \Pi$ $|A, A| = 0$, and we have $|A, B| > 0$ if $A \neq B$.*

(ii) *If $P \in [A, B]$, then $|A, P| \leq |A, B|$. If additionally $P \neq B$, then $|A, P| < |A, B|$.*

(iii) *If $A \neq C$ and B lies on the line AC but outside the segment $[A, C]$, then*

$$|A, B| + |B, C| > |A, C|.$$

(iv) *If $C \in [A, B$ is such that $|A, B| < |A, C|$, then $B \in [A, C]$.*

Proof.

(i) By A_4(iii) with $P = Q = A$ and any $R \in \Pi$, we have $|A, A| + |A, R| = |A, R|$, i.e. $x + y = y$ where $x = |A, A|$ and $y = |A, R|$. It follows that $x = 0$.

Next with $A \neq B$ let $l = AB$ and $\leq_l$ be the natural order on l for which $A \leq_l B$. Then we have

$$A \leq_l B, \ A \leq_l A, \ |A, A| = 0,$$

so that if we also had $|A, B| = 0$, then by the uniqueness part of A_4(iv) with $k = 0$, we would have $A = B$ and so have a contradiction. To avoid this we must have $|A, B| > 0$.

(ii) As $P \in [A, B]$, by A_4(iii) we have $|A, P| + |P, B| = |A, B|$. But by A_4(i) $|P, B| \geq 0$ and so $|A, P| \leq |A, B|$. If $P \neq B$, then by (i) of the present theorem $|P, B| > 0$ and so $|A, P| < |A, B|$.

(iii) As $B \notin [A, C]$ we have $B \neq A, B \neq C$ and so by 2.1.4 we have either $A \in [B, C]$ or $C \in [A, B]$. In the first of these $|B, A| + |A, C| = |B, C|$ by A_4(iii) and as $|A, B| = |B, A| > 0$ this gives $|A, C| < |B, C| < |A, B| + |B, C|$. In the second case we have $|A, C| + |C, B| = |A, B|$ by A_4(iii) and as $|C, B| = |B, C| > 0$, then $|A, C| < |A, B| < |A, B| + |B, C|$.

(iv) We have $A \neq B$ by definition of $[A, B$, and $A \neq C$ as $0 < |A, B| < |A, C|$ so that $0 < |A, C|$. We also have $B \neq C$ as $|A, B| < |A, C|$ combined with $B = C$ would give $|A, B| < |A, B|$, whereas once (A, B) is known $|A, B|$ is uniquely determined. We cannot have $A \in [B, C]$ as $C \in [A, B$. Then by 2.1.4 either $B \in [C, A]$ or $C \in [A, B]$. But by (ii) of the present result, if $C \in [A, B]$ we would have $|A, C| \leq |A, B|$. As this is ruled out by assumption, we must have $B \in [A, C]$.

Segments and half-lines have the following further properties:-

(i) *Let $l \in \Lambda$ be a line, $A \in l$ and $\leq_l$ a natural order on l. Then there are points B and C on l such that $A \leq_l B$ and $B \neq A$, and such that $C \leq_l A$ and $C \neq A$.*

(ii) *If $A \neq B$, there are points $X \in [A, B]$ such that $X \neq A$ and $X \neq B$.*

(iii) *If $[A, B = [C, D$ then $A = C$.*

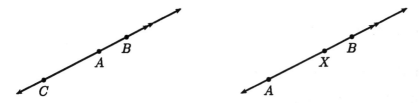

Figure 3.2.

Proof.

(i) By A_4(iv) with any $k > 0$, there is some $B \in l$ such that $A \leq_l B$ and $|A, B| = k$. As $|A, B| > 0$, we have $A \neq B$. Proof for the existence of C is similar.

(ii) Let $\leq_l$ be the natural order on $l = AB$ for which $A \leq_l B$. As $A \neq B$ we have $|A, B| > 0$ and then with any k such that $0 < k < |A, B|$, there is a point $X \in l$ such that $A \leq_l X$ and $|A, X| = k$. As $|A, X| \neq 0$, we have $A \neq X$; as $|A, X| < |A, B|$ then $X \neq B$. As $X \in [A, B$ and $|A, X| < |A, B|$, we have $X \in [A, B]$.

(iii) With the notation of (ii), $P \in [A, B$ if and only if $A \leq_l P$. Now $C \in [C, D = [A, B$ so $A \leq_l C$; similarly $A \leq_l D$.

CASE 1. Let $C \leq_l D$, so that $[C, D = \{Q \in l : C \leq_l Q\}$. As $A \in [A, B = [C, D$ we have $C \leq_l A$ and this combined with $A \leq_l C$ implies $A = C$.

CASE 2. Let $D \leq_l C$. Then $[C, D = \{Q \in l : Q \leq_l C\}$. By (i) of the present result there is an $X \in l$ such that $X \leq_l A$ and $X \neq A$. Then $X \leq_l A, A \leq_l C$ so $X \leq_l C$ and thus $X \in [C, D$. However $X \notin [A, B$ as otherwise we would have $A \leq_l X$ which combined with $X \leq_l A$ implies $X = A$ and involves a contradiction. Then $X \in [C, D$, $X \notin [A, B$ which contradicts the fact that $[A, B = [C, D$ and so this case cannot occur.

Figure 3.3.

3.2 MID-POINTS

3.2.1

If $A \neq B$ there is a unique point X on $l = AB$ such that $|A, X| = |X, B|$. In this in fact $X \in [A, B]$ and $X \neq A, X \neq B$.

Proof.

Existence. Let $\leq_l$ be the natural order on l for which $A \leq_l B$. With $k = \frac{1}{2}|A, B|$, by A_4(iv) there is a point X on l such that $A \leq_l X$ and $|A, X| = \frac{1}{2}|A, B|$. Clearly $X \in [A, B$. As $|A, B| > 0$ we have $|A, X| < |A, B|$; by 3.1.2 this implies that $X \in [A, B]$, $X \neq B$. By A_4(iii) $|A, X| + |X, B| = |A, B|$ and so $|X, B| = |A, B| - \frac{1}{2}|A, B| = \frac{1}{2}|A, B|$. Thus $|A, X| = |X, B|$ as required. We have already seen that $X \in [A, B]$ and $X \neq B$; as $|A, X| > 0$ we also have $X \neq A$.

Uniqueness. Suppose now that $Y \in l$ and $|A, Y| = |Y, B|$. Then Y cannot be A or B, as e.g. $Y = A$ implies that $|A, A| = |A, B|$, i.e. $0 = |A, B|$. Thus by 2.1.4 one of

$$Y \in [A, B], \ A \in [Y, B], \ B \in [A, Y],$$

holds. The second of these would imply $|Y, A| + |A, B| = |Y, B|$ and so $|Y, A| < |Y, B|$ as $|A, B| > 0$. The third of these would imply $|A, B| + |B, Y| = |A, Y|$ and so $|B, Y| < |A, Y|$. As these contradict our assumptions, we must have $Y \in [A, B]$. Then $|A, Y| + |Y, B| = |A, B|$ and as $|A, Y| = |Y, B|$ this implies that $|A, Y| = \frac{1}{2}|A, B|$. Then $A \leq_l X, A \leq_l Y$ and $|A, X| = |A, Y|$ so by the uniqueness in A_4(iv) we must have $X = Y$.

Definition.   Given any points $A, B \in \Pi$, we define the **mid-point** of A and B as follows: if $A = B$ then the mid-point is A; when $A \neq B$ the mid-point is the unique point X on the line AB such that $|A, X| = |X, B|$, which has just been guaranteed. We denote the mid-point of A and B by $\mathrm{mp}(A, B)$.

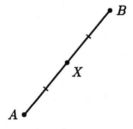

Figure 3.4. Mid-point of A and B.

Mid-points have the following properties:-

(i) *For all $A, B \in \Pi$, $\mathrm{mp}(A, B) = \mathrm{mp}(B, A)$.*

(ii) *For all $A, B \in \Pi$, $\mathrm{mp}(A, B) \in [A, B]$.*

(iii) *In all cases $\mathrm{mp}(A, A) = A$, and $\mathrm{mp}(A, B) \neq A$, $\mathrm{mp}(A, B) \neq B$ when $A \neq B$.*

(iv) *Given any points P and Q in Π, there is a unique point $R \in \Pi$ such that $Q = \mathrm{mp}(P, R)$.*

Proof.

(i) When $A \neq B$ this follows from the definition and A_4(ii); when $A = B$ it is immediate.

(ii) When $A \neq B$ this follows from the preparatory result. When $A = B$ it amounts to $A \in \{A\}$.

(iii) This follows from the definition and preparatory result.

(iv) *Existence.* If $Q = P$ we take $R = P$ and then $\mathrm{mp}(P, R) = \mathrm{mp}(P, P) = P = Q$. Suppose then that $P \neq Q$, let $l = PQ$ and let $\leq_l$ be the natural order on l under which $P \leq_l Q$. Take R on l so that $P \leq_l R$ and $|P, R| = 2|P, Q|$. Then P precedes both Q and R on l, while $|P, Q| = \frac{1}{2}|P, R|$. By our initial specification of X in the preparatory result we see that $Q = \mathrm{mp}(P, R)$.

Uniqueness. Suppose that also $Q = \mathrm{mp}(P, S)$. We again first take $Q = P$. Now in the preparatory result we had $X \neq A$, so that cannot be the situation here as $Q = P$; thus we must have $S = P$ and so $S = R$. Next suppose that $Q \neq P$; then we cannot have $S = P$, as we had $X \neq A$. Then $Q \in PS$, so by A_1 $S \in PQ$. In fact $Q \in [P, S]$ so as $P \leq_l Q$ we must have $P \leq_l S$; moreover $|P, R| = |P, S|$ as each is twice the distance $|P, Q|$. By the uniqueness in A_4(iv) we must then have $R = S$.

3.3   A RATIO RESULT

3.3.1

Let A, B, C be distinct collinear points, and write

$$\frac{|A, C|}{|A, B|} = r, \quad \frac{|A, C|}{|C, B|} = s.$$

Then if $C \in [A, B]$ we have

$$s = \frac{r}{1 - r}, \quad r = \frac{s}{1 + s}.$$

Proof. Let $|A, C| = x$, $|C, B| = y$. As $C \in [A, B]$ we have $|A, B| = x + y$. Then

$$\frac{x}{x + y} = r \quad \text{so that} \quad \frac{x + y}{x} = \frac{1}{r}.$$

Hence

$$\frac{y}{x} = \frac{1}{r} - 1 \quad \text{and so} \quad \frac{x}{y} = \frac{|A, C|}{|C, B|} = \frac{r}{1 - r}.$$

In turn

$$s = \frac{r}{1 - r} \quad \text{and so} \quad s - sr = r,$$

giving

$$s = r(1 + s) \quad \text{and thus} \quad r = \frac{s}{1 + s}.$$

3.4   THE CROSS-BAR THEOREM

3.4.1

THE CROSS-BAR THEOREM. *Let A, B, C be non-collinear points, $X \neq A$ any point on $[A, B$ and $Y \neq A$ any point on $[A, C$. If $D \neq A$ is any point in the interior region $\mathcal{IR}(|\underline{BAC})$, then $[A, D \cap [X, Y] \neq \emptyset$.*

Proof. If D is on $[A, B$ or $[A, C$ the result is clear, so we turn to other cases. By 3.1.2 there is a point $E \neq A$ such that $A \in [E, X]$. Thus X and E are on different sides of the line AC.

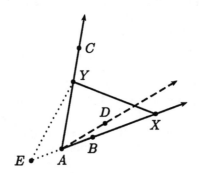

Figure 3.5. The Cross-Bar Theorem.

Then by 2.2.3(iv) every point of $[Y, E$ (other than Y) is on one side of AC, while every point of $[A, D$ (other than A) is on a different side of AC; thus $[A, D$ does not meet $[Y, E]$. Moreover the other points of the line AD are on one side of the line AB, while the points of $[E, Y$ (other than E) are on the other side of AB. On combining these two, we see that the line AD does not meet the side $[E, Y]$ of the triangle $[E, X, Y]$. As AD does meet the side $[E, X]$ of that triangle, we see by 2.4.3 that AD must meet the third side $[X, Y]$ of that triangle at some point F. As $F \in [X, Y] \subset \mathcal{IR}(|\underline{BAC})$, F must be on the part of AD in $\mathcal{IR}(|\underline{BAC})$, that is $F \in [A, D$.

3.5 DEGREE-MEASURE OF ANGLES

3.5.1 Axiom for degree-measure

Primitive Term. There is a function $|\ |°$ on the set of all wedge-angles and straight-angles, into **R**. Thus with each angle α, either a wedge-angle $\alpha = \angle BAC$ or a straight-angle with support $|\underline{BAC}$, there is associated a unique real number $|\alpha|°$, called its **degree-measure**.

AXIOM A$_5$. *Degree-measure $|\ |°$ of angles has the following properties:-*

(i) *In all cases $|\alpha|° \geq 0$;*

(ii) *if α is a straight-angle, then $|\alpha|° = 180$;*

(iii) *if $\angle BAC$ is a wedge-angle and the point $D \neq A$ lies in the interior region $\mathcal{IR}(|\underline{BAC})$, then*
$$|\angle BAD|° + |\angle DAC|° = |\angle BAC|°,$$
while if $|\underline{BAC}$ is a straight angle-support and $D \notin AB$, then
$$|\angle BAD|° + |\angle DAC|° = 180;$$

(iv) *if $B \neq A$, if $\mathcal{H}_1$ is a closed half-plane with edge AB and if the half-lines $[A, C$ and $[A, D$ in $\mathcal{H}_1$ are such that $|\angle BAC|° = |\angle BAD|°$, then $[A, D = [A, C$;*

(v) *if $B \neq A$, if $\mathcal{H}_1$ is a closed half-plane with edge AB and if $0 < k < 180$, then there is a half-line $[A, C$ in $\mathcal{H}_1$ such that $|\angle BAC|^\circ = k$.* |

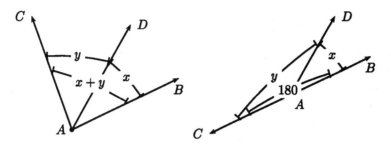

Figure 3.6. Addition of angle-measures.

COMMENT. The properties and proofs for degree-measure are quite like those for distance, with the role of interior regions analogous to that of segments. We note that A_5(i) is like A_4(i), A_5(iii) is like A_4(iii), A_5(iv) is like the uniqueness part of A_4(iv) and A_5(v) is like the existence part of A_4(iv). Wedge-angles $\angle BAD$ and $\angle DAC$ such as those in the second part of A_5(iii) are said to be **supplementary.**

3.5.2 Derived properties of degree-measure

Definition. For a wedge-angle $\angle BAC$, if we take a point $B_1 \neq A$ so that $A \in [B, B_1]$ and a point $C_1 \neq A$ so that $A \in [C, C_1]$, then $\angle B_1 AC_1$ is called the **opposite angle** of $\angle BAC$.

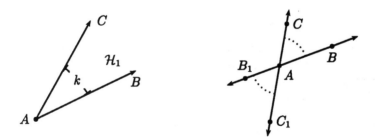

Figure 3.7. Laying off an angle.   Figure 3.8. Opposite angles at a vertex.

Degree-measure has the properties:-

(i) *The null-angle $\angle BAB$ has degree-measure 0.*

(ii) *For any non-null wedge-angle $\angle BAC$, we have $0 < |\angle BAC|^\circ < 180$.*

(iii) *If $\angle B_1 AC_1$ is the angle opposite to $\angle BAC$, then*

$$|\angle B_1 AC_1|^\circ = |\angle BAC|^\circ,$$

so that opposite angles have equal degree-measures.

Proof.

(i) Let C be a point not on AB. Then by A_5(iii) with $D = B$,

$$|\angle BAB|^\circ + |\angle BAC|^\circ = |\angle BAC|^\circ.$$

It follows that $|\angle BAB|^\circ = 0$.

(ii) Given any non-null wedge-angle $\angle BAC$, let $\mathcal{H}_1$ be the closed half-plane with edge AB in which C lies. If we had $|\angle BAC|^\circ = 0$, then we would have $|\angle BAC|^\circ = |\angle BAB|^\circ$ and so by A_5(iv) we would have $[A, C = [A, B$. This would imply that $\angle BAC$ is null, contrary to assumption. Then by A_5(i) $|\angle BAC|^\circ > 0$.

Now choose the point $E \neq A$ so that $A \in [B, E]$. Then by A_5(iii), as we have supplementary angles,

$$|\angle BAC|^\circ + |\angle CAE|^\circ = 180.$$

But $[A, E \neq [A, C$ as $\angle BAC$ is a wedge-angle, so $\angle CAE$ is not a null-angle. By the last paragraph we then have $|\angle CAE|^\circ > 0$ and it follows that $|\angle BAC|^\circ < 180$.

(iii) As $|\underline{BAB_1}, |\underline{CAC_1}$ are straight we have

$$|\angle BAC|^\circ + |\angle CAB_1|^\circ \ = \ 180,$$
$$|\angle CAB_1|^\circ + |\angle B_1 AC_1|^\circ \ = \ 180,$$

there being two pairs of supplementary angles. It follows that

$$|\angle BAC|^\circ + |\angle CAB_1|^\circ = |\angle CAB_1|^\circ + |\angle B_1 AC_1|^\circ,$$

from which we conclude by subtraction that $|\angle BAC|^\circ = |\angle B_1 AC_1|^\circ$.

Degree-measure has the further properties:-

(i) *If $\angle BAC$ is a wedge-angle and $D \neq A$ is in $\mathcal{IR}(|\underline{BAC})$, then $|\angle BAD|^\circ \leq |\angle BAC|^\circ$. If, further, $D \notin [A, C$ then $|\angle BAD|^\circ < |\angle BAC|^\circ$.*

(ii) *For non-collinear points A, B, C let $\mathcal{H}_1$ be the closed half-plane with edge AB in which C lies. If $D \neq A$ is in $\mathcal{H}_1$ and $|\angle BAD|^\circ \leq |\angle BAC|^\circ$, then $D \in \mathcal{IR}(|\underline{BAC})$.*

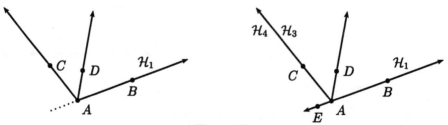

Figure 3.9.

Proof.

(i) As $D \in \mathcal{IR}(|\underline{BAC})$, by A_5(iii) $|\angle BAD|^\circ + |\angle DAC|^\circ = |\angle BAC|^\circ$. By A_5(i), $|\angle DAC|^\circ \geq 0$ so $|\angle BAD|^\circ \leq |\angle BAC|^\circ$.

If $D \notin [A, C$ then $\angle DAC$ is not a null-angle, so $|\angle DAC|^\circ > 0$ and hence $|\angle BAD|^\circ < |\angle BAC|^\circ$.

(ii) Let $E \neq A$ be such that $A \in [B, E]$. Let $\mathcal{H}_3, \mathcal{H}_4$ be the closed half-planes with common edge AC, with $B \in \mathcal{H}_3$ and $E \in \mathcal{H}_4$. Then

$$\begin{aligned} \mathcal{H}_1 &= \mathcal{H}_1 \cap \Pi = \mathcal{H}_1 \cap (\mathcal{H}_3 \cup \mathcal{H}_4) = (\mathcal{H}_1 \cap \mathcal{H}_3) \cup (\mathcal{H}_1 \cap \mathcal{H}_4) \\ &= \mathcal{IR}(|\underline{BAC}) \cup \mathcal{IR}(|\underline{EAC}). \end{aligned}$$

As $D \in \mathcal{H}_1$, then either $D \in \mathcal{IR}(|\underline{BAC})$ or $D \in \mathcal{IR}(|\underline{EAC})$.

Now suppose that $D \notin \mathcal{IR}(|\underline{BAC})$, so that $D \in \mathcal{IR}(|\underline{EAC})$ and $D \notin [A, C$. By A_5(iii),

$$|\angle EAD|° + |\angle DAC|° = |\angle EAC|°.$$

Hence by A_5(iii), as we have supplementary pairs of angles,

$$180 - |\angle BAD|° + |\angle DAC|° = 180 - |\angle BAC|°.$$

From this

$$|\angle BAC|° + |\angle DAC|° = |\angle BAD|°,$$

and as $|\angle DAC|° > 0$, we have $|\angle BAC|° < |\angle BAD|°$. This gives a contradiction with our hypothesis.

3.6 MID-LINE OF AN ANGLE-SUPPORT

3.6.1 Right-angles

Definition. Given any point $P \neq A$ of a line AB, by A_5(v) there is a half-line $[P, Q$ such that $|\angle APQ|° = 90$. Then $\angle APQ$ is called a **right-angle**. If $R \neq P$ is such that $P \in [A, R]$ then $\angle RPQ$ is also a right-angle. For $|\underline{APR}$ is a straight angle-support, so having supplementary angles,

$$|\angle APQ|° + |\angle QPR|° = 180.$$

As $|\angle APQ|° = 90$ it follows that $|\angle RPQ|° = 180 - 90 = 90$.

3.6.2 Perpendicular lines

Definition. If l, m are lines in Λ, we say that l is **perpendicular** m, written $l \perp m$, if l meets m at some point P and if $A \neq P$ is on l, and $Q \neq P$ is on m, then $\angle APQ$ is a right-angle.

COMMENT. In 3.6.1, we say that a **perpendicular PQ has been erected** to the line AB at the point P on it.

Perpendicularity has the following properties:-

(i) *If $l \perp m$, then $m \perp l$.*

(ii) *If $l \perp m$, then $l \neq m$ and $l \cap m \neq \emptyset$.*

Proof.
These follow immediately from the definition of perpendicularity.

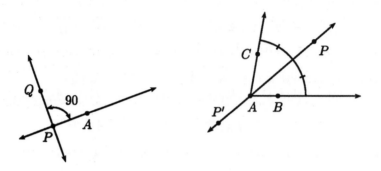

Figure 3.10. Perpendicular lines.   Mid-line of an angle-support.

3.6.3 Mid-lines

Given any angle-support $|\underline{BAC}$ such that $C \notin [A, B$, there is a unique line l such that $A \in l$ and for all $A \ne P \in l$, $|\angle BAP|° = |\angle PAC|°$.

Proof.

Existence.

This was already shown in 3.6.1 in the case when $|\underline{BAC}$ is straight, so we may assume that A, B, C are non-collinear.

By $A_5(v)$ and 3.5.2, as $0 < |\angle BAC|° < 180$ and so $0 < \frac{1}{2}|\angle BAC|° < 90$, there is a half-line $[A, P$ with P on the same side of AB as C is, such that $|\angle BAP|° = \frac{1}{2}|\angle BAC|°$. Then $[A, P \subset \mathcal{IR}(|\underline{BAC})$ by 3.5.2, so by $A_5(iii)$

$$|\angle BAP|° + |\angle PAC|° = |\angle BAC|°.$$

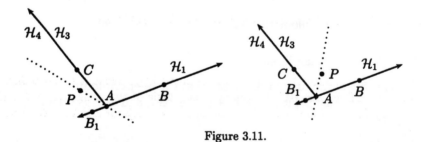

Figure 3.11.

It follows that

$$|\angle PAC|° = |\angle BAC|° - \frac{1}{2}|\angle BAC|° = \frac{1}{2}|\angle BAC|°$$

and so $|\angle BAP|° = |\angle PAC|°$.

If $P' \ne A$ is such that $A \in [P, P']$, then by $A_5(iii)$

$$|\angle BAP'|° = 180 - |\angle BAP|° = 180 - |\angle PAC|° = |\angle P'AC|°.$$

Uniqueness.

When $|\underline{BAC}$ is straight, by A_5(iii) $2|\angle BAP|^\circ = 180$ so $|\angle BAP|^\circ = 90$. By A_5(iv) this determines l uniquely. For the remainder we suppose then that we have a wedge-angle $\angle BAC$.

Let $\mathcal{H}_1, \mathcal{H}_2$ be the closed half-planes with common edge AB, with $C \in \mathcal{H}_1$, and $\mathcal{H}_3, \mathcal{H}_4$ be the closed half-planes with common edge AC, with $B \in \mathcal{H}_3$. Let $B_1 \neq A$ be such that $A \in [B, B_1]$.

Now if l contains a point $Q \neq A$ in $\mathcal{H}_4$ it will also contain a point $R \neq A$ of $\mathcal{H}_3$, so we may assume that l contains a point $P \neq A$ of $\mathcal{H}_3$. As

$$\begin{aligned} \mathcal{H}_1 &= \mathcal{H}_1 \cap \Pi = \mathcal{H}_1 \cap (\mathcal{H}_3 \cup \mathcal{H}_4) = (\mathcal{H}_1 \cap \mathcal{H}_3) \cup (\mathcal{H}_1 \cap \mathcal{H}_4) \\ &= \mathcal{IR}(|\underline{BAC}) \cup \mathcal{IR}(|\underline{B_1 AC}), \end{aligned}$$

we then have $P \in \mathcal{IR}(|\underline{BAC})$ or $P \in \mathcal{IR}(|\underline{B_1 AC})$.

We get a contradiction if l is either AB or AC. For if $l = AB$, then we have $|\angle BAP|^\circ = 0$, $|\angle PAC|^\circ > 0$. Similarly if $l = AC$.

We also get a contradiction if l contains a point $P \neq A$ in $\mathcal{IR}(|\underline{B_1 AC})$ which is not on AC. For then by 3.5.2 $|\angle B_1 AP|^\circ < |\angle B_1 AC|^\circ$, so that $180 - |\angle BAP|^\circ < 180 - |\angle BAC|^\circ$ and so $|\angle BAC|^\circ < |\angle BAP|^\circ$. It follows from 3.5.2 that $[A, C \subset \mathcal{IR}(|\underline{BAP})$ and so $|\angle BAC|^\circ + |\angle CAP|^\circ = |\angle BAP|^\circ$. But $|\angle BAC|^\circ > 0$ and so $|\angle CAP|^\circ < |\angle BAP|^\circ$, which gives a contradiction.

Thus l must contain a point $P \neq A$ in $\mathcal{IR}(|\underline{BAC})$. As then $|\angle BAP|^\circ + |\angle PAC|^\circ = |\angle BAC|^\circ$ and $|\angle BAP|^\circ = |\angle PAC|^\circ$, we must have $|\angle BAP|^\circ = \frac{1}{2}|\angle BAC|^\circ$ which determines $[A, P$ uniquely.

Definition. We define the **mid-line** or **bisector** of the angle-support $|\underline{BAC}$ as follows:- if $C \in [A, B$ then it is the line AB, and otherwise it is the unique line l just noted. We use the notation $\mathrm{ml}(|\underline{BAC})$for this.

3.7 DEGREE-MEASURE OF REFLEX ANGLES

3.7.1

Definition. Let α be a reflex angle with support $|\underline{BAC}$. We first suppose that $C \notin AB$, and as in 3.5.2 let $\angle B_1 AC_1$ be the opposite angle of the wedge-angle $\angle BAC$. Then $B_1 \notin AC$, $C_1 \notin AB$ and $\angle B_1 AC$ is the opposite angle for $\angle BAC_1$. By 3.5.2 we note that

$$180 + |\angle B_1 AC|^\circ = |\angle BAC_1|^\circ + 180,$$

and we define the degree-measure of α to be the common value of these:

$$|\alpha|^\circ = 180 + |\angle B_1 AC|^\circ = |\angle BAC_1|^\circ + 180.$$

Secondly, if $C \in [A, B$ so that α is a full-angle, we define $|\alpha|^\circ = 360$.

Then for each reflex-angle α, $|\alpha|^\circ$ is defined; by 3.5.2 it satisfies $180 < |\alpha|^\circ < 360$ unless α is a full-angle in which case $|\alpha|^\circ = 360$.

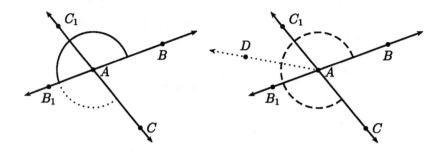

Figure 3.12. Measure of a reflex angle.

Let α be a non-full reflex-angle with support $|\underline{BAC}$ and take $B_1 \neq A$, $C_1 \neq A$ so that $A \in [B, B_1]$, $A \in [C, C_1]$. Let $[A, D \subset \mathcal{IR}(|\underline{B_1 AC_1})$ but $D \notin [A, C_1$, $D \notin [A, B_1$. Then

$$|\angle BAD|° + |\angle DAC|° = |\alpha|°.$$

Proof. As $[A, D \subset \mathcal{IR}(|\underline{B_1 AC_1})$, D is in the closed half-plane with edge AB in which C_1 lies, and also in the closed half-plane with edge AC in which B_1 lies. By 3.5.2, $|\angle B_1 AD|° < |\angle B_1 AC_1|°$ so by A_5(iii) $|\angle BAC_1|° < |\angle BAD|°$. By 3.5.2 then $[A, C_1 \subset \mathcal{IR}(|\underline{BAD})$, and by similar reasoning $[A, B_1 \subset \mathcal{IR}(|\underline{DAC})$. Then

$$
\begin{aligned}
|\angle BAD|° + |\angle DAC|° &= |\angle BAD|° + (|\angle DAB_1|° + |\angle B_1 AC|°) \\
&= (|\angle BAD|° + |\angle DAB_1|°) + |\angle B_1 AC|° \\
&= 180 + |\angle B_1 AC|° = |\alpha|°.
\end{aligned}
$$

COMMENT. We could use this last result to employ the measures of reflex-angles to a significant extent, but in fact do not do so until our full treatment of them in Chapter 9.

Exercises

3.1 If $B \in [A, C]$, then $|A, B| \leq |A, P| \leq |A, C|$ for all $P \in [B, C]$.

3.2 Let A, B, C be points of a line l, and $M = \text{mp}(A, B)$. If C is A or B, or if $C \in l \setminus [A, B]$, then $|C, A| + |C, B| = 2|C, M|$.

3.3 Let A, B, C be distinct points and $D = \text{mp}(B, C)$, $E = \text{mp}(C, A)$, $F = \text{mp}(A, B)$. Prove that D, E, F are distinct. If $A \notin BC$, show that neither E nor F belongs to BC.

3.4 If $A \neq B$, show that

$$\{P \in AB : |B, A| + |A, P| = |B, P|\}$$

is the half-line of AB with initial-point A which does not contain B, while

$$[A, B = \{P \in AB : |A, P| + |P, B| = |A, B| \text{ or } |A, B| + |B, P| = |A, P|\}.$$

3.5 Find analogues of 3.3.1 when $A \in [B, C]$ and when $B \in [C, A]$.

3.6 Show that if A, B, C, D are distinct collinear points such that $C \in [A, B]$, $B \in [A, D]$, and

$$\frac{|A, C|}{|C, B|} = \frac{|A, D|}{|D, B|},$$

then

$$\frac{1}{|A, C|} + \frac{1}{|A, D|} = \frac{2}{|A, B|}.$$

3.7 Show that if A, B, C are non-collinear points, and $P \neq A$ is a point of $\mathcal{IR}(|\underline{BAC})$, then

$$\mathcal{IR}(|\underline{BAP}) \cup \mathcal{IR}(|\underline{PAC}) = \mathcal{IR}(|\underline{BAC}), \ \mathcal{IR}(|\underline{BAP}) \cap \mathcal{IR}(|\underline{PAC}) = [A, P \,.$$

3.8 Show that if d is any positive real number and $|\ \ |$ is a distance function, then $d|\ \ |$ is also a distance function.

3.9 If α is the reflex angle with support $|\underline{{}^{!}AC}$ and β is the reflex angle with support $|\underline{BAF}$, show that if $[A, F \subset \mathcal{IR}(|\underline{BA}\ ?)$ then

$$|\alpha|^{\circ} + |\angle CAF|^{\circ} = |\beta|^{\circ}.$$

3.10 Prove that if $l = \text{ml}(|\underline{BAC})$ and $m = \text{ml}(|\underline{BAC_1})$ where $A \in [C, C_1]$, then $l \perp m$.

3.11 Suppose that B, C, B_1 and C_1 are points distinct from A and that $A \in [B, B_1]$, $A \in [C, C_1]$. Show that then $\text{ml}(|\underline{B_1 AC_1}) = \text{ml}(|\underline{BAC})$.

4

Congruence of triangles; parallel lines

COMMENT. In this chapter we deal with the notion of congruence of triangles, and make a start on the concept of parallelism of lines. As we have distance and angle-measure, we do not need special concepts of congruence of segments and congruence of angles, and we are able to *define* congruence of triangles instead of having it as a primitive term as in the traditional treatment. As a consequence there is a great gain in effectiveness and brevity.

4.1 PRINCIPLES OF CONGRUENCE

4.1.1 Congruence of triangles

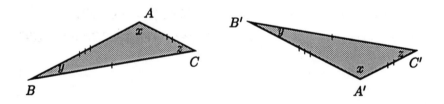

Figure 4.1. Congruent triangles.

Definition. Let T be a triangle with the vertices $\{A, B, C\}$ and T' a triangle with vertices $\{A', B', C'\}$. We say that T is **congruent to T' in the correspondence** $A \to A'$, $B \to B'$, $C \to C'$, if

$$|B, C| = |B', C'|, \quad |C, A| = |C', A'|, \quad |A, B| = |A', B'|,$$
$$|\angle BAC|° = |\angle B'A'C'|°, \quad |\angle CBA|° = |\angle C'B'A'|°, \quad |\angle ACB|° = |\angle A'C'B'|°.$$

We denote this by $T_{(A,B,C)} \equiv_{(A',B',C')} T'$.

47

We say that T is **congruent** to T', written $T \equiv T'$, if T is congruent to T' in at least one of the correspondences

$$(A,B,C) \;\rightarrow\; (A',B',C'), \; (A,B,C) \rightarrow (A',C',B'), \; (A,B,C) \rightarrow (B',C',A'),$$
$$(A,B,C) \;\rightarrow\; (B',A',C'), \; (A,B,C) \rightarrow (C',A',B'), \; (A,B,C) \rightarrow (C',B',A').$$

COMMENT. Originally, behind the concept of congruence lay the idea that T can be placed on T', fitting it exactly.

AXIOM A_6. *If triangles T and T', with vertices $\{A,B,C\}$ and $\{A',B',C'\}$, respectively, are such that*

$$|C,A| = |C',A'|, \; |A,B| = |A',B'|, \; |\angle BAC|^\circ = |\angle B'A'C'|^\circ,$$

then $T_{(A,B,C)\overset{\equiv}{\rightarrow}(A',B'C')} T'$. |

COMMENT. This is known as the SAS (side, angle, side) principle of congruence for triangles.

Triangles have the following properties:-

(i) *If in a triangle $[A,B,C]$, $|A,B| = |A,C|$ then $|\angle ABC|^\circ = |\angle ACB|^\circ$.*

(ii) *If in a triangle $[A,B,C]$, $|A,B| = |A,C|$ and D is the mid-point of B and C, then $AD \perp BC$.*

(iii) *If $B \neq C$, D is the mid-point of B and C, and $A \neq D$ is such that $AD \perp BC$, then $|A,B| = |A,C|$.*

(iv) *If $|\underline{BAC}$ is not straight, if $E \in [A,B \,, F \in [A,C$ are such that $|A,E| = |A,F| > 0$ and $G = mp(E,F)$, then $AG = ml(|\underline{BAC})$.*

Proof.

(i) Note that for the triangle T with vertices $\{A,B,C\}$, under the correspondence $(A,B,C) \rightarrow (A,C,B)$,

$$|A,B| = |A,C|, \; |A,C| = |A,B|, \; |\angle BAC|^\circ = |\angle CAB|^\circ,$$

so by the SAS principle $T_{(A,B,C)\overset{\equiv}{\rightarrow}(A,C,B)} T$. In particular $|\angle ABC|^\circ = |\angle ACB|^\circ$.

(ii) Note that if T_1, T_2 are the triangles with vertices $\{A,B,D\}, \{A,C,D\}$, respectively, then

$$|A,B| = |A,C|, \; |B,D| = |C,D|,$$

$$|\angle ABD|^\circ = |\angle ACD|^\circ,$$

so by the SAS principle, $T_1{}_{(A,B,D)\overset{\equiv}{\rightarrow}(A,C,D)} T_2$. In particular $|\angle ADB|^\circ = |\angle ADC|^\circ$. As $D \in [B,C]$, the sum of the degree-measures of these angles is 180 and so they must be right-angles.

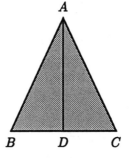

Figure 4.2. An isosceles triangle.

(iii) As $AD \perp BC$ we know that $A \notin BC$. If T_1, T_2 are the triangles with vertices $\{A, B, D\}, \{A, C, D\}$, respectively, then

$$|B, D| = |C, D|, \ |A, D| = |A, D|, \ |\angle BDA|^\circ = |\angle CDA|^\circ,$$

so by the SAS principle, $T_1 \ {}_{(A,B,D) \equiv (A,C,D)} \ T_2$. In particular $|A, B| = |A, C|$.

(iv) As in (ii), the triangles $[A, E, G], [A, F, G]$ are congruent, and so $|\angle EAG|^\circ = |\angle FAG|^\circ$.

Definition. A triangle is said to be **isosceles** if at least two of its sides have equal lengths.

If T, T' are triangles with vertices $\{A, B, C\}, \{A', B', C'\}$, respectively, for which

$$|B, C| = |B', C'|, \ |\angle CBA|^\circ = |\angle C'B'A'|^\circ, \ |\angle BCA|^\circ = |\angle B'C'A'|^\circ,$$

then $T \ {}_{(A,B,C) \equiv (A',B',C')} \ T'$.

Proof. Suppose that $|C', A'| \neq |C, A|$. Choose the point D' on the half-line $[C', A'$ such that $|C', D'| = |C, A|$. Then if T'' is the triangle with vertices $\{B', C', D'\}$, under the correspondence $(B, C, A) \to (B', C', D')$ we have

$$|B, C| = |B', C'|, \ |C, A| = |C', D'|, \ |\angle BCA|^\circ = |\angle B'C'D'|^\circ.$$

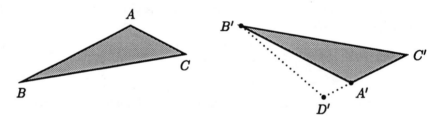

A

B'

C

C'

B

A'

D'

Figure 4.3.

Then by the SAS principle, $T \ {}_{(B,C,A) \equiv (B',C',D')} \ T''$. In particular

$$|\angle C'B'D'|^\circ = |\angle CBA|^\circ = |\angle C'B'A'|^\circ.$$

Then we have different wedge-angles $\angle C'B'A'$, $\angle C'B'D'$, laid off on the same side of $B'C'$ and having the same degree-measure. This gives a contradiction.

Thus $|C', A'| = |C, A|$, and as we also have

$$|C', B'| = |C, B|, \ |\angle B'C'A'|^\circ = |\angle BCA|^\circ,$$

by the SAS principle we have $T \ {}_{(B,C,A) \equiv (B',C',A')} \ T'$.

This is known as the ASA (angle, side, angle) principle of congruence.

If T and T' are triangles with vertices $\{A, B, C\}, \{A', B', C'\}$, respectively, for which

$$|B, C| = |B', C'|, \ |C, A| = |C', A'|, \ |A, B| = |A', B'|,$$

then $T \ {}_{(A,B,C) \equiv (A',B',C')} \ T'$.

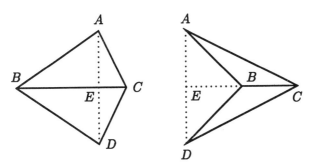

Figure 4.4. The SSS-principle of congruence.

Proof. Choose D on the opposite side of BC from A, so that $|\angle CBD|^\circ = |\angle C'B'A'|^\circ$ and $|B, D| = |B', A'|$. Let T'' be the triangle with vertices $\{B, C, D\}$. Then as $|B, C| = |B', C'|$, by the SAS principle

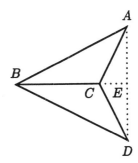

$$T'' \underset{(B,D,C) \overset{\equiv}{\rightarrow} (B',A',C')}{} T'.$$

Now $|B, A| = |B', A'| = |B, D|$ so we have an isosceles triangle and $|\angle BAD|^\circ = |\angle BDA|^\circ$. Similarly $|\angle CAD|^\circ = |\angle CDA|^\circ$.

Note that A and D are on different sides of BC, so a point E of $[A, D]$ is on BC.

CASE 1. Let $E \in [B, C]$. Then $[A, D \subset \mathcal{IR}(|\underline{BAC})$ and $[D, A \in \mathcal{IR}(|\underline{BDC})$. It follows that

$$|\angle BAC|^\circ = |\angle BAD|^\circ + |\angle DAC|^\circ = |\angle BDA|^\circ + |\angle ADC|^\circ = |\angle BDC|^\circ.$$

CASE 2. Let $B \in [E, C]$. Then $[A, B \subset \mathcal{IR}(|\underline{DAC})$ and $[D, B \in \mathcal{IR}(|\underline{ADC})$. It follows that

$$|\angle BAC|^\circ = |\angle DAC|^\circ - |\angle DAB|^\circ = |\angle ADC|^\circ - |\angle ADB|^\circ = |\angle BDC|^\circ.$$

CASE 3. Let $C \in [B, E]$. Then $[A, C \subset \mathcal{IR}(|\underline{BAD})$ and $[D, C \in \mathcal{IR}(|\underline{BDA})$. It follows that

$$|\angle BAC|^\circ = |\angle BAD|^\circ - |\angle DAC|^\circ = |\angle BDA|^\circ - |\angle ADC|^\circ = |\angle BDC|^\circ.$$

Now combining the cases, by the SAS principle, as

$$|A, B| = |D, B|, \ |A, C| = |D, C|, \ |\angle BAC|^\circ = |\angle BDC|^\circ,$$

we have $T \underset{(A,B,C) \overset{\equiv}{\rightarrow} (D,B,C)}{} T''$. But $T'' \underset{(D,B,C) \overset{\equiv}{\rightarrow} (A',B',C')}{} T'$ so $T \underset{(A,B,C) \overset{\equiv}{\rightarrow} (A',B',C')}{} T'$. This is known as the SSS(side, side, side) principle of congruence for triangles.

4.2   ALTERNATE ANGLES, PARALLEL LINES

4.2.1   Alternate angles

Let A, B, C be non-collinear points, and take $D \neq C$ so that $C \in [A, D]$. Then $|\angle BCD|^\circ > |\angle CBA|^\circ$.

Proof. Let $E = mp(B, C)$ and choose F so that $E = mp(A, F)$. Then if T_1, T_2 are the triangles with vertices $\{A, B, E\}$, $\{F, C, E\}$, respectively, by the SAS principle of congruence $T_1 \underset{(A,B,E) \cong (F,C,E)}{\equiv} T_2$. In particular,

$$|\angle EBA|^\circ = |\angle ECF|^\circ, \text{ i.e. } |\angle CBA|^\circ = |\angle BCF|^\circ.$$

But $[C, F \subset \mathcal{IR}(|\underline{BCD})$ as E, and so F, is in the closed half-plane with edge AC in which B lies, and D and F are on the opposite side of BC from A. Also $F \notin AD$ as $F \in AD$ would imply that $E = C$. Then by 3.5.2 $|\angle BCF|^\circ < |\angle BCD|^\circ$.

COROLLARY. In the theorem let $G \neq C$ be such that $C \in [B, G]$. Then $|\angle ACG|^\circ > |\angle ABC|^\circ$.

Proof. This follows immediately as $\angle ACG$ and $\angle BCD$ are opposite angles.

COMMENT. If D and H are on opposite sides of BC, then $\angle CBH$ and $\angle BCD$ are known as *alternate angles* . This last result implies that if alternate angles $\angle CBH$ and $\angle BCD$ are equal in measure, then CD and BH cannot meet at some point A.

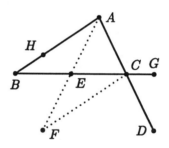

Figure 4.5. Result on alternate angles.

Given any line l and any point $P \notin l$, there is a line m which contains P and is such that $l \cap m = \emptyset$.

Proof. Take any points $A, B \in l$ and lay off an angle $\angle APQ$ on the opposite side of AP from B, so that $|\angle APQ|^\circ = |\angle PAB|^\circ$. Than by the last result the line PQ does not meet l. In this $\angle APQ$ and $\angle PAB$ are alternate angles which are equal in measure.

4.2.2   Parallel lines

Definition. If l and m are lines in Λ, we say that l is **parallel** to m, written $l \parallel m$, if $l = m$ or $l \cap m = \emptyset$.

Parallelism has the following properties:-

(i) $l \parallel l$ for all $l \in \Lambda$;

(ii) If $l \parallel m$ then $m \parallel l$;

(iii) Given any line $l \in \Lambda$ and any point $P \in \Pi$, there is at least one line m which contains P and is such that $l \parallel m$.

(iv) *If the lines l and m are both perpendicular to the line n, then l and m are parallel to each other.*

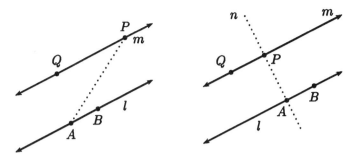

Figure 4.6. Parallel lines.

Proof.
(i) and (ii) follow immediatly from the definition, while (iii) follows from 4.2.1.

(iv) As perpendicular lines form right-angles with each other at some point, l must meet n at some point A, and m must meet n at some point P such that if B is any other point of l and Q is any point of m on the other side of n from B, then $|\angle PAB|^\circ = 90$, $|\angle APQ|^\circ = 90$. Then, as there are alternate angles equal in measure, by 4.2.1 $l \parallel m$.

4.3   PROPERTIES OF TRIANGLES AND HALF-PLANES

4.3.1   Side-angle relationships; the triangle inequality

If A, B, C are non-collinear points and $|A, B| > |B, C|$, then $|\angle ACB|^\circ > |\angle BAC|^\circ$, so that in a triangle a greater angle is opposite a longer side.

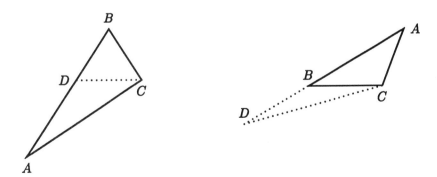

Figure 4.7. Angle opposite longer side.   Figure 4.8. The triangle inequality.

Proof. Choose $D \in [B, A$ so that $|B, D| = |B, C|$. Then $D \in [B, A]$ as $|B, D| < |B, A|$. Now $|\angle ACB|^\circ > |\angle DCB|^\circ$ as $[C, D \subset \mathcal{IR}(|\underline{BCA})$, and $|\angle DCB|^\circ = |\angle BDC|^\circ$ by 4.1.1. But $|\angle BDC|^\circ > |\angle DAC|^\circ$ by 4.2.1, so

$$|\angle ACB|° > |\angle DCB|° = |\angle BDC|° > |\angle DAC|°.$$

Hence $|\angle ACB|° > |\angle DAC|°$ and $\angle DAC = \angle BAC$ as $D \in [B, A]$.

COROLLARY. *If A, B, C are non-collinear points and $|\angle ACB|° > |\angle BAC|°$, then $|A, B| > |B, C|$, so that in a triangle a longer side is opposite a greater angle.*

Proof. For if $|A, B| \le |B, C|$, we have $|\angle ACB|° \le |\angle BAC|°$ by 4.1.1 and this result.

THE TRIANGLE INEQUALITY. *If A, B, C are non-collinear points, then $|C, A| < |A, B| + |B, C|$.*

Proof. Take a point D so that $B \in [A, D]$ and $|B, D| = |B, C|$. As $[C, B \subset \mathcal{IR}(|\underline{ACD})$ we have $|\angle DCA|° > |\angle DCB|°$. But $|\angle DCB|° = |\angle CDB|°$ by 4.1.1, so by our last result, $|A, D| > |A, C|$. However $|A, D| = |A, B| + |B, D|$ as $B \in [A, D]$, and the result follows.

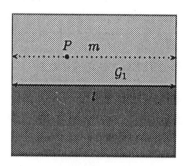

Figure 4.9.

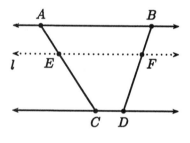

Figure 4.10.

4.3.2 Properties of parallelism

Let $l \in \Lambda$ be a line, G_1 an open half-plane with edge l and P a point of G_1. If m is a line such that $P \in m$ and $l \parallel m$, then $m \subset G_1$.

Proof. As $P \notin l, P \in m$ we have $l \ne m$. Then as $l \parallel m$ we have $l \cap m = \emptyset$. Thus there cannot be a point of m on l. Neither can there be a point Q of m in G_2, the other open half-plane with edge l. For then we would have $[P, Q] \cap l \ne \emptyset$ and so a point R of m would be on l, as $[P, Q] \subset PQ = m$.

Let AB, CD be distinct lines and l distinct from and parallel to both. If l meets $[A, C]$ in a point E, then l meets $[B, D]$ in a point F.

Proof. By the Pasch property applied to $[A, B, C]$ as l does not meet $[A, B]$ it meets $[B, C]$ at some point G. Then by the Pasch property applied to $[B, C, D]$, as l does not meet $[C, D]$ it meets $[B, D]$ in some point F.

4.3.3 Dropping a perpendicular

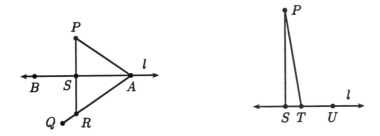

Figure 4.11. Dropping a perpendicular.

Given any line $l \in \Lambda$ and any point $P \notin l$, there is a unique line m such that $P \in m$ and $l \perp m$.

Proof.

Existence. Let A, B be distinct points of l. Take a point Q on the opposite side of l from P and such that $|\angle BAQ|° = |\angle BAP|°$. Also take $R \in [A, Q$ so that $|A, R| = |A, P|$. As P and R are on opposite sides of l, $[P, R]$ meets l in a point S.

We first suppose that $A \notin PR$ so that $A \neq S$. Then $[A, P, S]$ and $[A, R, S]$ are congruent by the SAS-principle, so in particular $|\angle ASP|° = |\angle ASR|°$. As $S \in [P, R]$ it follows that these are right-angles and so $PR \perp l$.

In the second case suppose that $A \in PR$ so that $A = S$. Then $S \in [P, R]$ and by construction $|\angle BSR|° = |\angle BSP|°$. Again these are right-angles so $PR \perp l$.

Uniqueness. Suppose that there are distinct points $S, T \in l$ such that $PS \perp l$, $PT \perp l$. Choose $U \neq T$ so that $T \in [S, U]$. Then $|\angle UTP|° = |\angle USP|° = 90$ and this contradicts 4.2.1.

COMMENT. We refer to this last as **dropping a perpendicular** from P to l.

Let A, B, C be non-collinear points such that $AB \perp AC$ and let D be the foot of the perpendicular from A to BC. Then $D \in [B, C]$, $D \neq B$, $D \neq C$.

Proof.   By 4.2.1, in a right-angled triangle each of the other two angles have degree-measure less than 90. By 4.3.1 it then follows that the side opposite the right-angle is longer than each of the other sides.   It follows that $|B, D| < |A, B| < |B, C|$. By a similar argument $|C, D| < |B, C|$.

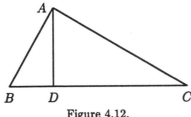

Figure 4.12.

We cannot then have $B \in [C, D]$ as that would imply $|C, B| \leq |C, D|$, and similarly we cannot have $C \in [B, D]$ with as that would imply $|B, C| \leq |B, D|$. Hence $D \in [B, C]$, $D \neq B$, $D \neq C$.

4.3.4   Projection and axial symmetry

Definition. For any line $l \in \Lambda$ we define a function $\pi_l : \Pi \to l$ by specifying that for all $P \in \Pi$, $\pi_l(P)$ is the foot of the perpendicular from P to l. We refer to π_l as **projection to the line** l.

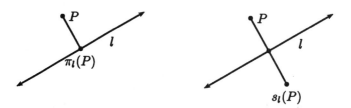

Figure 4.13. Projection to the line l.   Axial symmetry in the line l.

Definition. For any line $l \in \Lambda$ we define a function $s_l : \Pi \to \Pi$ by specifying that for all $P \in \Pi$, $s_l(P)$ is the point Q such that

$$\pi_l(P) = mp(P,Q).$$

We refer to s_l as **axial symmetry in the line** l.

Let $\mathcal{H}_1, \mathcal{H}_2$ be closed half-planes with common edge l, let $P_1 \in \mathcal{H}_1 \setminus l$ and $P_2 = s_l(P_1)$. Then, for all $P \in \mathcal{H}_1$, $|P, P_1| \leq |P, P_2|$.

Proof. If $P \in l$, then $|P, P_1| = |P, P_2|$, by 4.1.1 when $P \notin P_1 P_2$, and as $P = mp(P_1, P_2)$ otherwise.

When $P \in \mathcal{G}_1 = \mathcal{H}_1 \setminus l$ we suppose first that $P \notin P_1 P_2$. Then $[P, P_2]$ meets l in a point Q and we have

$$|P, P_2| = |P, Q| + |Q, P_2| = |P, Q| + |Q, P_1|.$$

Now we cannot have $Q \in [P, P_1]$ as $[P, P_1] \subset \mathcal{G}_1$ and $Q \in l$. Thus either $Q \notin PP_1$ or $Q \in PP_1 \setminus [P, P_1]$. We then have $|P, Q| + |Q, P_1| > |P, P_1|$ by 4.3.1 and 3.1.2. For the case when $P \in P_1 P_2$, we denote by R the point of intersection of $P_1 P_2$ and l, so that $R = mp(P_1, P_2)$.

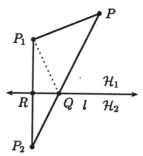

Figure 4.14. Distance and half-planes.

Now $P \in [R, P_1]$ so either $P \in [R, P_1]$ or $P_1 \in [R, P]$. In the first of these cases we have

$$|P_1, P| < |P_1, R| = |R, P_2| < |P, P_2|,$$

as $R \in [P, P_2]$. In the second case we have $|P, P_1| < |P, R| < |P, P_2|$ as $R \in [P, P_2]$.

Exercises

4.1 If $D \neq A$ is in $[A, B, C]$ but not in $[B, C]$, then $|B, D| + |D, C| < |B, A| + |A, C|$ and $|\angle BDC|^\circ > |\angle BAC|^\circ$.

4.2 There is an AAS-principle of congruence that if

$$|\angle BAC|^\circ = |\angle EDF|^\circ, \ |\angle CBA|^\circ = |\angle FED|^\circ,$$
$$|B, C| = |E, F|,$$

then the triangles $[A, B, C], [D, E, F]$ are congruent. [Hint. Suppose that $|\angle BCA|^\circ < |\angle EFD|^\circ$; lay off an angle $\angle BCG$ equal in magnitude to $\angle EFD$ and with G on the same side of BC as A is; then $[C, G$ meets $[A, B]$ at a point H; also $[H, B, C] \equiv [D, E, F]$ and in particular $|\angle BHC|^\circ = |\angle EDF|^\circ = |\angle BAC|^\circ$; deduce a contradiction and then apply the ASA-principle.]

4.3 There is an ASS-principle of congruence for right-angled triangles, that if $BC \perp BA$, $EF \perp ED$, $|C, A| = |F, D|$, $|A, B| = |D, E|$, then $[A, B, C] \equiv [D, E, F]$. [Hint. Take C' so that $E \in [F, C']$ and $|E, C'| = |B, C|$.]

4.4 If $P \in \mathrm{ml}(|\underline{BAC})$ and $Q = \pi_{AB}(P)$, $R = \pi_{AC}(P)$, then $|P, Q| = |P, R|$. Conversely, if $P \in \mathcal{IR}(|\underline{BAC})$ and $|P, Q| = |P, R|$ where $Q = \pi_{AB}(P)$, $R = \pi_{AC}(P)$, then $P \in \mathrm{ml}(|\underline{BAC})$.

4.5 In triangles $[A, B, C], [D, E, F]$ let

$$|A, B| = |D, E|, \ |A, C| = |D, F|, \ |\angle BAC|^\circ > |\angle EDF|^\circ.$$

Then $|B, C| > |E, F|$. [Hint. Lay off the angle $\angle BAG$ with $|\angle BAG|^\circ = |\angle EDF|^\circ$ and with G on the same side of AB as C is. If $G \in BC$ proceed; if $G \notin BC$, let $K = \mathrm{mp}(G, C)$ and show that $[A, K$ meets $[B, C]$ in a point H.]

4.6 If $AB \parallel AC$, then $AB = AC$.

4.7 Let $\mathcal{H}_1$ be a closed half-plane with edge l, let $P \in \mathcal{H}_1$ and let $O = \pi_l(P)$. Then if m is any line in Λ such that $O \in m$, we must have $\pi_m(P) \in \mathcal{H}_1$.

5

The parallel axiom; Euclidean geometry

COMMENT. The effect of introducing any axiom is to narrow things down, and depending on the final axiom still to be taken, we can obtain two quite distinct well-known types of geometry. By introducing our final axiom, we confine ourselves to the familiar school geometry, which is known as **Euclidean geometry**.

5.1 THE PARALLEL AXIOM

5.1.1 Uniqueness of a parallel line

We saw in 4.2 that given any line l and any point $P \notin l$ there is at least one line m such that $P \in m$ and $l \parallel m$. We now assume that there is only one such line ever.

AXIOM A_7. *Given any line $l \in \Lambda$ and any point $P \notin l$, there is at most one line m such that $P \in m$ and $l \parallel m$.* |

COMMENT. By 4.2 and A_7, given any line $l \in \Lambda$ and any point $P \in \Pi$, there is a unique line m through P which is parallel to l.

Let $l \in \Lambda$, $P \in \Pi$ and $n \in \Lambda$ be such that $l \neq n$, $P \in n$ and $l \parallel n$. Let A and B be any distinct points of l and R a point of n such that R and B are on opposite sides of AP. Then $|\angle APR|° = |\angle PAB|°$, so that for parallel lines alternate angles must have equal degree-measures.
    Proof. Let m be the line PQ in 4.2.1 such that $|\angle APQ|° = |\angle BAP|°$. Then $l \parallel m$. As m and n both contain P and l is parallel to both of them, by A_7 we have $m = n$, so that $R \in [P, Q$ and so $|\angle APR|° = |\angle APQ|°$. Thus $|\angle APR|° = |\angle APQ|° = |\angle PAB|°$.
    Let l, n be distinct parallel lines, A, $B \in l$ and P, $T \in n$ be such that B and T are on the one side of AP, and $S \neq P$ be such that $P \in [A, S]$. Then the angles $\angle BAP$, $\angle TPS$ have equal degree-measures.
    Proof. Choose $R \neq P$ so that $P \in [T, R]$. Then $R \in n$ and B and R are on opposite sides of AP, so that $\angle BAP$, $\angle APR$ are alternate angles and so have

equal degree-measures. But $\angle APR$ and $\angle TPS$ are opposite angles and so have equal degree-measures. Hence $|\angle BAP|° = |\angle TPS|°$.

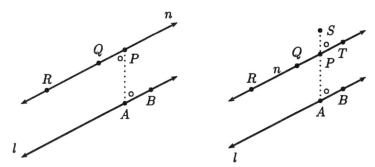

Figure 5.1. Alternate angles.   Corresponding angles.

We call such angles $\angle BAP$, $\angle TPS$ **corresponding angles for a transversal**.

If lines l, m, n are such that $l \parallel m$ and $m \parallel n$, then $l \parallel n$.

Proof. If $l = n$, the result is trivial as $l \parallel l$, so suppose $l \neq n$. If l is not parallel to n, then l and n will meet at some point P, and then we will have distinct lines l and n, both containing P and both parallel to m, which gives a contradiction by A_7. Thus parallelism is a transitive relation. Combined with the properties in 4.2.2 this makes it an equivalence relation.

If lines are such that $l \perp n$ and $l \parallel m$, then $m \perp n$.

Proof. As l is perpendicular to n they must meet at some point A. As $l \parallel m$, we cannot have $m \parallel n$, as by transitivity that would imply $l \parallel n$. Thus m meets n in some point P, and if we choose B on l, Q on m on opposite sides of n, then we have $|\angle APQ|° = |\angle PAB|°$ as these are alternate angles for parallel lines. Hence $|\angle APQ|° = 90$ and $m \perp n$.

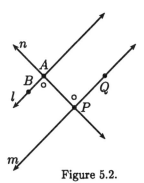

Figure 5.2.

5.2 PARALLELOGRAMS

5.2.1 Parallelograms and rectangles

Definition. Let points A, B, C, D be such that no three of them are collinear and $AB \parallel CD$, $AD \parallel BC$. Let $\mathcal{H}_1$ be the closed half-plane with edge AB in which C lies; as $CD \parallel AB$ then, by 4.3.2, $D \in \mathcal{H}_1$. Similarly let $\mathcal{H}_3$ be the closed half-plane with edge BC in which A lies; as $AD \parallel BC$, then $D \in \mathcal{H}_3$. Thus $D \in \mathcal{H}_1 \cap \mathcal{H}_3 = \mathcal{IR}(|ABC)$ and so by the cross-bar theorem $[A, C]$ meets $[B, D$ in some point T, which is unique as $AC = BD$ would imply $B \in AC$. Similarly $C \in \mathcal{IR}(|BAD)$ so T is on $[B, D]$. Thus $[A, C] \cap [B, D] \neq \emptyset$ so as in 2.4.4 a convex quadrilateral $[A, B, C, D]$ can be defined, and in this case it is called a **parallelogram**. The terminology of 2.4.4 then applies.

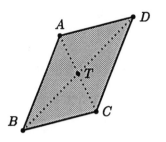 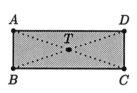

Figure 5.3. A parallelogram.   A rectangle.

Definition If $[A, B, C, D]$ is a parallelogram in which $AB \perp AD$, then, as $AB \parallel CD$, by 5.1.1 we have $AD \perp CD$. Thus if two adjacent side-lines of a parallelogram are perpendicular, each pair of adjacent side-lines are perpendicular; we call such a parallelogram a **rectangle**.

Parallelograms have the following properties:-

(i) *Opposite sides of a parallelogram have equal lengths.*

(ii) *The point of intersection of the diagonals of a parallelogram is the mid-point of each diagonal.*

Proof.

(i) With the notation above for a parallelogram, the triangles with vertices $\{A, B, D\}$ and $\{C, D, B\}$ are congruent in the correspondence $(A, B, D) \to (C, D, B)$ by the ASA principle. First note that $|B, D| = |D, B|$. Secondly note that $AB \parallel CD$ and A and C are on opposite sides of BD so that $\angle ABD$ and $\angle CDB$ are alternate angles, and hence $|\angle ABD|° = |\angle CDB|°$. Finally $AD \parallel BC$, and A and C are on opposite sides of BD, so that $\angle ADB$ and $\angle CBD$ are alternate angles and hence $|\angle ADB|° = |\angle CBD|°$. It follows that $|A, B| = |C, D|$, $|A, D| = |B, C|$.

(ii) Let T be the point of intersection of the diagonals. Then the triangles $[A, B, T]$, $[C, D, T]$ are congruent by the ASA principle, as

$$|A, B| = |C, D|, \ |\angle ABT|° = |\angle CDT|°, \ |\angle BAT|° = |\angle DCT|°.$$

It follows that $|A, T| = |C, T|$, $|B, T| = |D, T|$.

5.2.2   Sum of measures of wedge-angles of a triangle

If A, B, C are non-collinear points, then

$$|\angle CAB|° + |\angle ABC|° + |\angle BCA|° = 180.$$

Thus the sum of the degree-measures of the wedge-angles of a triangle is equal to 180.

Proof. Let l be the line through A which is parallel to BC. If m is the line through B which is parallel to AC, then we cannot have $l \parallel m$ as we would then have $l \parallel m$, $m \parallel AC$ which would imply $l \parallel AC$; we would then have $BC \parallel l$, $l \parallel AC$ and so $BC \parallel AC$; thus as $BC \cap AC \neq \emptyset$ we would have $BC = AC$; this would make A, B, C collinear and contradict our assumption.

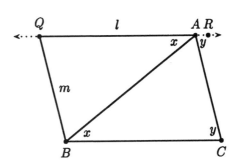

Figure 5.4. Angles of a triangle.

Thus m meets l at some point, Q say. Then $[A, C, B, Q]$ is a parallelogram and $[A, B]$, $[Q, C]$ meet at a point T. Now Q is on the opposite side of AB from C, so that $\angle CBA$ and $\angle BAQ$ are alternate angles and so $|\angle CBA|° = |\angle BAQ|°$. Moreover $[A, B \subset \mathcal{IR}(|CAQ)$ and so $|\angle CAB|° + |\angle BAQ|° = |\angle CAQ|°$.

Choose $R \neq A$ so that $A \in [Q, R]$. Then $R \in l$ and R is on the opposite side of AC from Q. But $BQ \parallel AC$ so B and Q are on the same side of AC, and hence B and R are on opposite sides of AC. Then $\angle BCA$ and $\angle CAR$ are alternate angles, so $|\angle BCA|° = |\angle CAR|°$. Thus

$$
\begin{aligned}
(|\angle CAB|° + |\angle CBA|°) + |\angle BCA|° &= (|\angle CAB|° + |\angle BAQ|°) + |\angle BCA|° \\
&= |\angle CAQ|° + |\angle CAR|° \\
&= 180.
\end{aligned}
$$

COROLLARY. *If the points A, B, C are non-collinear, and $D \neq C$ is chosen so that $C \in [B, D]$, then $|\angle ACD|° = |\angle BAC|° + |\angle CBA|°$. Thus the degree-measure of an exterior wedge-angle of a triangle is equal to the sum of the degree-measures of the two remote wedge-angles of the triangle.*

Proof. For each of these is equal to $180 - |\angle ACB|°$, as $C \in [B, D]$.

5.3   RATIO RESULTS FOR TRIANGLES

5.3.1   Lines parallel to one side-line of a triangle

Let A, B, C be non-collinear points, and with $l = AB$, $m = AC$, let $\leq_l$, $\leq_m$ be natural orders such that $A \leq_l B$, $A \leq_m C$. Let D_1, D_2, D_3 be points of AB such that $A \leq_l D_1 \leq_l D_2 \leq_l D_3 \leq_l B$ and $|D_1, D_2| = |D_2, D_3|$, so that D_2 is the mid-point of D_1 and D_3. Then the lines through D_1, D_2 and D_3 which are all parallel to BC, will meet AC in points E_1, E_2, E_3, respectively, such that $A \leq_m E_1 \leq_m E_2 \leq_m E_3 \leq_m C$ and $|E_1, E_2| = |E_2, E_3|$.

Proof. By Pasch's property in 2.4.3 applied to the triangle $[A, B, C]$, the lines through $[D_1, D_2, D_3]$ which are parallel to BC will meet $[A, C]$ in points E_1, E_2, E_3, respectively. By Pasch's property applied to $[A, D_3, E_3]$, since $D_2 \in [A, D_3]$ and the lines through D_2 and D_3 parallel to BC are parallel to each other, $E_2 \in [A, E_3]$. By Pasch's property applied to $[A, D_2, E_2]$, since $D_1 \in [A, D_2]$ and the lines through D_1 and D_2 parallel to BC are parallel to each other, $E_1 \in [A, E_2]$. It remains to show that E_2 is equidistant from E_1 and E_3.

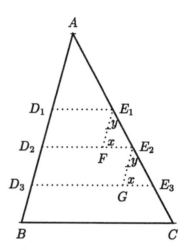

Figure 5.5. Transversals to parallel lines.

By Pasch's property applied to $[A, D_2, E_2]$, since $E_1 \in [A, E_2]$ the line through E_1 which is parallel to $AB = AD_2$ will meet $[D_2, E_2]$ in a point F. By Pasch's property applied to $[A, D_3, E_3]$, since $E_2 \in [A, E_3]$ the line through E_2 which is parallel to $AB = AD_3$ will meet $[D_3, E_3]$ in a point G.

Let T_1, T_2 be the triangles with vertices $\{E_1, F, E_2\}, \{E_2, G, E_3\}$, respectively. Our objective is to show that

$$T_1 \underset{(E_1,F,E_2) \rightrightarrows (E_2,G,E_3)}{} T_2.$$

Now $D_1 E_1 \parallel D_2 F$, $D_1 D_2 \parallel E_1 F$, so $[D_1, D_2, F, E_1]$ is a parallelogram, and so by 5.2.1 $|D_1, D_2| = |E_1, F|$. Similarly $D_2 E_2 \parallel D_3 G$, $D_2 D_3 \parallel E_2 G$ so $[D_2, D_3, G, E_2]$ is a parallelogram, and so $|D_2, D_3| = |E_2, G|$. But $|D_1, D_2| = |D_2, D_3|$ and hence $|E_1, F| = |E_2, G|$.

Let $\mathcal{H}_1$ be the closed half-plane with edge AC in which B lies. Then $[A, B] \subset \mathcal{H}_1$, so $D_2, D_3 \in \mathcal{H}_1$. Then $[D_2, E_2], [D_3, E_3] \subset \mathcal{H}_1$, so $F, G \in \mathcal{H}_1$. Then F and G are on the one side of the line AC, and as $D_2 E_2 \parallel D_3 E_3$ and $E_2 \in [E_3, E_1]$, the angles $\angle FE_2 E_1, \angle GE_3 E_2$ are corresponding angles for parallel lines and so have equal degree-measures. Thus $|\angle FE_2 E_1|^\circ = |\angle GE_3 E_2|^\circ$.

By transitivity $E_1 F \parallel E_2 G$ as both are parallel to AB, F and G are on the one side of AC, and $E_2 \in [E_1, E_3]$, so the angles $\angle FE_1 E_2$ and $\angle GE_2 E_3$ are corresponding angles for parallel lines and so have equal degree-measures. Thus $|\angle FE_1 E_2|^\circ = |\angle GE_2 E_3|^\circ$.

As

$$|\angle FE_2 E_1|^\circ = |\angle GE_3 E_2|^\circ, \ |\angle FE_1 E_2|^\circ = |\angle GE_2 E_3|^\circ,$$

by 5.2.2 $|\angle E_1 FE_2|^\circ = |\angle E_2 GE_3|^\circ$. Thus

$$|E_1, F| = |E_2, G|, \ |\angle FE_1 E_2|^\circ = |\angle GE_2 E_3|^\circ, \ |\angle E_1 FE_2|^\circ = |\angle E_2 GE_3|^\circ,$$

so by the ASA principle, the triangles T_1, T_2 are congruent in the correspondence $(E_1, F, E_2) \to (E_2, G, E_3)$. It follows that $|E_1, E_2| = |E_2, E_3|$.

Let A, B, C be non-collinear points and let $P \in [A, B$ and $Q \in [A, C$ be such that $PQ \parallel BC$. Then

$$\frac{|A, P|}{|A, B|} = \frac{|A, Q|}{|A, C|}.$$

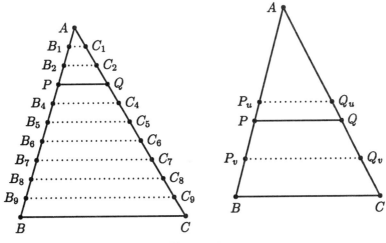

Figure 5.6.

Proof. We assume first that $P \in [A, B]$. Within this first case, we suppose initially that

$$\frac{|A, P|}{|A, B|} = \frac{s}{t},$$

where s and t are positive whole numbers with $s < t$, so that s/t is an arbitrary rational number between 0 and 1. For $0 \le j \le t$ let B_j be the point on $[A, B$ such that

$$\frac{|A, B_j|}{|A, B|} = \frac{j}{t},$$

so that $B_0 = A$, $B_t = B$ and $B_s = P$. If $AB = l$ and $\le_l$ is the natural order for which $A \le_l B$, then $A \le_l B_{j-1} \le_l B_j \le_l B_{j+1} \le_l B$ and $|B_{j-1}, B_j| = |B_j, B_{j+1}|$. If $AC = m$ and $\le_m$ is the natural order for which $A \le_m C$, then by the last result applied with $(D_1, D_2, D_3) = (B_{j-1}, B_j, B_{j+1})$, for $1 \le j \le t - 1$ the line through B_j which is parallel to BC will meet AC in a point C_j such that $A \le_m C_{j-1} \le_m C_j \le_m C_{j+1} \le_m C$ and $|C_{j-1}, C_j| = |C_j, C_{j+1}|$.

It follows that, for $0 \le j \le t$, $|A, C_j| = j|A, C_1|$ and so as $C_t = C$,

$$\frac{|A, C_j|}{|A, C|} = \frac{j|A, C_1|}{t|A, C_1|} = \frac{j}{t}.$$

In particular, as $C_s = Q$, it follows that

$$\frac{|A, Q|}{|A, C|} = \frac{s}{t} = \frac{|A, P|}{|A, B|}.$$

Still within the first case, now suppose that

$$\frac{|A,P|}{|A,B|} = x, \quad \frac{|A,Q|}{|A,C|} = y,$$

where x is an irrational number with $0 < x < 1$. If u is any positive rational number less than x, and P_u is a point chosen on $[A,B]$ so that

$$\frac{|A,P_u|}{|A,B|} = u,$$

then the line through P_u which is parallel to BC will meet $[A,C]$ in a point Q_u such that

$$\frac{|A,Q_u|}{|A,C|} = u.$$

Similarly if v is any rational number such that $x < v < 1$, and P_v is a point chosen on $[A,B]$ so that

$$\frac{|A,P_v|}{|A,B|} = v,$$

then the line through P_v which is parallel to BC will meet $[A,C]$ in a point Q_v such that

$$\frac{|A,Q_v|}{|A,C|} = v.$$

As $|A,P_u| < |A,P| < |A,P_v|$ we have $P \in [P_u, P_v]$. It follows by 4.3.2 that $Q \in [Q_u, Q_v]$ and so $u < y < v$. Thus for all rational u and v such that $u < x < v$ we have $u < y < v$. It follows that $x = y$.

This completes the first case. For the second case note that if $P \notin [A,B]$ we have $B \in [A,P]$. Then by the first case

$$\frac{|A,B|}{|A,P|} = \frac{|A,C|}{|A,Q|},$$

so the reciprocals of these are equal.

5.3.2   Similar triangles

Let A,B,C and A',B',C' be two sets of non-collinear points such that

$$|\angle BAC|^\circ = |\angle B'A'C'|^\circ, \; |\angle CBA|^\circ = |\angle C'B'A'|^\circ, \; |\angle ACB|^\circ = |\angle A'C'B'|^\circ.$$

Then

$$\frac{|B',C'|}{|B,C|} = \frac{|C',A'|}{|C,A|} = \frac{|A',B'|}{|A,B|}.$$

Thus if the degree-measures of the angles of one triangle are equal, respectively, to the degree-measures of the angles of a second triangle, then the ratios of the lengths of corresponding sides of the two triangles are equal.

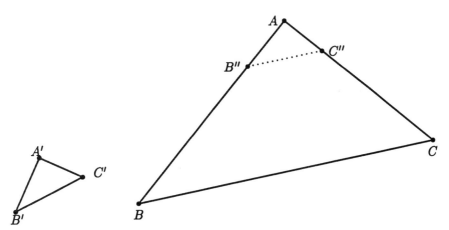

Figure 5.7. Similar triangles.

Proof. Choose $B'' \in [A, B$ and $C'' \in [A, C$ so that $|A, B''| = |A', B'|$, $|A, C''| = |A', C'|$. Then as $|\angle B'' AC''|° = |\angle BAC|° = |\angle B'A'C'|°$, by the SAS principle we see that the triangles $[A, B'', C''], [A', B', C']$ are congruent in the correspondence $(A, B'', C'') \to (A', B', C')$. In particular $|\angle AB''C''|° = |\angle A'B'C'|°$ and so $|\angle AB''C''|° = |\angle ABC|°$. These are corresponding angles in the sense of 5.1.1, so $B''C'' \parallel BC$ and then by 5.3.1

$$\frac{|A, B''|}{|A, B|} = \frac{|A, C''|}{|A, C|},$$

so

$$\frac{|A', B'|}{|A, B|} = \frac{|A', C'|}{|A, C|}.$$

By a similar argument on taking a triangle $[B, E, F]$ which is congruent to $[B', C', A']$, we have

$$\frac{|B', C'|}{|B, C|} = \frac{|B', A'|}{|B, A|}.$$

COMMENT. Triangles like these, which have the degree-measures of corresponding angles equal and so the ratios of the lengths of corresponding sides are equal, are said to be **similar** in the correspondence $(A, B, C) \to (A', B', C')$.

Let A, B, C and A', B', C' be two sets of non-collinear points such that

$$\frac{|A', B'|}{|A, B|} = \frac{|A', C'|}{|A, C|}, \quad |\angle B'A'C'|° = |\angle BAC|°.$$

Then the triangles are similar.

Proof. Choose $B'' \in [A, B$, $C'' \in [A, C$ so that $|A, B''| = |A', B'|$, $|A, C''| = |A', C'|$. Then as $|\angle B'A'C'|° = |\angle BAC|° = |\angle B''AC''|°$, by the SAS principle we see that the triangles $[A', B', C'], [A, B'', C'']$ are congruent. We note that

$$\frac{|A, B''|}{|A, B|} = \frac{|A, C''|}{|A, C|}.$$

Now the line through B'' which is parallel to BC will meet $[A, C$ in a point D such that

$$\frac{|A, B''|}{|A, B|} = \frac{|A, D|}{|A, C|}.$$

Hence

$$\frac{|A, C''|}{|A, C|} = \frac{|A, D|}{|A, C|},$$

from which it follows that $|A, D| = |A, C''|$. As $C'', D \in [A, C$ we then have $D = C''$ and so $B''C'' \parallel BC$. Thus the degree-measures of the angles of $[A, B, C]$ are equal to those of the corresponding angles of $[A, B'', C'']$ and so in turn to those of the corresponding angles in $[A', B', C']$.

5.4 PYTHAGORAS' THEOREM, c.550B.C.

5.4.1

PYTHAGORAS' THEOREM. *Let A, B, C be non-collinear points such that $AB \perp AC$. Then*

$$|B, C|^2 = |C, A|^2 + |A, B|^2.$$

Proof. Let D be the foot of the perpendicular from A to BC; then by 4.3.3 D is between B and C. The triangles $[D, B, A]$, $[A, B, C]$ are similar as $|\angle ADB|° = |\angle CAB|° = 90$, $|\angle DBA|° = |\angle ABC|°$, and then by 5.2.2 $|\angle BAD|° = |\angle BCA|°$. Then by the last result

$$\frac{|A, B|}{|B, C|} = \frac{|B, D|}{|A, B|},$$

so that $|A, B|^2 = |B, D||B, C|$. By a similar argument applied to the triangles $[D, C, A]$, $[A, B, C]$ we get that $|A, C|^2 = |D, C||B, C|$. Then by addition, as $D \in [B, C]$,

$$|A, B|^2 + |A, C|^2 = (|B, D| + |D, C|)|B, C| = |B, C|^2.$$

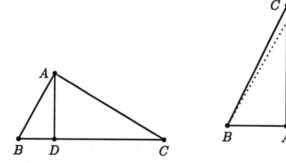

Figure 5.8. Pythagoras' theorem.   Figure 5.9. Impossible figure for converse.

CONVERSE of PYTHAGORAS' THEOREM. *Let A, B, C be non-collinear points such that*

$$|B, C|^2 = |C, A|^2 + |A, B|^2.$$

Then $\angle BAC$ is a right-angle.

Proof. Choose the point E so that $|A, C| = |A, E|$, E is on the same side of AB as C is, and $\angle BAE$ is right-angle. By Pythagoras' theorem,

$$|B, E|^2 = |A, E|^2 + |A, B|^2 = |C, A|^2 + |A, B|^2 = |B, C|^2.$$

Thus $|B, E| = |B, C|$, and the lengths of the sides of the triangle $[B, A, C]$ are equal to those of $[B, A, E]$. By the SSS principle, $[B, A, C] \equiv [B, A, E]$. In particular $|\angle BAC|° = |\angle BAE|°$ and this latter is a right-angle by construction. In fact $E = C$.

NOTE. In a right-angled triangle, the side opposite the right- angle is known as the **hypotenuse**

5.5   MID-LINES AND TRIANGLES

5.5.1   Harmonic ranges

Let A, B, C be non-collinear points such that $|A, B| > |A, C|$. Take $D \neq A$ so that $A \in [B, D]$. Then the mid-lines of $|\underline{BAC}$ and $|\underline{CAD}$ meet BC at points E, F, respectively, such that $\{E, F\}$ divide $\{B, C\}$ internally and externally in the same ratio.

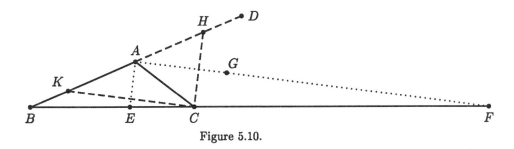

Figure 5.10.

Proof. By the cross-bar theorem the mid-line of $|\underline{BAC}$ meets $[B, C]$ in a point E. Let G be a point of the mid-line of $|\underline{CAD}$, on the same side of AB as C is. We cannot have $AG \parallel BC$ as that would imply

$$|\angle BCA|° = |\angle CAG|° = |\angle GAD|° = |\angle CBA|°,$$

and this in turn would imply that $|A, B| = |A, C|$, contrary to hypothesis. Then AG meets BC in some point F.

Take $H \in [A, D$ so that $|A, H| = |A, C|$. Then $|\angle AHC|° = |\angle ACH|°$. We have that

$$
\begin{aligned}
|\angle BAC|° &= |\angle AHC|° + |\angle ACH|°, & |\angle AHC|° &= |\angle ACH|°, \\
|\angle BAC|° &= |\angle BAE|° + |\angle EAC|°, & |\angle BAE|° &= |\angle EAC|°.
\end{aligned}
$$

It follows that $|\angle EAC|° = |\angle ACH|°$, and as E, H are on opposite sides of AC this implies that $AE \parallel HC$. It then follows that

$$\frac{|B,E|}{|E,C|} = \frac{|B,A|}{|A,H|} = \frac{|B,A|}{|A,C|}.$$

Next choose $K \in [A, B$ so that $|A, K| = |A, C|$. Then $|A, K| < |A, B|$ so $K \in [A, B]$. Now

$$|\angle HAC|° = |\angle AKC|° + |\angle ACK|°, \quad |\angle AKC|° = |\angle ACK|°,$$
$$|\angle HAC|° = |\angle HAG|° + |\angle GAC|°, \quad |\angle HAG|° = |\angle GAC|°.$$

It follows that $|\angle GAC|° = |\angle ACK|°$. But H, K are on opposite sides of AC, H, G are on the same side, and so G, K are on opposite sides. This implies that $AG \parallel KC$. Now AG meets BC at F, and $K \in [A, B]$ so $C \in [B, F]$. It follows that

$$\frac{|B,F|}{|F,C|} = \frac{|B,A|}{|A,K|} = \frac{|B,A|}{|A,C|}.$$

On combining the two results, we then have

$$\frac{|B,E|}{|E,C|} = \frac{|B,F|}{|F,C|}.$$

NOTE. We also refer to the mid-line of $|\underline{CAD}$ above as the **external bisector** of $|\underline{BAC}$. When $\{E, F\}$ divide $\{B, C\}$ internally and externally in the same ratio, we say that (B, C, E, F) form a **harmonic range**.

Let (A, B, C, D) be a harmonic range and $S \notin AB$. Let the line through C, parallel to SD, meet SA at G and SB at H. Then C is the mid-point of G and H.

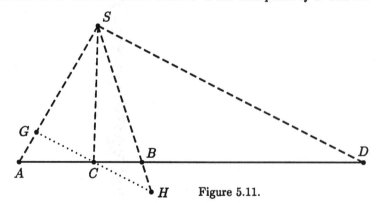

Figure 5.11.

Proof. We are given

$$\frac{|A,C|}{|C,B|} = \frac{|A,D|}{|D,B|},$$

so

$$\frac{|A,C|}{|A,D|} = \frac{|C,B|}{|D,B|}.$$

As $GC \parallel SD$ the triangles $[A, D, S]$ and $[A, C, G]$ are similar, so

$$\frac{|A, C|}{|A, D|} = \frac{|G, C|}{|S, D|}.$$

In the similar triangles $[B, C, H]$ and $[B, D, S]$,

$$\frac{|B, C|}{|B, D|} = \frac{|C, H|}{|S, D|}.$$

Then

$$\frac{|G, C|}{|S, D|} = \frac{|C, H|}{|S, D|}.$$

It follows that $|G, C| = |C, H|$.

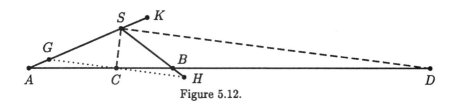

Figure 5.12.

Let (A, B, C, D) be a harmonic range, $S \notin AB$ and $K \neq S$ be such that $S \in [A, K]$. Suppose that $CS \perp DS$. Then CS and DS are the mid-lines of $|ASB$ and $|BSK$.

Proof. Let the line through C, parallel to DS meet SA at G and SB at H. Then C is the mid-point of G and H. Also $CS \perp SD$, $SD \parallel GH$ so $SC \perp GH$. It follows that the triangles $[G, C, S]$ and $[H, C, S]$ are congruent by the SAS-principle. In particular $|\angle GSC|° = |\angle HSC|°$ and so SC is the mid-line of $|ASB$. But also $|\angle CGS|° = |\angle CHS|°$ and in fact the triangle $[S, G, H]$ is isosceles. Now $\angle CGS$ and $\angle DSK$ are corresponding angles and $\angle CHS$ and $\angle DSH$ are alternate angles. It follows that $|\angle DSK|° = |\angle DSH|°$ and so the mid-line of $|BSK$ is SD.

5.6   AREA OF TRIANGLES, AND CONVEX QUADRILATERALS AND POLYGONS

5.6.1   Area of a triangle

Let A, B, C be non-collinear points, and $D \in BC$, $E \in CA$, $F \in AB$ points such that $AD \perp BC$, $BE \perp CA$, $CF \perp AB$. Then

$$|A, D||B, C| = |B, E||C, A| = |C, F||A, B|.$$

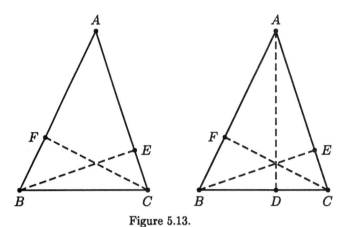

Figure 5.13.

Proof.

The triangles $[A, B, E]$ and $[A, C, F]$ are similar in the correspondence $(A, B, E) \rightarrow (A, C, F)$, as $\angle BAE = \angle CAF$ is in both, $|\angle AEB|^\circ = |\angle AFC|^\circ = 90$, and then by 5.2.2 $|\angle ABE|^\circ = |\angle ACF|^\circ$. By 5.3.2

$$\frac{|B, E|}{|C, F|} = \frac{|A, B|}{|C, A|}.$$

On cross multiplication,

$$|B, E||C, A| = |C, F||A, B|.$$

By a similar argument, we can show that $|A, D||B, C|$ is equal to these.

Definition. With the notation of the last result, the **area** of the triangle $[A, B, C]$, denoted by $\Delta[A, B, C]$, is the common value of:

$$\tfrac{1}{2}|A, D||B, C|, \ \tfrac{1}{2}|B, E||C, A|, \ \tfrac{1}{2}|C, F||A, B|.$$

Area of triangles has the following properties:-

(i) *If $P \in [B, C]$ is distinct from B and C, then*

$$\Delta[A, B, P] + \Delta[A, P, C] = \Delta[A, B, C].$$

(ii) *If $[A, B, C, D]$ is a convex quadrilateral, then*

$$\Delta[A, B, D] + \Delta[C, B, D] = \Delta[B, C, A] + \Delta[D, C, A].$$

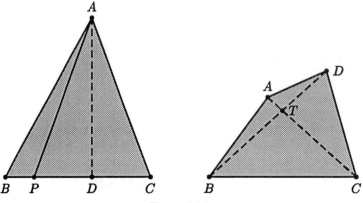

Figure 5.14.

Proof.

(i) For D is the foot of the perpendicular from the vertex A to the opposite side-line in each of the triangles $[A, B, P]$ and $[A, P, C]$, so with $p_1 = |A, D|$ we have

$$\Delta[A, B, P] = \tfrac{1}{2}p_1|B, P|, \quad \Delta[A, P, C] = \tfrac{1}{2}p_1|P, C|,$$

and the sum of these is

$$\tfrac{1}{2}p_1(|B, P| + |P, C|) = \tfrac{1}{2}p_1|B, C|,$$

as $P \in [B, C]$.

(ii) As in 5.2.1 denote by T the point which $[A, C]$ and $[B, D]$ have in common. Then by (i) above,

$$\Delta[A, B, D] + \Delta[C, B, D] = (\Delta[A, B, T] + \Delta[A, D, T]) + (\Delta[C, B, T] + \Delta[C, D, T])$$

$$\Delta[A, B, C] + \Delta[A, D, C] = (\Delta[A, B, T] + \Delta[C, B, T]) + (\Delta[A, D, T] + \Delta[C, D, T])$$

and these are clearly equal.

5.6.2   Area of a convex quadrilateral

Definition. We define **the area of the convex quadrilateral** $[A, B, C, D]$ to be $\Delta[A, B, D] + \Delta[C, B, D]$, and denote it by $\Delta[A, B, C, D]$.

If $[A, B, C, D]$ is a rectangle, then

$$\Delta[A, B, C, D] = |A, B||B, C|,$$

that is the area is equal to the product of the lengths of two adjacent sides.

Proof. For $\Delta[A, B, D] = \tfrac{1}{2}|A, B||A, D|$, $\Delta[C, B, D] = \tfrac{1}{2}|D, C||B, C|$. As by 5.2.1 $|D, C| = |A, B|$ and $|B, C| = |A, D|$, the result follows by addition.

5.6.3   Area of a convex polygon

Definition. For an integer $n \geq 3$ let $P_1, P_2, \ldots, P_n$ be n points such that no three of them are collinear. Writing also $P_{n+1} = P_1$, for each integer j such that $1 \leq j \leq n$ let $\mathcal{H}_{2j-1}$, $\mathcal{H}_{2j}$ be the closed half-planes with common edge the line $P_j P_{j+1}$, and suppose that all the points P_k lie in $\mathcal{H}_{2j-1}$ in each case. Then the intersection $\bigcap_{j=1}^{n} \mathcal{H}_{2j-1}$ is called a **convex polygon**. The intersection of the corresponding open half-planes is called the **interior** of the convex polygon. The notation for convex quadrangles is extended to convex polygons in a straightforward way.

Consider a convex polygonal region with sides $[P_1, P_2], [P_2, P_3], \ldots, [P_n, P_1]$. Let a point U interior to the polygon be joined by segments to the vertices. Then

$$\sum_{j=1}^{n-1} \Delta[U, P_j, P_{j+1}] + \Delta[U, P_n, P_1] = \sum_{j=2}^{n-1} \Delta[P_1, P_j, P_{j+1}].$$

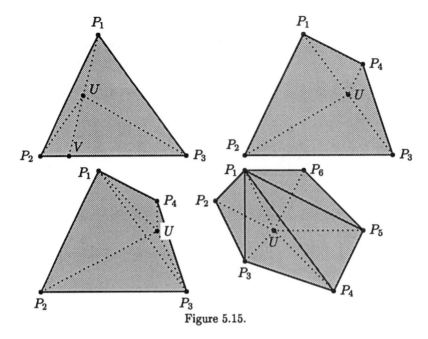

Figure 5.15.

Proof.

CASE 1. We first take the case of a triangle so that $n = 3$. Now $[P_1, U$ will meet $[P_2, P_3]$ in a point V. Then by 5.6.1

$$\Delta[U, P_1, P_2] + \Delta[U, P_2, P_3] + \Delta[U, P_3, P_1]$$
$$= \Delta[U, P_1, P_2] + \{\Delta[U, P_2, V] + \Delta[U, V, P_3]\} + \Delta[U, P_3, P_1]$$
$$= \{\Delta[U, P_1, P_2] + \Delta[U, P_2, V]\} + \{\Delta[U, V, P_3] + \Delta[U, P_3, P_1]\}$$
$$= \Delta[P_1, P_2, V] + \Delta[V, P_3, P_1] = \Delta[P_1, P_2, P_3].$$

CASE 2. Secondly we take the case of a convex quadrilateral so that $n = 4$. Suppose first that $U \in [P_1, P_3]$. Then by 5.6.1 used twice,

$$\{\Delta[U, P_1, P_2] + \Delta[U, P_2, P_3]\} + \{\Delta[U, P_3, P_4] + \Delta[U, P_4, P_1]\}$$
$$= \Delta[P_1, P_2, P_3] + \Delta[P_1, P_3, P_4].$$

Suppose next that $U \notin [P_1, P_3]$. Then U is interior to $[P_1, P_2, P_3]$ or $[P_1, P_3, P_4]$, say $U \in [U_1, P_3, P_4]$. Then by 5.6.1

$$\Delta[U, P_1, P_2] + \Delta[U, P_2, P_3] = \Delta[P_1, P_2, P_3] + \Delta[U, P_1, P_3],$$

so

$$\Delta[U, P_1, P_2] + \Delta[U, P_2, P_3] + \Delta[U, P_3, P_4] + \Delta[U, P_4, P_1]$$
$$= \Delta[P_1, P_2, P_3] + \{\Delta[U, P_1, P_3] + \Delta[U, P_3, P_4] + \Delta[U, P_4, P_1]\}$$
$$= \Delta[P_1, P_2, P_3] + \Delta[P_1, P_3, P_4]$$

by CASE 1.

CASE 3. We now suppose that the result holds, for some $n \geq 4$, for any convex polygonal region with n sides. Then for that n consider any convex polygonal region with $n+1$ sides, $[P_1, P_2]$, $[P_2, P_3]$, $\ldots$, $[P_n, P_{n+1}]$, $[P_{n+1}, P_1]$. As $n + 1 \geq 5$, $[P_1, P_2, P_3]$ and $[P_1, P_n, P_{n+1}]$ have only P_1 in common, so U cannot be in both. Suppose that $U \notin [P_1, P_2, P_3]$. By 5.6.1

$$\Delta[U, P_1, P_2] + \Delta[U, P_2, P_3] = \Delta[P_1, P_2, P_3] + \Delta[U, P_1, P_3].$$

Hence as U is interior to the polygon with n sides $[P_1, P_3]$, $[P_3, P_4]$, $\ldots$, $[P_n, P_{n+1}]$, $[P_{n+1}, P_1]$,

$$\sum_{j=1}^{n} \Delta[U, P_j, P_{j+1}] + \Delta[U, P_{n+1}, P_1]$$

$$= \Delta[P_1, P_2, P_3] + \Delta[U, P_1, P_3] + \sum_{j=3}^{n} \Delta[U, P_j, P_{j+1}] + \Delta[U, P_{n+1}, P_1]$$

$$= \Delta[P_1, P_2, P_3] + \sum_{j=3}^{n} \Delta[P_1, P_j, P_{j+1}] = \sum_{j=2}^{n} \Delta[P_1, P_j, P_{j+1}].$$

If instead $U \in [P_1, P_n, P_{n+1}]$ we get the same conclusion by similar reasoning. The result now follows by induction on n.

Definition. The **area** of the polygonal region in the present section is defined to be the sum of the areas of the triangles involved.

Exercises

5.1 Opposite wedge-angles in a parallelogram have equal degree-measures.

5.2 If two adjacent sides of a rectangle have equal lengths, then all the sides have equal lengths. Such a rectangle is called a **square**.

5.3 If the diagonals of a parallelogram have equal lengths, it must be a rectangle.

5.4 If the diagonal lines of a rectangle are perpendicular, it must be a square.

5.5 Let A, B, C be non-collinear points and let $P \in [A, B$ and $Q \in [A, C$ be such that
$$\frac{|A, P|}{|A, B|} = \frac{|A, Q|}{|A, C|}.$$
Then $PQ \parallel BC$.

5.6 Let $AB \perp AC$ and let $D = \text{mp}(B, C)$. Prove that $|D, A| = |D, B| = |D, C|$.

5.7 Let A, B, C be non-collinear points and for $A \in [B, P$ and $A \in [C, Q$ let $PQ \parallel BC$. Show that then
$$\frac{|A, P|}{|A, B|} = \frac{|A, Q|}{|A, C|}.$$

5.8 Suppose that A, B, C are non-collinear points with $|A, B| > |A, C|$ and let $D = \pi_{BC}(A)$. Prove that then

$$|A, B|^2 - |A, C|^2 = |B, D|^2 - |C, D|^2.$$

5.9 Suppose that A, B, C are non-collinear points and D is the mid-point of B and C. Prove that then

$$|A, B|^2 + |A, C|^2 = 2|B, D|^2 + 2|A, D|^2.$$

[Hint. Consider the foot of the perpendicular from A to BC.]

5.10 Show that the AAS-principle of congruence in Ex.4.2 can be deduced from 5.2.2 and the ASA-principle.

5.11 Show that the AAS-principle of congruence for right-angled triangles in Ex.4.3 can be deduced from Pythagoras' theorem and the SSS-principle.

5.12 For $C \notin AB$, suppose that m is the line through C which is parallel to AB. Prove that for any point $D \notin AB$ the line AD meets m in a unique point E. When, additionally, $D \in \mathcal{IR}(|\underline{BAC})$ then E is on $[A, D$ and is also on $m \cap \mathcal{IR}(|\underline{BAC})$.

5.13 In a triangle $[A, B, C]$, let $|A, B| > |A, C|$. Let $D \in [A, B$ be such that $|A, D| = |A, C|$. Prove that then

$$2|\angle BCD|^\circ = |\angle ACB|^\circ - |\angle CBA|^\circ.$$

6

Cartesian coordinates; applications

COMMENT. Hitherto we have confined ourselves to synthetic or pure geometrical arguments aided by a little algebra, and traditionally this is continued with. This is a difficult process because of the scarcity of manipulations, operations and transformations to aid us. The main difficulties in synthetic proofs are locational, to show that points are where the diagrams suggest they should be, and in making sure that all possible cases are covered.

For ease and efficiency we now introduce coordinates, and hence thoroughgoing algebraic methods. These not only enable us to deal with the concepts already introduced but also to elaborate on them in an advantageous way.

In Chapter 6 we do the basic coordinate geometry of lines, segments, half-lines and half-planes. The only use we make of angles here is to deal with perpendicularity.

6.1 FRAME OF REFERENCE, CARTESIAN COORDINATES

6.1.1

Definition. A couple or ordered pair $\mathcal{F} = ([O,I\,, [O,J\,)$ of half-lines such that $OI \perp OJ$, will be called a **frame of reference** for Π. With it, as standard notation, we shall associate the pair of closed half-planes $\mathcal{H}_1, \mathcal{H}_2$, with common edge OI, and with $J \in \mathcal{H}_1$, and the pair of closed half-planes $\mathcal{H}_3, \mathcal{H}_4$, with common edge OJ, and with $I \in \mathcal{H}_3$. We refer to $\mathcal{Q}_1 = \mathcal{H}_1 \cap \mathcal{H}_3$, $\mathcal{Q}_2 = \mathcal{H}_1 \cap \mathcal{H}_4$, $\mathcal{Q}_3 = \mathcal{H}_2 \cap \mathcal{H}_4$ and $\mathcal{Q}_4 = \mathcal{H}_2 \cap \mathcal{H}_3$, respectively, as **the first, second, third** and **fourth quadrants** of $\mathcal{F}$. We refer to OI and OJ as the **axes** and to O as the **origin**.

Given any point Z in Π, (rectangular) Cartesian coordinates for Z are defined as follows. Let U be the foot of the perpendicular from Z to OI and V the foot of the perpendicular from Z to OJ. We let

$$x = \begin{cases} |O, U|, & \text{if } Z \in \mathcal{H}_3, \\ -|O, U|, & \text{if } Z \in \mathcal{H}_4, \end{cases} \quad \text{and} \quad y = \begin{cases} |O, V|, & \text{if } Z \in \mathcal{H}_1, \\ -|O, V|, & \text{if } Z \in \mathcal{H}_2. \end{cases}$$

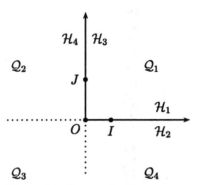

Figure 6.1. Frame of reference.   Figure 6.2. Cartesian coordinates.

Then the ordered pair (x, y) are called **Cartesian coordinates** for Z, relative to $\mathcal{F}$. We denote this in symbols by $Z \equiv_{\mathcal{F}} (x, y)$, but when $\mathcal{F}$ is fixed and can be understood, we relax this notation to $Z \equiv (x, y)$.

Cartesian coordinates have the following properties:-

(i) *If $Z \in \mathcal{Q}_1$, then $x \geq 0$, $y \geq 0$; if $Z \in \mathcal{Q}_2$, then $x \leq 0$, $y \geq 0$; if $Z \in \mathcal{Q}_3$, then $x \leq 0$, $y \leq 0$; if $Z \in \mathcal{Q}_4$, then $x \geq 0$, $y \leq 0$.*

(ii) *If $Z_1 \equiv (x_1, y_1)$, $Z_2 \equiv (x_2, y_2)$ and*

$$U_1 = \pi_{OI}(Z_1), \ V_1 = \pi_{OJ}(Z_1), \ U_2 = \pi_{OI}(Z_2), \ V_2 = \pi_{OJ}(Z_2),$$

then $|U_1, U_2| = \pm(x_2 - x_1)$, $|V_1, V_2| = \pm(y_2 - y_1)$.

(iii) *If $Z_1 \equiv (x_1, y_1)$, $Z_2 \equiv (x_2, y_2)$, then*

$$|Z_1, Z_2| = \sqrt{(x_2 - x_1)^2 + (y_2 - y_1)^2}.$$

(iv) *If $Z_1 \equiv (x_1, y_1)$, $Z_2 \equiv (x_2, y_2)$ and $Z_3 \equiv (x_3, y_3)$ where*

$$x_3 = \tfrac{1}{2}(x_1 + x_2), \ y_3 = \tfrac{1}{2}(y_1 + y_2),$$

then $Z_3 = \mathrm{mp}(Z_1, Z_2)$.

(v) *Let $\leq_l$ be the natural order on $l = OI$ under which $O \leq_l I$. If $x_1 < x_2$, $U_1 \equiv (x_1, 0)$ and $U_2 \equiv (x_2, 0)$, then $U_1 \leq_l U_2$.*

Proof.
(i) This is clear from the definition of coordinates.
(ii) For if $Z_1, Z_2 \in \mathcal{H}_3$ we have

$$|O, U_2| = x_1, \ |O, U_2| = x_2,$$

and so as $U_1, U_2 \in [O, I$,

$$|U_1, U_2| = \pm(x_2 - x_1)$$

according as $U_1 \in [O, U_2]$ or $U_2 \in [O, U_1]$. Similarly if $Z_1, Z_2 \in \mathcal{H}_3$, we have

$$|O, U_1| = -x_1, \ |O, U_2| = -x_2$$

and
$$|U_1, U_2| = \pm[-x_2 - (-x_1)]$$

according as $U_1 \in [O, U_2]$ or $U_2 \in [O, U_1]$. Finally if $Z_1 \in \mathcal{H}_3$, $Z_2 \in \mathcal{H}_4$ then

$$|O, U_1| = x_1, \quad |O, U_2| = -x_2$$

and $O \in [U_1, U_2]$ so that
$$|U_1, U_2| = x_1 + (-x_2);$$

similarly if $Z_1 \in \mathcal{H}_4$, $Z_2 \in \mathcal{H}_3$.

That $|V_1, V_2| = \pm(y_2 - y_1)$ can be shown in the same way.

(iii) Now the lines through Z_1 parallel to OI and through Z_2 parallel to OJ are perpendicular to each other, and so meet in a unique point Z_4. Clearly $\pi_{OI}(Z_4) = \pi_{OI}(Z_2) = U_2$ so Z_2 and Z_4 have the same first coordinate, x_2; $\pi_{OJ}(Z_4) = \pi_{OJ}(Z_1) = V_1$ so Z_1 and Z_4 have the same second coordinate, y_1. Thus Z_4 has coordinates (x_2, y_1). If the points Z_1, Z_2, Z_4 are not collinear, then by Pythagoras' theorem

$$|Z_1, Z_2|^2 = |Z_1, Z_4|^2 + |Z_4, Z_2|^2;$$

if they are collinear we must have $Z_1 = Z_4$ or $Z_2 = Z_4$ and this identity is trivially true. But $|Z_1, Z_4| = |U_1, U_2|$ as $[Z_1, Z_4, U_2, U_1]$ is a rectangle, or else $Z_1 = Z_4$ and $U_1 = U_2$, or $Z_1 = U_1$, $Z_4 = U_2$. Similarly $|Z_2, Z_4| = |V_1, V_2|$. Thus we have the distance formula

$$\begin{aligned} |Z_1, Z_2|^2 &= |U_1, U_2|^2 + |V_1, V_2|^2 \\ &= (x_2 - x_1)^2 + (y_2 - y_1)^2, \end{aligned}$$

which expresses the distance $|Z_1, Z_2|$ in terms of the coordinates of Z_1 and Z_2.

(iv) If $Z_1 = Z_2$, then $x_2 = x_1$, $y_2 = y_1$ so that $x_3 = x_1$, $y_3 = y_1$. Thus $Z_3 = Z_1 = \text{mp}(Z_1, Z_1)$, as required.

Suppose then that $Z_1 \neq Z_2$. Note that

$$|Z_1, Z_3|^2 = \left[\frac{x_1 + x_2}{2} - x_1\right]^2 + \left[\frac{y_1 + y_2}{2} - y_1\right]^2 = \left[\frac{x_2 - x_1}{2}\right]^2 + \left[\frac{y_2 - y_1}{2}\right]^2$$

and so $|Z_1, Z_3| = \frac{1}{2}|Z_1, Z_2|$. Similarly

$$|Z_3, Z_2|^2 = \left[\frac{x_1 + x_2}{2} - x_2\right]^2 + \left[\frac{y_1 + y_2}{2} - y_2\right]^2 = \left[\frac{x_1 - x_2}{2}\right]^2 + \left[\frac{y_1 - y_2}{2}\right]^2$$

and so $|Z_3, Z_2| = \frac{1}{2}|Z_1, Z_2|$. Then

$$|Z_1, Z_3| + |Z_3, Z_2| = |Z_1, Z_2|.$$

It follows by 3.1.2 and 4.3.1 that $Z_3 \in [Z_1, Z_2] \subset Z_1 Z_2$. As $|Z_1, Z_3| = |Z_3, Z_2|$ it then follows that $Z_3 = \text{mp}(Z_1, Z_2)$.

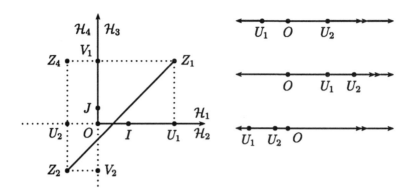

Figure 6.3. The distance formula.   Order of points on the x-axis.

(v) By 2.1.4 at least one of

$$\text{(a) } O \in [U_1, U_2], \text{ (b) } U_1 \in [O, U_2], \text{ (c) } U_2 \in [O, U_1],$$

holds.

In (a), U_1 and U_2 are in different half-lines with end-point O. We cannot have $U_1 \in [O, I$ as then we would have $x_1 \geq 0$, $x_2 \leq 0$, a contradiction. Thus $U_2 \in [O, I$ so that $U_1 \leq_l O$, $O \leq_l U_2$ and thus $U_1 \leq_l U_2$.

In (b) we cannot have $U_1 \leq_l O$. For then we would have $U_2 \leq_l O$ and

$$|O, U_1| = -x_1, \quad |O, U_2| = -x_2.$$

As $U_1 \in [O, U_2]$ we have $|O, U_1| \leq |O, U_2|$ which yields $-x_1 \leq -x_2$ and so $x_1 \geq x_2$, a contradiction. Hence $O \leq_l U_1$ and so as $U_1 \in [O, U_2]$, $U_1 \leq_l U_2$.

In (c) we cannot have $O \leq_l U_1$. For then we would have $O \leq_l U_2$ and so

$$|O, U_1| = x_1, \quad |O, U_2| = x_2.$$

As $U_2 \in [O, U_1]$ we have $|O, U_2| \leq |O, U_1|$, so that $x_2 \leq x_1$, a contradiction. Hence $U_1 \leq_l O$ so $U_1 \leq_l U_2$.

6.2   ALGEBRAIC NOTE ON LINEAR EQUATIONS

6.2.1

It is convenient to note here some results on solutions of two simultaneous linear equations in two unknowns.

(a) If

$$a_{1,1}a_{2,2} - a_{1,2}a_{2,1} \neq 0, \tag{6.2.1}$$

then the pair of simultaneous equations

$$
\begin{aligned}
a_{1,1}x + a_{1,2}y &= k_1, \\
a_{2,1}x + a_{2,2}y &= k_2,
\end{aligned}
\tag{6.2.2}
$$

has precisely one solution pair (x, y), and that is given by

$$(x, y) = \left(\frac{a_{2,2}k_1 - a_{1,2}k_2}{a_{1,1}a_{2,2} - a_{1,2}a_{2,1}}, \frac{a_{1,1}k_2 - a_{2,1}k_1}{a_{1,1}a_{2,2} - a_{1,2}a_{2,1}} \right). \qquad (6.2.3)$$

(b) If

$$(a_{1,1}, a_{1,2}) \neq (0,0) \quad \text{and} \quad (a_{2,1}, a_{2,2}) \neq (0,0), \qquad (6.2.4)$$

and

$$a_{1,1}a_{2,2} - a_{1,2}a_{2,1} = 0, \qquad (6.2.5)$$

then there is some $j \neq 0$ such that

$$a_{2,1} = ja_{1,1}, \ a_{2,2} = ja_{1,2}. \qquad (6.2.6)$$

(c) If (6.2.4) holds, then for the system (6.2.2) of simultaneous equations to have either no, or more than one, solution pair (x, y) it is necessary and sufficient that (6.2.5) hold.

Note in particular that when (6.2.4) holds, for the pair of homogeneous linear equations

$$\begin{aligned} a_{1,1}x + a_{1,2}y &= 0, \\ a_{2,1}x + a_{2,2}y &= 0, \end{aligned} \qquad (6.2.7)$$

to have a solution (x, y) other than the obvious one $(0,0)$, it is necessary and sufficient that (6.2.5) hold.

6.3   CARTESIAN EQUATION OF A LINE

6.3.1

Given any line $l \in \Lambda$, there are numbers a, b and c, with the case $a = b = 0$ excluded, such that $Z \equiv (x, y) \in l$ if and only if

$$ax + by + c = 0.$$

Proof. Take any point $Z_2 \equiv (x_2, y_2) \notin l$ and let $Z_3 \equiv (x_3, y_3) = s_l(Z_2)$. Then $Z_2 \neq Z_3$. Now l is the perpendicular bisector of $[Z_2, Z_3]$, so by 4.1.1 $Z \in l$ if and only if $|Z, Z_2| = |Z, Z_3|$. As these are both non-negative, this is the case if and only if $|Z, Z_2|^2 = |Z, Z_3|^2$. By 6.1.1 this happens if and only if

$$(x - x_2)^2 + (y - y_2)^2 = (x - x_3)^2 + (y - y_3)^2.$$

This simplifies to

$$2(x_3 - x_2)x + 2(y_3 - y_2)y + x_2^2 + y_2^2 - x_3^2 - y_3^2 = 0.$$

On writing

$$a = 2(x_3 - x_2), \ b = 2(y_3 - y_2), \ c = x_2^2 + y_2^2 - x_3^2 - y_3^2,$$

we see that $Z \equiv (x, y) \in l$ if and only if $ax + by + c = 0$. Now $a = b = 0$ corresponds to $x_2 = x_3$, $y_2 = y_3$, which is ruled out as $Z_2 \neq Z_3$.

COROLLARY. *Let* $Z_0 \equiv (x_0, y_0)$, $Z_1 \equiv (x_1, y_1)$ *be distinct points and* $Z \equiv (x, y)$. *Then* $Z \in Z_0 Z_1$ *if and only if*

$$-(y_1 - y_0)(x - x_0) + (x_1 - x_0)(y - y_0) = 0.$$

Proof. By the theorem, there exist numbers a, b, c, with the case $a = b = 0$ excluded, such that $Z \in Z_0 Z_1$ if and only if $ax + by + c = 0$. As $Z_0, Z_1 \in Z_0 Z_1$ we then have

$$ax_0 + by_0 + c = 0,$$
$$ax_1 + by_1 + c = 0.$$

We subdivide into two cases as follows.

CASE 1. Let $x_0 \neq x_1$. We rewrite our equations as

$$ax_1 + c = -by_1,$$
$$ax_0 + c = -by_0,$$

and regard these as equations in the unknowns a and c. As $x_1 - x_0 \neq 0$, we note that by 6.2.1 we must have

$$a = \frac{-b(y_1 - y_0)}{x_1 - x_0}, \quad c = \frac{-b(x_1 y_0 - x_0 y_1)}{x_1 - x_0}.$$

Note that $b \neq 0$, as $b = 0$ would imply $a = 0$ here. On inserting these values for a and c above we see that $Z \in l$ if and only if

$$\frac{-b(y_1 - y_0)}{x_1 - x_0} x + by + \frac{-b(x_1 y_0 - x_0 y_1)}{x_1 - x_0} = 0,$$

and so as $b/(x_1 - x_0) \neq 0$, if and only if

$$-(y_1 - y_0)x + (x_1 - x_0)y - x_1 y_0 + x_0 y_1 = 0.$$

This is equivalent to the stated equation.

CASE 2. Let $y_0 \neq y_1$. We rewrite our equations as

$$by_1 + c = -ax_1,$$

$$by_0 + c = -ax_0,$$

and note that, as $y_1 - y_0 \neq 0$, by 6.2.1 we must have

$$b = \frac{-a(x_1 - x_0)}{y_1 - y_0}, \quad c = \frac{-a(y_1 x_0 - y_0 x_1)}{y_1 - y_0}.$$

Note that $a \neq 0$, as $a = 0$ would imply $b = 0$ here. On inserting these values for b and c above we see that $Z \in l$ if and only if

$$ax + \frac{-a(x_1 - x_0)}{y_1 - y_0}y + \frac{-a(y_1 x_0 - y_0 x_1)}{y_1 - y_0} = 0,$$

and so as $-a/(y_1 - y_0) \neq 0$, if and only if

$$-(y_1 - y_0)x + (x_1 - x_0)y - x_1 y_0 + x_0 y_1 = 0.$$

This is equivalent to the stated equation.

Now either CASE 1 or CASE 2 (or both) must hold, as otherwise we have $x_0 = x_1$, $y_0 = y_1$ and so $Z_0 = Z_1$, contrary to what is given.

Definition. If $l \in \Lambda$ and $l = \{Z \equiv (x,y) : ax + by + c = 0\}$, we call $ax + by + c = 0$ a **Cartesian equation** of l relative to $\mathcal{F}$, and we write $l \equiv_{\mathcal{F}} ax + by + c = 0$. When $\mathcal{F}$ can be understood we relax this to $l \equiv ax + by + c = 0$.

Let $l \in \Lambda$ be a line, with Cartesian equation
(i)
$$ax + by + c = 0.$$

Then l also has
(ii)
$$a_1 x + b_1 y + c_1 = 0,$$

as an equation if and only if
(iii)
$$a_1 = ja, \; b_1 = jb, \; c_1 = jc,$$

for some $j \neq 0$.

Proof.

Necessity. Suppose first that l can be expressed in each of the forms (i) and (ii) above. We subdivide into four cases as follows.

CASE 1. Suppose that $a \neq 0$, $b \neq 0$ and $c \neq 0$. Then we note from (i) that the points $A \equiv (-c/a, 0)$ and $B \equiv (0, -c/b)$ are in l, and are in fact the only points of l in either OI or OJ, as A is the only point with $y = 0$ and B is the only point with $x = 0$.

We now note that none of a_1, b_1, c_1 can be equal to 0. For if $a_1 = 0$, by (ii) we would have $y = -c_1/b_1$ for all points Z in l; this would make l parallel to OI and give a contradiction. Similarly $b_1 = 0$ would imply that $x = -c_1/a_1$ for all points Z in l, making l parallel to OJ and again giving a contradiction. Moreover if $c_1 = 0$, by (ii) we would have that $O \in l$, again a contradiction.

We note from (ii) that the points $A_1 \equiv (-c_1/a_1, 0)$, $B_1 \equiv (0, -c_1/b_1)$ are in l and are in fact the only points of l in either OI or OJ. Thus we must have $A_1 = A$, $B_1 = B$ and so

$$-\frac{c_1}{a_1} = -\frac{c}{a}, \quad -\frac{c_1}{b_1} = -\frac{c}{b}.$$

Thus

$$\frac{a_1}{a} = \frac{b_1}{b} = \frac{c_1}{c},$$

and if we denote the common value of these by j, we have $j \neq 0$ and (iii).

CASE 2. Suppose that $a = 0$. Then $b \neq 0$ and by (i) for every $Z \in l$ we have $y = -c/b$, so that l contains B and is parallel to OI; when $c \neq 0$, l has no point in common with OI, and when $c = 0$, l coincides with OI. Now we must have $a_1 = 0$, as otherwise l would meet OI in the unique point A_1, and that would give a contradiction. Then $b_1 \neq 0$ and for every $Z \in l$ we have $y = -c_1/b_1$, so that l contains B_1 and is parallel to OI. Thus we must have

$$-\frac{c_1}{b_1} = -\frac{c}{b}.$$

When $c = 0$, this implies that $c_1 = 0$, so that if we take $j = b_1/b$, we have satisfied (iii). When $c \neq 0$, we must have that $b_1/b = c_1/c$, and if we take j to be the common value of these we have (iii) again.

CASE 3. Suppose that $b = 0$. This is treated similarly to CASE 2.

CASE 4. Finally suppose that $a \neq 0$, $b \neq 0$ and $c = 0$. Then by (i) we see that $O \in l$ and then by (ii) we must have $c_1 = 0$. We see from (i) that $C \equiv (1, -a/b)$ is in l, and on using this information in (ii) we find that $a_1 + b_1(-a/b) = 0$. This implies that $a_1/a = b_1/b$, and if we take j to be the common value of these, we must have (iii).

This establishes the necessity of (iii).

Sufficiency. Suppose now that (iii) holds. Then $a_1 x + b_1 y + c_1 = j(ax + by + c)$ and as $j \neq 0$ we have $a_1 x + b_1 y + c_1 = 0$ if and only if $ax + by + c = 0$.

6.4 PARAMETRIC EQUATIONS OF A LINE

6.4.1

Let l be a line with Cartesian equation $ax + by + c = 0$.

(i) *If $Z_0 \equiv (x_0, y_0)$ is in l, then*

$$l = \{Z \equiv (x, y) : x = x_0 + bt, \ y = y_0 - at, \ (t \in \mathbf{R})\}.$$

(ii) *If $Z_1 \equiv (x_1, y_1) = (x_0 + b, y_0 - a)$ and $\leq_l$ is the natural order on l for which $Z_0 \leq_l Z_1$, then for $Z_2 \equiv (x_0 + bt_2, y_0 - at_2)$, $Z_3 \equiv (x_0 + bt_3, y_0 - at_3)$ we have $t_2 \leq t_3$ if and only if $Z_2 \leq_l Z_3$.*

(iii) *If $Z_1 \equiv (x_1, y_1) = (x_0 + b, y_0 - a)$, then*

$$[Z_0, Z_1] = \{Z \equiv (x, y) : x = x_0 + bt, \ y = y_0 - at, \ (0 \leq t \leq 1)\}.$$

(iv) *With Z_1 as in (ii),*

$$[Z_0, Z_1 = \{Z \equiv (x, y) : x = x_0 + bt, \ y = y_0 - at, \ (t \geq 0)\}.$$

Proof.
(i) If $Z \in l$ then $ax + by + c = 0$, $ax_0 + by_0 + c = 0$, so that

$$b(y - y_0) = -a(x - x_0). \tag{6.4.1}$$

When $b \neq 0$, let us define t by $t = (x - x_0)/b$; then by (6.4.1) we must have, $y - y_0 = -at$. Thus

$$x = x_0 + bt, \ y = y_0 - at, \tag{6.4.2}$$

for some $t \in \mathbf{R}$.

When $b = 0$ then $a \neq 0$, and by (6.4.1) we must have $x = x_0$. If we define t by $t = (y - y_0)/(-a)$, then we have (6.4.2) for some $t \in \mathbf{R}$.

Conversely suppose that (6.4.2) holds for any $t \in \mathbf{R}$. Then

$$ax + by + c = a(x_0 + bt) + b(y_0 - at) + c = ax_0 + by_0 + c = 0.$$

(ii) We first suppose that l is not perpendicular to $m = OI$, so that $b \neq 0$. We recall that Z_0, Z_1 are distinct points on l for which $Z_0 \leq_l Z_1$. Let $\leq_m$ be the natural order on m for which $O \leq_m I$. Let $U_0 = \pi_m(Z_0)$, $U_1 = \pi_m(Z_1)$ so that $U_0 \equiv (x_0, 0)$, $U_1 \equiv (x_0 + b, 0)$.

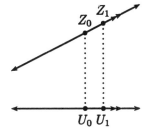 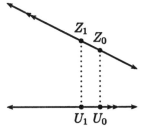

Figure 6.4. Direct correspondence.   Indirect correspondence.

If $b > 0$, then $x_0 < x_0 + b$ and so by 6.1.1 $U_0 \leq_m U_1$. In this case we say that the correspondence between $\leq_l$ and $\leq_m$ is **direct** . If $b < 0$ then $x_0 + b < x_0$ and so $U_1 \leq_m U_0$. In this case we say that the correspondence between $\leq_l$ and $\leq_m$ is **indirect**. In what follows we assume that $b > 0$ so that the correspondence between $\leq_l$ and $\leq_m$ is direct. The other case can be covered by replacing $\leq_m$ by $\geq_m$ in the following.

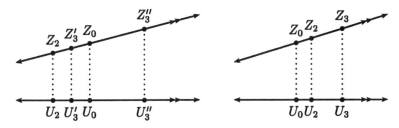

Suppose now that $Z_2 \leq_l Z_3$; we wish to show that $U_2 \leq_m U_3$ where $U_2 = \pi_m(Z_2)$, $U_3 = \pi_m(Z_3)$. We subdivide into three cases.

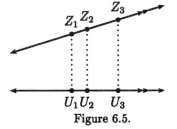

Figure 6.5.

CASE 1. Suppose that $Z_2 \leq_l Z_0$. Then $Z_2 \leq_l Z_0 \leq_l Z_1$ so that $Z_0 \in [Z_2, Z_1]$. Then by 4.3.2, $U_0 \in [U_2, U_1]$. As $U_0 \leq_m U_1$, we then have $U_2 \leq_m U_0$. There are now two possibilities, that $Z_3 \leq_l Z_0$ or that $Z_0 \leq_l Z_3$. In the first of these subcases, $Z_3 \in [Z_2, Z_0]$ so $U_3 \in [U_2, U_0]$. As $U_2 \leq_m U_0$ we then have $U_2 \leq_m U_3$. In the second of these subcases we have $Z_0 \in [Z_2, Z_3]$ so $U_0 \in [U_2, U_3]$. As $U_2 \leq_m U_0$ we have $U_0 \leq_m U_3$ so $U_2 \leq_m U_3$.

CASE 2. Suppose that $Z_0 \leq_l Z_2 \leq_l Z_1$. Then $Z_2 \in [Z_0, Z_1]$ so $U_2 \in [U_0, U_1]$. As $U_0 \leq_m U_1$ then $U_0 \leq_m U_2 \leq_m U_1$. Now $Z_2 \in [Z_0, Z_3]$ so $U_2 \in [U_0, U_3]$. As $U_0 \leq_m U_2$ it follows that $U_2 \leq_m U_3$.

CASE 3. Suppose that $Z_1 \leq_l Z_2$. Then $Z_1 \in [Z_0, Z_2]$ so that $U_1 \in [U_0, U_2]$. As $U_0 \leq_m U_1$ we then have $U_1 \leq_m U_2$. Then $Z_2 \in [Z_1, Z_3]$ so $U_2 \in [U_1, U_3]$. As $U_1 \leq_m U_2$ we have $U_2 \leq_m U_3$.

Now continuing with all three cases, we note that $U_2 \equiv (x_0 + bt_2, 0)$, $U_3 \equiv (x_0 + bt_3, 0)$ and as $U_2 \leq_m U_3$ by 6.1.1 we have $x_0 + bt_2 \leq x_0 + bt_3$. As $b > 0$ this implies that $t_2 \leq t_3$.

We also have that $t_2 \leq t_3$ implies $Z_2 \leq_l Z_3$. For otherwise $Z_3 \leq_l Z_2$ and so by the above $t_3 \leq t_2$, which gives a contradiction unless $Z_2 = Z_3$.

When l is perpendicular to OI we use π_{OJ} instead of π_m. By a similar argument we reach the same conclusion.

(iii) This follows directly from (ii) of the present theorem. It can also be proved as follows. Note that in (6.4.2) $t = 0$ gives Z_0 and $t = 1$ gives Z_1. Then for $Z \equiv (x, y)$ with x and y as in (6.4.2), by 6.1.1 we have

$$|Z_0, Z| = \sqrt{(x - x_0)^2 + (y - y_0)^2} = \sqrt{(bt)^2 + (-at)^2} = |t|\sqrt{b^2 + a^2},$$
$$|Z_1, Z| = \sqrt{(bt - t)^2 + (a - at)^2} = \sqrt{(t - 1)^2(b^2 + a^2)} = |t - 1|\sqrt{b^2 + a^2},$$
$$|Z_0, Z_1| = \sqrt{b^2 + (-a)^2} = \sqrt{b^2 + a^2}.$$

Thus when $t < 0$,

$$|Z_0, Z| = (-t)\sqrt{b^2 + a^2}, \quad |Z_1, Z| = (1 - t)\sqrt{b^2 + a^2},$$

and so $|Z, Z_0| + |Z_0, Z_1| = |Z, Z_1|$; thus by 3.1.2 and (i) above, $Z_0 \in [Z, Z_1]$, $Z_0 \neq Z$, $Z \neq Z_1$.

When $0 \leq t \leq 1$,

$$|Z_0, Z| = t\sqrt{b^2 + a^2}, \quad |Z, Z_1| = (1 - t)\sqrt{b^2 + a^2},$$

and so $|Z_0, Z| + |Z, Z_1| = |Z_0, Z_1|$; thus $Z \in [Z_0, Z_1]$.

When $t > 1$,

$$|Z_0, Z| = t\sqrt{b^2 + a^2}, \quad |Z_1, Z| = (t - 1)\sqrt{b^2 + a^2},$$

and so $|Z_0, Z_1| + |Z_1, Z| = |Z_0, Z|$; thus $Z_1 \in [Z_0, Z]$ and $Z \neq Z_0$, $Z \neq Z_1$.

These combined show that the values of t for which $0 \leq t \leq 1$ are those for which $Z \in [Z_0, Z_1]$.

(iv) This follows directly from (ii) of the present theorem. It can also be proved as follows. As in the proof of (iii) above, we see that the values of t for which $t \geq 0$ are those for which $Z \in [Z_0, Z_1$.

COROLLARY. Let $Z_0 \equiv (x_0, y_0)$ and $Z_1 \equiv (x_1, y_1)$ be distinct points. Then the following hold:-

(i)

$$Z_0 Z_1 = \{Z \equiv (x, y) : x = x_0 + t(x_1 - x_0), \ y = y_0 + t(y_1 - y_0), \ t \in \mathbf{R}\}.$$

(ii) Let $\leq_l$ be the natural order on $l = Z_0 Z_1$ for which $Z_0 \leq_l Z_1$. Let

$$\begin{aligned} Z_2 &\equiv (x_0 + t_2(x_1 - x_0), y_0 + t_2(y_1 - y_0)), \\ Z_3 &\equiv (x_0 + t_3(x_1 - x_0), y_0 + t_3(y_1 - y_0)). \end{aligned}$$

Then we have $t_2 \leq t_3$ if and only if $Z_2 \leq_l Z_3$.

(iii)

$$[Z_0, Z_1] = \{Z \equiv (x, y) : x = x_0 + t(x_1 - x_0), \ y = y_0 + t(y_1 - y_0), \ 0 \leq t \leq 1\}.$$

(iv)

$$[Z_0, Z_1 = \{Z \equiv (x, y) : x = x_0 + t(x_1 - x_0), \ y = y_0 + t(y_1 - y_0), \ t \geq 0\}.$$

Proof. By 6.3.1, in the above we can take $a = -(y_1 - y_0)$, $b = x_1 - x_0$ and the conclusions follow immediately.

NOTE. We refer to

$$x = x_0 + bt, \ y = y_0 - at, \ (t \in \mathbf{R})$$

in 6.4.1 as **parametric equations** of the line l, and t as the **parameter** of the point $Z \equiv (x, y)$.

6.5 PERPENDICULARITY AND PARALLELISM OF LINES

6.5.1

Let $l \equiv ax + by + c = 0$, $m \equiv a_1 x + b_1 y + c_1 = 0$.

(i) Then $l \perp m$ if and only if
$$aa_1 + bb_1 = 0. \tag{6.5.1}$$

(ii) Also $l \parallel m$ if and only if
$$ab_1 - a_1 b = 0. \tag{6.5.2}$$

Proof.
(i) Suppose that $l \perp m$. Then l meets m in a unique point which we denote by Z_0. By 6.4.1 $Z_1 \equiv (x_0 + b, y_0 - a)$ is a point of l and similarly $Z_2 \equiv (x_0 + b_1, y_0 - a_1)$ is a point of m. Now by Pythagoras' theorem $|Z_0, Z_1|^2 + |Z_0, Z_2|^2 = |Z_1, Z_2|^2$ and so by 6.1.1
$$[b^2 + (-a)^2] + [b_1^2 + (-a_1)^2] = (b - b_1)^2 + (a_1 - a)^2.$$

This simplifies to (6.5.1).

Conversely suppose that (6.5.1) holds. Then we cannot have (6.5.2) as well. For if we did, on multiplying (6.5.1) by a and (6.5.2) by b we would find that

$$a^2 a_1 + abb_1 = 0, \quad -b^2 a_1 + abb_1 = 0,$$

so that $(a^2 + b^2) a_1 = 0$, and hence as $(a, b) \neq (0, 0)$, $a_1 = 0$. Similarly

$$aba_1 + b^2 b_1 = 0, \quad -aba_1 + a^2 b_1 = 0,$$

so that $b_1 = 0$ as well, giving a contradiction. We now search for a point of intersection of l and m, and so consider solving for (x, y) the simultaneous equations

$$ax + by = -c, \quad a_1 x + b_1 y = -c_1.$$

As $ab_1 - a_1 b \neq 0$, by 6.2.1 these will have a unique solution, yielding a point which we shall denote by $Z_0 \equiv (x_0, y_0)$. Then by 6.4.1

$$\begin{aligned} l &= \{Z \equiv (x, y) : x = x_0 + bt, \ y = y_0 - at, \ t \in \mathbf{R}\}, \\ m &= \{Z \equiv (x, y) : x = x_0 + b_1 t, \ y = y_0 - a_1 t, \ t \in \mathbf{R}\}. \end{aligned}$$

We choose $Z_1 \in l$, $Z_2 \in m$ as above, and from (6.5.1) find that $|Z_0, Z_1|^2 + |Z_0, Z_2|^2 = |Z_1, Z_2|^2$. By 6.4.1 we can conclude that $l \perp m$.

(ii) By 6.2.1 the equations $ax + by + c = 0$, $a_1 x + b_1 y + c_1 = 0$ have either no solution or more than one if and only if (6.5.2) holds.

Alternatively, by (i) above we have $l \parallel m$ if and only if there is some $(a_2, b_2) \neq (0, 0)$ such that

$$aa_2 + bb_2 = 0, \quad a_1 a_2 + b_1 b_2 = 0.$$

But the equations

$$au + bv = 0, \quad a_1 u + b_1 v = 0,$$

have a solution (u, v) other than $(0, 0)$ if and only if $ab_1 - a_1 b = 0$. Thus (6.5.2) is a condition for l and m to be parallel.

COROLLARY.

(i) The lines $Z_1 Z_2$ and $Z_3 Z_4$ are perpendicular if and only if

$$(y_2 - y_1)(y_4 - y_3) + (x_2 - x_1)(x_4 - x_3) = 0.$$

(ii) These lines are parallel if and only if

$$-(y_2 - y_1)(x_4 - x_3) + (y_4 - y_3)(x_2 - x_1) = 0.$$

6.6   PROJECTION AND AXIAL SYMMETRY

6.6.1

Let $l \equiv ax + by + c = 0$ *and* $Z_0 \equiv (x_0, y_0)$. *Then*

(i)
$$|Z_0, \pi_l(Z_0)| = \frac{|ax_0 + by_0 + c|}{\sqrt{a^2 + b^2}}.$$

(ii)
$$\pi_l(Z_0) \equiv \left(x_0 - \frac{a}{a^2 + b^2}(ax_0 + by_0 + c),\ y_0 - \frac{b}{a^2 + b^2}(ax_0 + by_0 + c)\right).$$

(iii)
$$s_l(Z_0) \equiv \left(x_0 - \frac{2a}{a^2 + b^2}(ax_0 + by_0 + c),\ y_0 - \frac{2b}{a^2 + b^2}(ax_0 + by_0 + c)\right).$$

Proof. Let m be the line such that $l \perp m$ and $Z_0 \in m$. Then as $l \perp m$, by 6.5.1 we will have $m \equiv -bx + ay + c_1 = 0$ for some c_1, and as $Z_0 \in m$ we have $c_1 = bx_0 - ay_0$. To find the coordinates (x, y) of $\pi_l(Z_0)$ we need to solve simultaneously the equations

$$ax + by = -c,\quad -bx + ay = -bx_0 + ay_0.$$

As for (i) we shall then go on to apply 6.1.1 it is $(x - x_0)^2$ and $(y - y_0)^2$ that we shall actually use, and it is easier to work directly with these. We rewrite the equations as

$$\begin{aligned} a(x - x_0) + b(y - y_0) &= -(ax_0 + by_0 + c), \\ -b(x - x_0) + a(y - y_0) &= 0. \end{aligned}$$

Now on squaring each of these and adding, we find that

$$(a^2 + b^2)[(x - x_0)^2 + (y - y_0)^2] = (ax_0 + by_0 + c)^2.$$

The conclusion (i) now readily follows.

For (ii) we solve these equations, obtaining

$$x - x_0 = -\frac{a}{a^2 + b^2}(ax_0 + by_0 + c),\ y - y_0 = -\frac{b}{a^2 + b^2}(ax_0 + by_0 + c)).$$

For (iii) we recall that if $s_l(Z_0) \equiv (x_1, y_1)$ and $\pi_l(Z_0) \equiv (x_2, y_2)$, then as $mp(Z_0, s_l(Z_0)) = \pi_l(Z_0)$ we have $x_1 + x_0 = 2x_2$, $y_1 + y_0 = 2y_2$. Now x_2 and y_2 are given by (ii) of the present theorem, and the result follows.

6.6.2   Formula for area of a triangle

Let $Z_1 \equiv_{\mathcal{F}} (x_1, y_1)$, $Z_2 \equiv_{\mathcal{F}} (x_2, y_2)$ *and* $Z_3 \equiv_{\mathcal{F}} (x_3, y_3)$ *be non- collinear points. Then the area* $\Delta[Z_1, Z_2, Z_3]$ *is equal to* $|\delta_{\mathcal{F}}(Z_1, Z_2, Z_3)|$ *where*

$$\begin{aligned} \delta_{\mathcal{F}}(Z_1, Z_2, Z_3) &= \tfrac{1}{2}[x_1(y_2 - y_3) - y_1(x_2 - x_3) + x_2 y_3 - x_3 y_2] \\ &= \tfrac{1}{2}\det\begin{pmatrix} x_1 & y_1 & 1 \\ x_2 & y_2 & 1 \\ x_3 & y_3 & 1 \end{pmatrix}. \end{aligned}$$

Proof. By 6.3.1 $Z_2 Z_3 \equiv$
$-(y_3 - y_2)(x - x_2) + (x_3 - x_2)(y - y_2) = 0$, so by 6.6.1
$|Z_1, \pi_{Z_2 Z_3}(Z_1)|$ is equal to
$\frac{|-(y_3 - y_2)(x_1 - x_2) + (x_3 - x_2)(y_1 - y_2)|}{\sqrt{(y_3 - y_2)^2 + (x_3 - x_2)^2}}$.

But $\Delta[Z_1, Z_2, Z_3] = \frac{1}{2}|Z_2, Z_3||Z_1, \pi_{Z_2 Z_3}(Z_1)|$, and the denominator above is equal to $|Z_2, Z_3|$. Hence the area is equal to half the numerator.

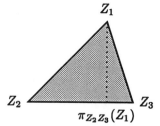

Figure 6.6. Area of a triangle.

6.6.3   Inequalities for closed half-planes

Let $l \equiv ax + by + c = 0$. Then the sets

$$\{Z \equiv (x,y) : ax + by + c \le 0\}, \qquad (6.6.1)$$
$$\{Z \equiv (x,y) : ax + by + c \ge 0\}, \qquad (6.6.2)$$

are the closed half-planes with common edge l.

  Proof. Let $Z_1 \equiv (x_1, y_1)$ be a point not in l, and let $s_l(Z_1) = Z_2 \equiv (x_2, y_2)$. Let $Z \equiv (x, y)$. Then as in 6.3.1, $Z \in l$ if and only if $|Z, Z_1|^2 = |Z, Z_2|^2$, and this occurs when $(x - x_1)^2 + (y - y_1)^2 = (x - x_2)^2 + (y - y_2)^2$, which simplifies to

$$2(x_2 - x_1)x + 2(y_2 - y_1)y + x_1^2 + y_1^2 - x_2^2 - y_2^2 = 0.$$

This is an equation for l and so by 6.3.1 there is some $j \ne 0$ such that

$$ax + by + c = j \left[2(x_2 - x_1)x + 2(y_2 - y_1)y + x_1^2 + y_1^2 - x_2^2 - y_2^2 \right].$$

By 4.3.4 the sets

$$\{Z \equiv (x,y) : 2(x_2 - x_1)x + 2(y_2 - y_1)y + x_1^2 + y_1^2 - x_2^2 - y_2^2 \le 0\}, \quad (6.6.3)$$
$$\{Z \equiv (x,y) : 2(x_2 - x_1)x + 2(y_2 - y_1)y + x_1^2 + y_1^2 - x_2^2 - y_2^2 \ge 0\}, \quad (6.6.4)$$

are the closed half-planes with edge l, as they correspond to $|Z, Z_1| \le |Z, Z_2|$ and $|Z, Z_1| \ge |Z, Z_2|$, respectively. But when $j > 0$, (6.6.1) and (6.6.3) coincide as do (6.6.2) and (6.6.4), while when $j < 0$, (6.6.1) and (6.6.4) coincide as do (6.6.2) and (6.6.3).

6.7   COORDINATE TREATMENT OF HARMONIC RANGES

6.7.1   New parametrization of a line

As in 6.4.1, if $Z_1 \equiv (x_1, y_1)$, $Z_2 \equiv (x_2, y_2)$, $Z \equiv (x, y)$ where $x = x_1 + t(x_2 - x_1)$, $y = y_1 + t(y_2 - y_1)$, then $Z \in Z_1 Z_2$ and

$$
\begin{aligned}
|Z_1, Z|^2 &= [t(y_2 - y_1)]^2 + [t(y_2 - y_1)]^2 = t^2 |Z_1, Z_2|^2, \\
|Z, Z_2|^2 &= [(1 - t)(x_2 - x_1)]^2 + [(1 - t)(y_2 - y_1)]^2 = (1 - t)^2 |Z_1, Z_2|^2, \\
\frac{|Z_1, Z|}{|Z, Z_2|} &= \left| \frac{t}{1 - t} \right|.
\end{aligned}
$$

Accordingly, if we write $\frac{t}{1-t} = \lambda$ where $\lambda \neq 0$ and so have $t = \frac{\lambda}{1+\lambda}$, we have

$$\frac{|Z_1, Z|}{|Z, Z_2|} = |\lambda|.$$

Thus Z divides (Z_1, Z_2) in the ratio $|\lambda| : 1$.

Changing our notation slightly, if we denote by $Z_3 \equiv (x_3, y_3)$ the point with

$$x_3 = x_1 + \frac{\lambda}{1+\lambda}(x_2 - x_1) = \frac{1}{1+\lambda}x_1 + \frac{\lambda}{1+\lambda}x_2,$$

$$y_3 = y_1 + \frac{\lambda}{1+\lambda}(y_2 - y_1) = \frac{1}{1+\lambda}y_1 + \frac{\lambda}{1+\lambda}y_2,$$

then Z_3 divides (Z_1, Z_2) in the ratio $|\lambda| : 1$. Consequently if we denote by $Z_4 \equiv (x_4, y_4)$ the point with

$$x_4 = \frac{1}{1+\lambda'}x_1 + \frac{\lambda'}{1+\lambda'}x_2, \ y_4 = \frac{1}{1+\lambda'}y_1 + \frac{\lambda'}{1+\lambda'}y_2,$$

where $\lambda' = -\lambda$, so that

$$x_4 = \frac{1}{1-\lambda}x_1 - \frac{\lambda}{1-\lambda}, \ y_4 = \frac{1}{1-\lambda}y_1 - \frac{\lambda}{1-\lambda}y_2,$$

then Z_4 also divides (Z_1, Z_2) in the ratio $|-\lambda| : 1 = |\lambda| : 1$.

Now $\lambda = \frac{t}{1-t}$ and if we write $-\lambda = \frac{s}{1-s}$ we have Z_4 in the original format,

$$x_4 = x_1 + s(x_2 - x_1), \ y_4 = y_1 + s(y_2 - y_1).$$

Then

$$\frac{t}{1-t} = -\frac{s}{1-s}$$

so that

$$s = \frac{\frac{1}{2}t}{t - \frac{1}{2}}.$$

Thus

$$s - \frac{1}{2} = \frac{\frac{1}{2}t}{t - \frac{1}{2}} - \frac{1}{2} = \frac{\frac{1}{2}t - \frac{1}{2}t + \frac{1}{4}}{t - \frac{1}{2}} = \frac{1}{4}\frac{1}{t - \frac{1}{2}}.$$

Hence

$$\left(s - \tfrac{1}{2}\right)\left(t - \tfrac{1}{2}\right) = \tfrac{1}{4}, \ \left|(s - \tfrac{1}{2})(t - \tfrac{1}{2})\right| = \tfrac{1}{4}. \tag{6.7.1}$$

Then we have three possibilities,

(a) $\left|t - \tfrac{1}{2}\right| < \tfrac{1}{2}, \ \left|s - \tfrac{1}{2}\right| > \tfrac{1}{2},$

(b) $\left|s - \tfrac{1}{2}\right| < \tfrac{1}{2}, \ \left|t - \tfrac{1}{2}\right| > \tfrac{1}{2},$

(c) $\left|s - \tfrac{1}{2}\right| = \tfrac{1}{2}, \ \left|t - \tfrac{1}{2}\right| = \tfrac{1}{2}.$

In (a) we have $-\tfrac{1}{2} < t - \tfrac{1}{2} < \tfrac{1}{2}$ and either $s - \tfrac{1}{2} < -\tfrac{1}{2}$ or $s - \tfrac{1}{2} > \tfrac{1}{2}$. Hence $0 < t < 1$ and either $s < 0$ or $s > 1$. It follows that $Z_3 \in [Z_1, Z_2]$, $Z_4 \notin [Z_1, Z_2]$.

The situation in (b) is like that in (a) with the roles of t and s, and so of Z_3 and Z_4 interchanged.

In (c) $-\frac{1}{2} = t - \frac{1}{2}$ or $t - \frac{1}{2} = \frac{1}{2}$, so either $t = 0$ or $t = 1$. Similarly either $s = 0$ or $s = 1$. We rule out the case of $t = 1$ as then λ would be undefined, and we rule out the case of $s = 1$ as then $-\lambda$ would be undefined. What remains is $t = s = 0$ and we excluded this by taking $\lambda \neq 0$; it would imply that $Z_3 = Z_4 = Z_1$.

Thus just one of Z_3, Z_4 is in the segment $[Z_1, Z_2]$ and the other is on the line $Z_1 Z_2$ but outside this segment. Hence Z_3 and Z_4 divide $\{Z_1, Z_2\}$ internally and externally in the same ratio. We recall that we then call (Z_1, Z_2, Z_3, Z_4) a harmonic range.

We note above that there can be no solution for s if $t = \frac{1}{2}$; thus there is no corresponding Z_4 when Z_3 is the mid-point Z_0 of Z_1 and Z_2. Similarly there can be no solution for t if $s = \frac{1}{2}$; thus there is no corresponding Z_3 when Z_4 is Z_0.

6.7.2 Interchange of pairs of points

If the points Z_3 and Z_4 divide $\{Z_1, Z_2\}$ internally and externally in the same ratio, then it turns out that the points Z_1 and Z_2 also divide $\{Z_3, Z_4\}$ internally and externally in the same ratio.

Proof. For we had

$$x_3 = \frac{1}{1+\lambda}x_1 + \frac{\lambda}{1+\lambda}x_2, \; y_3 = \frac{1}{1+\lambda}y_1 + \frac{\lambda}{1+\lambda}y_2,$$

$$x_4 = \frac{1}{1-\lambda}x_1 - \frac{\lambda}{1-\lambda}x_2, \; y_4 = \frac{1}{1-\lambda}y_1 - \frac{\lambda}{1-\lambda}y_2.$$

Then

$$(1+\lambda)x_3 = x_1 + \lambda x_2,$$
$$(1-\lambda)x_4 = x_1 - \lambda x_2.$$

By addition and subtraction, we find that

$$x_1 = \frac{1+\lambda}{2}x_3 + \frac{1-\lambda}{2}x_4,$$

$$x_2 = \frac{1+\lambda}{2\lambda}x_3 + \frac{\lambda-1}{2\lambda}x_4,$$

and by a similar argument,

$$y_1 = \frac{1+\lambda}{2}y_3 + \frac{1-\lambda}{2}y_4,$$

$$y_2 = \frac{1+\lambda}{2\lambda}y_3 + \frac{\lambda-1}{2\lambda}y_4.$$

If we define μ by

$$\frac{1}{1+\mu} = \frac{1+\lambda}{2},$$

so that

$$\mu = \frac{1-\lambda}{1+\lambda}, \qquad \frac{\mu}{1+\mu} = \frac{1-\lambda}{2},$$

then

$$x_1 = \frac{1}{1+\mu}x_3 + \frac{\mu}{1+\mu}x_4, \ y_1 = \frac{1}{1+\mu}y_3 + \frac{\mu}{1+\mu}y_4.$$

If we define μ' by

$$\frac{1}{1+\mu'} = \frac{1+\lambda}{2\lambda},$$

so that

$$\mu' = \frac{\lambda - 1}{1+\lambda}, \qquad \frac{\mu'}{1+\mu'} = \frac{\lambda - 1}{2\lambda},$$

then

$$x_2 = \frac{1}{1+\mu'}x_3 + \frac{\mu'}{1+\mu'}x_4, \ y_2 = \frac{1}{1+\mu'}y_3 + \frac{\mu'}{1+\mu'}y_4.$$

As $\mu' = -\mu$, this shows that Z_1 and Z_2 divide $\{Z_3, Z_4\}$ internally and externally in the same ratio.

6.7.3   Distances from mid-point

Let Z_0 be the mid-point of distinct points Z_1 and Z_2. Then points $Z_3, Z_4 \in Z_1 Z_2$ divide $\{Z_1, Z_2\}$ internally and externally in the same ratio if and only if Z_3 and Z_4 are on the one side of Z_0 on the line $Z_1 Z_2$ and

$$|Z_0, Z_3||Z_0, Z_4| = \tfrac{1}{4}|Z_1, Z_2|^2.$$

Proof. We have $Z_0 \equiv (x_0, y_0)$ where $x_0 = \frac{1}{2}(x_1 + x_2)$, $y_0 = \frac{1}{2}(y_1 + y_2)$. Then

$$x_3 - x_0 = (t - \tfrac{1}{2})(x_2 - x_1), \quad y_3 - y_0 = (t - \tfrac{1}{2})(y_2 - y_1),$$
$$x_4 - x_0 = (s - \tfrac{1}{2})(x_2 - x_1), \quad y_4 - y_0 = (s - \tfrac{1}{2})(y_2 - y_1),$$

and so

$$|Z_0, Z_3||Z_0, Z_4| = |(t - \tfrac{1}{2})(s - \tfrac{1}{2})||Z_1, Z_2|^2.$$

By (6.7.1) Z_3, Z_4 divide $\{Z_1, Z_2\}$ internally and externally in the same ratio if and only if $(s - \tfrac{1}{2})(t - \tfrac{1}{2}) = \tfrac{1}{4}$. This is equivalent to having $|(s - \tfrac{1}{2})(t - \tfrac{1}{2})| = \tfrac{1}{4}$ and $(s - \tfrac{1}{2})(t - \tfrac{1}{2}) > 0$. The latter is equivalent to having either $s - \tfrac{1}{2} > 0$ and $t - \tfrac{1}{2} > 0$, or $s - \tfrac{1}{2} < 0$ and $t - \tfrac{1}{2} < 0$, so that Z_3 and Z_4 are on the one side of Z_0 on the line $Z_1 Z_2$.

6.7.4   Distances from end-point

Let $\{Z_3, Z_4\}$ divide $\{Z_1, Z_2\}$ internally and externally in the same ratio with $Z_2 \in [Z_1, Z_4]$. Then

$$\frac{1}{2}\left(\frac{1}{|Z_1, Z_3|} + \frac{1}{|Z_1, Z_4|}\right) = \frac{1}{|Z_1, Z_2|}.$$

Proof. We have as before

$$x_3 = x_1 + \frac{\lambda}{1+\lambda}(x_2 - x_1), \ y_3 = y_1 + \frac{\lambda}{1+\lambda}(y_2 - y_1),$$

$$x_4 = x_1 + \frac{\lambda}{\lambda - 1}(x_2 - x_1), \quad y_4 = y_1 + \frac{\lambda}{\lambda - 1}(y_2 - y_1),$$

Now $\lambda/(\lambda - 1) > 1$ and so $\lambda > 1$. Hence $\frac{1}{2} < \lambda/(1 + \lambda) < 1$, and so $Z_3 \in [Z_1, Z_2]$. Thus Z_2, Z_3 and Z_4 are on the one side of Z_1 on the line $Z_1 Z_2$. Then

$$\frac{|Z_1, Z_3|}{|Z_1, Z_2|} = \frac{\lambda}{\lambda + 1}, \quad \frac{|Z_1, Z_4|}{|Z_1, Z_2|} = \frac{\lambda}{\lambda - 1},$$

so that

$$\frac{|Z_1, Z_2|}{|Z_1, Z_3|} = \frac{\lambda + 1}{\lambda}, \quad \frac{|Z_1, Z_2|}{|Z_1, Z_4|} = \frac{\lambda - 1}{\lambda},$$

and so

$$\frac{|Z_1, Z_2|}{|Z_1, Z_3|} + \frac{|Z_1, Z_2|}{|Z_1, Z_4|} = \frac{\lambda + 1}{\lambda} + \frac{\lambda - 1}{\lambda} = 2.$$

Hence

$$\frac{1}{2}\left(\frac{1}{|Z_1, Z_3|} + \frac{1}{|Z_1,, Z_4|}\right) = \frac{1}{|Z_1, Z_2|}.$$

This is expressed by saying that $|Z_1, Z_2|$ is the **harmonic mean** of $|Z_1, Z_3|$ and $|Z_1, Z_4|$.

6.7.5   Construction for a harmonic range

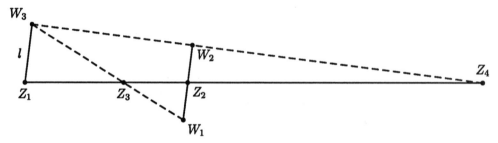

Figure 6.7.

Let Z_1, Z_2, Z_3 be distinct collinear points with Z_3 not the mid-point of Z_1 and Z_2. Take any points W_1 and W_2, not on $Z_1 Z_2$, so that Z_2 is the mid-point of W_1 and W_2. Let l be the line through Z_1 which is parallel to $W_1 W_2$ and let W_3 be the point in which $W_1 Z_3$ meets l, with Z_4 the point in which $W_2 W_3$ meets $Z_1 Z_2$. Then (Z_1, Z_2, Z_3, Z_4) is a harmonic range.

Proof. Without loss of generality we may take the x-axis to be the line $Z_1 Z_2$ and so take coordinates

$$Z_1 \equiv (x_1, 0), \quad Z_2 \equiv (x_2, 0), \quad Z_3 \equiv (x_3, 0), \quad Z_4 \equiv (x_4, 0),$$

and $W_1 \equiv (u_1, v_1)$, $W_2 \equiv (2x_2 - u_1, -v_1)$. The lines l and $W_1 Z_3$ have equations

$$(u_1 - x_2)y = v_1(x - x_1), \quad (u_1 - x_3)y = v_1(x - x_3),$$

respectively, and so W_3 has coordinates

$$u_3 = \frac{x_3(u_1 - x_2) - x_1(u_1 - x_3)}{x_3 - x_2}, \quad v_3 = v_1 \frac{x_3 - x_1}{x_3 - x_2}.$$

On the forming the equation of $W_2 W_3$ and finding where it meets $Z_1 Z_2$ we obtain

$$x_4 = \frac{-x_3(x_1 + x_2) + 2x_1 x_2}{x_1 + x_2 - 2x_3},$$

from which it follows that

$$x_4 - x_1 = \frac{(x_1 - x_2)(x_3 - x_1)}{x_1 + x_2 - 2x_3}, \quad x_2 - x_4 = \frac{(x_1 - x_2)(x_3 - x_2)}{x_1 + x_2 - 2x_3}.$$

From these we see that

$$\frac{x_4 - x_1}{x_2 - x_4} = -\frac{x_3 - x_1}{x_2 - x_3}.$$

Exercises

6.1 Suppose that Z_1, Z_2, Z_3 are non-collinear points and $Z_5 = \text{mp}\{Z_3, Z_1\}$, $Z_6 = \text{mp}\{Z_1, Z_2\}$. Show that if $|Z_2, Z_5| = |Z_3, Z_6|$, then $|Z_3, Z_1| = |Z_1, Z_2|$.[Hint. Select a frame of reference to simplify the calculations.]

6.2 Let l_1, l_2 be distinct intersecting lines and Z_0 a point not on either of them. Show that there are unique points $Z_1 \in l_1$, $Z_2 \in l_2$ such that Z_0 is the mid-point of Z_1 and Z_2.

6.3 Suppose that Z_1, Z_2, Z_3 are non-collinear points. Show that the points $Z \equiv (x, y)$, the perpendicular distances from which to the lines $Z_1 Z_2, Z_1 Z_3$ are equal, are those the coordinates of which satisfy

$$\frac{-(y_2 - y_1)(x - x_1) + (x_2 - x_1)(y - y_1)}{\sqrt{(x_2 - x_1)^2 + (y_2 - y_1)^2}}$$
$$\pm \frac{-(y_3 - y_1)(x - x_1) + (x_3 - x_1)(y - y_1)}{\sqrt{(x_3 - x_1)^2 + (y_3 - y_1)^2}} = 0.$$

Show that if $x = (1 - t)x_2 + tx_3$, $y = (1 - t)y_2 + ty_3$ then Z lies on the line with equation

$$\frac{-(y_2 - y_1)(x - x_1) + (x_2 - x_1)(y - y_1)}{\sqrt{(x_2 - x_1)^2 + (y_2 - y_1)^2}}$$
$$+ \frac{-(y_3 - y_1)(x - x_1) + (x_3 - x_1)(y - y_1)}{\sqrt{(x_3 - x_1)^2 + (y_3 - y_1)^2}} = 0,$$

if and only

$$t = \frac{|Z_1, Z_2|^2}{|Z_1, Z_2|^2 + |Z_2, Z_3|^2}.$$

Deduce that this latter line is the mid-line of $|Z_2 Z_1 Z_3$.

6.4 If the fixed triangle $[Z_1, Z_2, Z_3]$ is isosceles, with $|Z_1, Z_2| = |Z_1, Z_3|$, and Z is a variable point on the side $[Z_2, Z_3]$, show that the sum of the perpendicular distances from Z to the lines $Z_1 Z_2$ and $Z_1 Z_3$ is constant.[Hint. Select a frame of reference to simplify the calculations.]

6.5 Let $[A, B, C, D]$ be a parallelogram, $E = \text{mp}\{C, D\}$, $F = \text{mp}\{A, B\}$, and let AE and CF meet BD at G and H, respectively. Prove that $AE \parallel CF$ and $|D, G| = |G, H| = |H, B|$.

7

Circles; their basic properties

Hitherto our sets have involved lines and half-planes, and specific subsets of these. Now we introduce circles and study their relationships to lines. We do not do this just to admire the circles, and to behold their striking properties of symmetry. They are the means by which we control angles, and simplify our work on them.

7.1 INTERSECTION OF A LINE AND A CIRCLE

7.1.1

Definition. If O is any point of the plane Π and k is any positive real number, we call the set $C(O; k)$ of all points X in Π which are at a distance k from O, i.e. $C(O; k) = \{X \in \Pi : |O, X| = k\}$, the **circle** with *centre* O and length of radius k. If $X \in C(O; k)$ the segment $[O, X]$ is called a **radius** of the circle. Any point U such that $|O, U| < k$ is said to be an **interior point** for this circle. Any point V such that $|O, V| > k$ is said to be an **exterior point** for this circle.

For every circle $C(O; k)$ and line l, one of the following holds:-

(i) $l \cap C(O; k) = \{P\}$ *for some point P, in which case every point of $l \setminus \{P\}$ is exterior to the circle.*

(ii) $l \cap C(O; k) = \{P, Q\}$ *for some points P and Q, with $P \neq Q$, in which case every point of $[P, Q] \setminus \{P, Q\}$ is interior to the circle, and every point of $PQ \setminus [P, Q]$ is exterior to the circle.*

(iii) $l \cap C(O; k) = \emptyset$, *in which case every point of l is exterior to the circle.*

Proof. Let $M = \pi_l(O)$, and let m be the line which contains M and is perpendicular to l, so that $O \in m$.

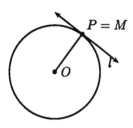

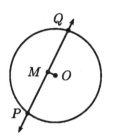

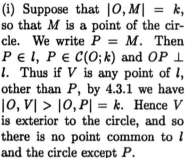

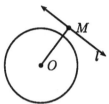

Figure 7.1.

(i) Suppose that $|O, M| = k$, so that M is a point of the circle. We write $P = M$. Then $P \in l$, $P \in \mathcal{C}(O; k)$ and $OP \perp l$. Thus if V is any point of l, other than P, by 4.3.1 we have $|O, V| > |O, P| = k$. Hence V is exterior to the circle, and so there is no point common to l and the circle except P.

(ii) Suppose that $|O, M| < k$, so that M is interior to the circle. Then $k^2 - |O, M|^2 > 0$ so that its square root can be extracted as a positive real number. By A_4(iv) choose $P \in l$ so that $|M, P| = \sqrt{k^2 - |O, M|^2}$. There is also a point $Q \in l$ on the other side of M from P and such that $|M, Q| = |M, P|$. Clearly M is the mid-point of P and Q.

When $M = O$, this gives $|O, P| = k$ so that $P \in \mathcal{C}(O; k)$. By 2.1.3 any point $X \neq P$ of the half-line $[O, P$ must satisfy either $X \in [O, P]$ or $P \in [O, X]$. If $X \in [O, P]$ then by 3.1.2 $|O, X| < |O, P| = k$, so that X is interior to the circle. On the other hand if $P \in [O, X]$ then $|O, X| > |O, P| = k$, and so X is an exterior point for the circle. Moreover Q is also on the circle and similar results hold when $X \in [O, Q$.

When $M \neq O$, we have $MP = l$, $MO = m$, so that $MP \perp MO$ and then by Pythagoras' theorem

$$|O, P|^2 = |O, M|^2 + |M, P|^2 = |O, M|^2 + [k^2 - |O, M|^2] = k^2;$$

thus again $|O, P| = k$, so that P is on the circle. By 2.1.3 any point $X \neq P$ of the half-line $[M, P$ must satisfy either $X \in [M, P]$ or $P \in [M, X]$. If $X \in [M, P]$, then by 3.1.2 $|M, X| < |M, P|$; when $X = M$, clearly X an interior point; when $X \neq M$, by Pythagoras' theorem this gives

$$|O, X|^2 = |O, M|^2 + |M, X|^2 < |O, M|^2 + |M, P|^2 = k^2,$$

so that $|O, X| < k$ and so again X is interior to the circle. If on the other hand $P \in [M, X]$, while still $X \neq P$, then by 3.1.2 $|M, X| > |M, P|$; by Pythagoras' theorem we have

$$|O, X|^2 = |O, M|^2 + |M, X|^2 > |O, M|^2 + |M, P|^2 = k^2,$$

so that $|O, X| > k$ and so X is exterior to the circle. Thus the points of $[M, P] \setminus \{P\}$ are interior to the circle, and the points of $([M, P) \setminus [M, P]$ are exterior to the circle.

Similar results hold when $X \in [M,Q$, that is the points of $[M,Q] \setminus \{Q\}$ are interior to the circle while the points of $([M,Q) \setminus [M,Q]$ are exterior to the circle. But $M \in [P,Q]$ so that $[P,M] \cup [M,Q] = [P,Q]$, $[M,P \cup [M,Q = PQ$ and so we can take these results together. Thus the points of $[P,Q]$, other than P and Q, are interior to the circle, and the points of $PQ \setminus [P,Q]$ are exterior to the circle, leaving just the points P and Q of the line $l = PQ$ in the circle.

(iii) Suppose that $|O,M| > k$, so that M is exterior to the circle. Then $M \in l$ and $OM \perp l$. If $X \in l, X \neq M$, then by 4.3.1, $|O,X| > |O,M| > k$, so that X is exterior to the circle.

Definition. If l is a line such that $l \cap \mathcal{C}(O;k) = \{P\}$ for a point P, then l is called a **tangent** to $\mathcal{C}(O;k)$ at P, and P is called the **point of contact.** If $l \cap \mathcal{C}(O;k) = \{P,Q\}$ for distinct points P and Q, then l is called a **secant** for $\mathcal{C}(O;k)$ and the segment $[P,Q]$ is called a **chord** of the circle; when $O \in l = PQ$, the chord $[P,Q]$ is called a **diameter** of the circle; in that case $O = \mathrm{mp}(P,Q)$. If $l \cap \mathcal{C}(O;k) = \emptyset$, then l is called a **non-secant** line for the circle.

NOTE. By the above every point of a tangent to a circle, other than the point of contact, is an exterior point. If $[P,Q]$ is a chord, every point of the chord other than its end-points P and Q is an interior point, while every point of $PQ \setminus [P,Q]$ is exterior. Every point of a non-secant line is an exterior point.

7.2 PROPERTIES OF CIRCLES

7.2.1

Circles have the following properties:-

(i) *If $[Q,S]$ is a diameter of the circle $\mathcal{C}(O;k)$ and P any point of the circle other than Q and S, then $PQ \perp PS$.*

(ii) *If points P,Q,S are such that $PQ \perp PS$, then P is on a circle with diameter $[Q,S]$.*

(iii) *If P is any point of the circle $\mathcal{C}(O;k)$, $[Q,S]$ is any diameter and $U = \pi_{QS}(P)$, then $U \in [Q,S]$ and $|Q,U| \leq 2k$.*

(iv) *If Q is a point of a circle with centre O and l is the tangent to the circle at Q, then every point of the circle lies in the closed half-plane with edge l in which O lies.*

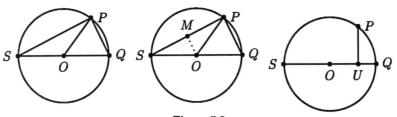

Figure 7.2.

Proof.

(i) By 5.2.2

$$|\angle OSP|^\circ + |\angle SPO|^\circ + |\angle POS|^\circ = 180, \quad |\angle OQP|^\circ + |\angle QPO|^\circ + |\angle POQ|^\circ = 180.$$

But by 4.1.1,

$$|\angle OSP|^\circ = |\angle SPO|^\circ, \quad |\angle OQP|^\circ = |\angle QPO|^\circ,$$

and so

$$2|\angle SPO|^\circ + 2|\angle QPO|^\circ + |\angle POS|^\circ + |\angle POQ|^\circ = 360.$$

Now $O \in [Q, S]$ so $|\angle POS|^\circ + |\angle POQ|^\circ = 180$, and as $[P, O \subset \mathcal{IR}(|\underline{QPS})$ we have $|\angle SPO|^\circ + |\angle OPQ|^\circ = |\angle QPS|^\circ$. Thus $|\angle QPS|^\circ = 90$.

(ii) Let O be the mid-point of Q and S and through O draw the line parallel to PQ. It will meet $[P, S]$ in a point M. Then by 5.3.1 M is the mid-point of P and S. But $PQ \perp PS$ and $PQ \parallel MO$ so by 5.1.1 $MO \perp PS$. Then $[O, P, M] \equiv [O, S, M]$ by the SAS principle of congruence. It follows that $|O, P| = |O, S|$.

(iii) If $P \notin QR$, then by (i) of the present theorem and 4.3.3 $U \in [Q, S]$. If $P \in QS$ then U is either Q or S and so $U \in [Q, S]$. Then by 3.1.2 $|Q, U| \le |Q, S|$. But as $O = \mathrm{mp}(Q, S)$, by 3.2.1 $|Q, O| = \frac{1}{2}|Q, S|$, and so $|Q, S| = 2k$.

(iv) Let $[Q, S]$ be the diameter containing Q and $\mathcal{H}_1$ the closed half-plane with edge l which contains O. Then by 2.2.3 every point of $[Q, O$ lies in $\mathcal{H}_1$. If P is any point of the circle and $U = \pi_{QO}(P)$ then by (iii) above $U \in [Q, S] \subset [Q, O$ so $U \in \mathcal{H}_1$. But $l \perp QS$, $UP \perp QS$ so $UP \parallel l$. Then by 4.3.2 $P \in \mathcal{H}_1$.

7.2.2   Equation of a circle

Let $Z_0 \equiv (x_0, y_0)$ and $k > 0$. Then $Z \equiv (x, y)$ is on $\mathcal{C}(Z_0; k)$ if and only if

$$(x - x_0)^2 + (y - y_0)^2 = k^2.$$

Proof. This is immediate by the distance formula in 6.1.1.

7.2.3   Circle through three points

Given any three non-collinear points A, B and C, there is a unique circle which passes through them.

Proof. Let l and m be the perpendicular bisectors of $[B, C]$ and $[C, A]$, respectively. Then if we had $l \parallel m$ we would have $l \parallel m$, $m \perp CA$ and so $l \perp CA$ by 5.1.1; this would yield $BC \perp l$, $CA \perp l$ and so $BC \parallel CA$ by 4.2.2(iv). This would make the points A, B, C collinear and so give a contradiction.

Thus l must meet m in a unique point, D say. Then by 4.1.1(iii) D is equidistant from B and C as it is on l, and it is equidistant from C and A as it is on m. Thus the circle with centre D and length of radius $|D, A|$ passes through A, B and C.

Conversely, suppose that a circle passes through A, B and C. Then by 4.1.1(ii) its centre must be on l and on m and so it must be D. The length of radius then must be $|D, A|$.

COROLLARY. *Two distinct circles cannot have more than two points in common.*

7.3 FORMULA FOR MID-LINE OF AN ANGLE-SUPPORT

7.3.1

COMMENT. We now start to prepare the ground for our treatment of angles. Earlier on we found that mid-points have a considerable role. Now we shall find that mid-lines of angle-supports, dealt with in 3.6, have a prominent role as well. Given any angle-support $|\underline{BAC}$, if we take any number $k > 0$ there are unique points P_1 and P_2 on $[A, B$ and $[A, C$ respectively, such that $|A, P_1| = k$, $|A, P_2| = k$. Thus P_1 and P_2 are the points of $[A, B$ and $[A, C$ on the circle $\mathcal{C}(A; k)$. Then $|\underline{BAC} = |\underline{P_1 A P_2}$ and it is far more convenient to work with the latter form. We first prove a result which will enable us to deal with the mid-lines of angle-supports by means of Cartesian coordinates.

With a frame of reference $\mathcal{F} = ([O, I , [O, J),$ let $P_1, P_2 \in \mathcal{C}(O; 1)$ be such that $P_1 \equiv_{\mathcal{F}} (a_1, b_1), P_2 \equiv_{\mathcal{F}} (a_2, b_2)$. Then the mid-line l of $|\underline{P_1 O P_2}$ has equation

$$(b_1 + b_2)x - (a_1 + a_2)y = 0$$

when P_1 and P_2 are not diametrically opposite, and equation $a_1 x + b_1 y = 0$ when they are.

Proof. When P_1 and P_2 are not diametrically opposite, their mid-point M is not O and we have $l = OM$. As M has coordinates $(\frac{1}{2}(a_1 + a_2), \frac{1}{2}(b_1 + b_2))$, the line OM has equation $(b_1 + b_2)x - (a_1 + a_2)y = 0$. When P_1 is diametrically opposite to P_2, l is the line through O which is perpendicular to OP and this has the given equation.

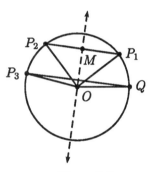

Figure 7.3.

With the notation of the last result, let $Q \equiv_{\mathcal{F}} (1, 0)$ and $s_l(Q) = P_3$ where $P_3 \equiv_{\mathcal{F}} (a_3, b_3)$. Then

$$a_3 = \frac{(a_1 + a_2)^2 - (b_1 + b_2)^2}{(a_1 + a_2)^2 + (b_1 + b_2)^2}, \quad b_3 = \frac{2(a_1 + a_2)(b_1 + b_2)}{(a_1 + a_2)^2 + (b_1 + b_2)^2},$$

when P_1 and P_2 are not diametrically opposite, and

$$a_3 = b_1^2 - a_1^2, \quad b_3 = -2a_1 b_1,$$

when they are.

Proof. For $l \equiv ax + by + c = 0$, we recall from 6.6.1 that

$$s_l(Z_0) \equiv \left(x_0 - \frac{2a}{a^2 + b^2}(ax_0 + by_0 + c), y_0 - \frac{2b}{a^2 + b^2}(ax_0 + by_0 + c) \right).$$

When P_1 and P_2 are not diametrically opposite, $l \equiv (b_1 + b_2)x - (a_1 + a_2)y = 0$. Thus for it $x_0 = 1$, $y_0 = 0$, $a = b_1 + b_2$, $b = -(a_1 + a_2)$, $c = 0$ and so

$$a_3 = \frac{(a_1 + a_2)^2 - (b_1 + b_2)^2}{(a_1 + a_2)^2 + (b_1 + b_2)^2}, \quad b_3 = \frac{2(a_1 + a_2)(b_1 + b_2)}{(a_1 + a_2)^2 + (b_1 + b_2)^2}.$$

When P_1 and P_2 are diametrically opposite, $l \equiv a_1x + b_1y = 0$. Thus for it $x_0 = 1$, $y_0 = 0$, $a = a_1$, $b = b_1$, $c = 0$, so we have

$$a_3 = b_1^2 - a_1^2, \ b_3 = -2a_1b_1,$$

as $a_1^2 + b_1^2 = 1$.

7.4 POLAR PROPERTIES OF A CIRCLE

7.4.1 Tangents from an exterior point

Let P be a point exterior to a circle C. Then two tangents to the circle pass through P. Their points of contact are equidistant from P.

Proof. Let the circle have centre O and length of radius a. Let $|O, P| = b$, so that $b > a$. Choose the point $U \in [O, P$ so that $x = |O, U| = a^2/b$. As $b > a$, then $x < a < b$ so $U \in [O, P]$. Erect a perpendicular to OP at U and mark off on it a distance

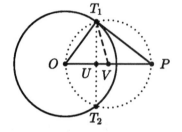

T_1

T_2

Figure 7.4.

$$y = |U, T_1| = a\sqrt{1 - \left(\frac{a}{b}\right)^2}.$$

Then, by Pythagoras' Theorem,

$$|O, T_1|^2 = |O, U|^2 + |U, T_1|^2 = x^2 + y^2 = \frac{a^4}{b^2} + a^2 - \frac{a^4}{b^2} = a^2,$$

so that $T_1 \in C$.

Let V be the mid-point of O and P, so that $V \in [O, P]$ and $|O, V| = \frac{b}{2}$. Then

$$|U, V| = \pm(|O, V| - |O, U|) = \pm(\tfrac{1}{2}b - x).$$

Again by Pythagoras' Theorem,

$$|V, T_1|^2 = |U, V|^2 + |U, T_1|^2 = (\tfrac{1}{2}b - x)^2 + y^2$$
$$= \left(\tfrac{1}{2}b - \frac{a^2}{b}\right)^2 + a^2\left(1 - \frac{a^2}{b^2}\right) = \tfrac{1}{4}b^2.$$

Thus T_1 is on the circle C_1 with centre V and radius length $\frac{b}{2}$. Note that C_1 also passes through O and P. Then $\angle OT_1P$ is an angle in a semi-circle of C_1, so that by 7.2.1 it is a right-angle. Thus by 7.1.1 PT_1 is a tangent to C at T_1.

By a similar argument, if we take T_2 so that U is the mid-point of T_1 and T_2, then PT_2 is also a tangent to C at T_2. We note that T_1 and T_2 are both on the line which is perpendicular to OP at the point U.

By Pythagoras' theorem

$$|P, T_1|^2 = |O, P|^2 - |O, T_1|^2 = |O, P|^2 - |O, T_2|^2 = |P, T_2|^2,$$

and so $|P, T_1| = |P, T_2|$.

There cannot be a third tangent PT_3 as then T_3 would be on C and C_1, whereas by 7.2.3 these circles have only two points in common.

7.4.2   The power property of a circle

For a fixed circle $C(O; k)$ and fixed point $P \notin C(O; k)$, let a variable line l through P meet $C(O; k)$ at R and S. Then the product of distances $|P, R||P, S|$ is constant. When P is exterior to the circle,

$$|P, R||P, S| = |P, T_1|^2,$$

where T_1 is the point of contact of a tangent from P to the circle.

Proof. By the distance formula $Z \equiv (x, y)$ is on $C(O; k)$ if and only if $x^2 + y^2 - k^2 = 0$. If $P \equiv (x_0, y_0)$ and l has Cartesian equation $ax + by + c = 0$, by 6.4.1 points Z on l have parametric equations of the form $x = x_0 + bt$, $y = y_0 - at$ $(t \in \mathbb{R})$. Now l also has Cartesian equation

$$\frac{a}{\sqrt{a^2 + b^2}}x + \frac{b}{\sqrt{a^2 + b^2}}y + \frac{c}{\sqrt{a^2 + b^2}} = 0.$$

Thus as we we may replace a and b by $a/\sqrt{a^2 + b^2}$ and $b/\sqrt{a^2 + b^2}$, without loss of generality we may assume that $a^2 + b^2 = 1$. Then the point Z on the line lies on the circle if $(x_0 + bt)^2 + (y_0 - at)^2 - k^2 = 0$, that is if

$$t^2 + 2(bx_0 - ay_0)t + x_0^2 + y_0^2 - k^2 = 0.$$

If t_1, t_2 are the roots of this equation, then $t_1 t_2 = x_0^2 + y_0^2 - k^2$. As for R and S we have

$$x_1 = x_0 + bt_1,\ y_1 = y_0 - at_1,\ x_2 = x_0 + bt_2,\ y_2 = y_0 - at_2,$$

so $|P, R| = |t_1|$, $|P, S| = |t_2|$. Thus

$$|P, R||P, S| = |t_1 t_2| = |x_0^2 + y_0^2 - k^2|,$$

which is constant.

When P is exterior to the circle, the roots of the quadratic equation are equal if

$$(bx_0 - ay_0)^2 = x_0^2 + y_0^2 - k^2,$$

and the repeated root is given by $t = -(bx_0 - ay_0)$. Then for a point of intersection T_1 of the line and circle, we have for the coordinates of T_1

$$x = x_0 - (bx_0 - ay_0)b,\quad y = y_0 + (bx_0 - ay_0)a.$$

Hence

$$|P, T_1|^2 = (x - x_0)^2 + (y - y_0)^2 = (bx_0 - ay_0)^2 = x_0^2 + y_0^2 - k^2 = |O, P|^2 - k^2.$$

It is also easy to give a synthetic proof as follows. We first take P interior to the circle. Let M be the mid-point of R and S so that M is the foot of the perpendicular from O to RS. Then P is in either $[R, M]$ or $[M, S]$; we suppose that $P \in [R, M]$.

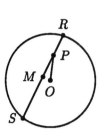
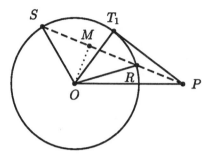

Figure 7.5.

Then

$$|P,R||P,S| = (|M,R| - |P,M|)(|M,S| + |P,M|) = |M,R|^2 - |P,M|^2$$
$$= |M,R|^2 - \left(|P,O|^2 - |O,M|^2\right) = \left(|M,R|^2 + |O,M|^2\right) - |P,O|^2$$
$$= |O,R|^2 - |P,O|^2 = k^2 - |P,O|^2,$$

and this is fixed.

We continue with the case where P is exterior to the circle, and may suppose that $|P,R| < |P,S|$, as otherwise we can just interchange the points R and S. As P is outside the circle, by 7.1.1 it is outside the segment $[R,S]$ on the line RS. Then we have

$$|P,R||P,S| = (|P,M| - |M,R|)(|P,M| + |M,S|) = |P,M|^2 - |M,R|^2$$
$$= \left(|P,O|^2 - |O,M|^2\right) - |M,R|^2 = |P,O|^2 - \left(|O,M|^2 + |M,R|^2\right)$$
$$= |P,O|^2 - |O,R|^2 = |P,O|^2 - |O,T_1|^2 = |P,T_1|^2.$$

7.4.3　A harmonic range

Let T_1 and T_2 be the points of contact of the tangents from an exterior point P to a circle C with centre O. If a line l through P cuts C in the points R and S, and cuts T_1T_2 in Q, then P and Q divide $\{R,S\}$ internally and externally in the same ratio.

Proof. We use the notation of 7.4.1 and first recall that T_1T_2 cuts OP at right-angles at a point U. Then, by 7.2.1(ii), the circle C_2 on $[O,Q]$ as a diameter passes through U. We let M be the mid-point of R and S; then by 4.1.1 $OM \perp MQ$, and so M also lies on the circle C_2.

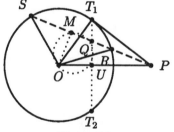

Figure 7.6.

We have $|P,R||P,S| = |P,T_1|^2$ by 7.4.2, $|P,T_1|^2 = |P,U||P,O|$ by the proof of Pythagoras' theorem in 5.4.1, and $|P,U||P,O| = |P,Q||P,M|$, by the 7.4.2 applied to the circle C_2. On combining these we have $|P,R||P,S| = |P,Q||P,M|$.

We cannot have $l \perp OP$ as that would make $l \parallel T_1T_2$ whereas l meets T_1T_2. Then, with the notation of 7.4.1, l is not a tangent to C_1 at P so, by 7.1.1 l must meet C_1 at a point H. We are supposing that l is not either of PT_1, PT_2 and so H is not T_1 or T_2. We let K be the foot of the perpendicular from H to OP. Then by 4.3.3 $K \in [P,O]$ and by the proof of Pythagoras' theorem in 5.4.1 $|P,H|^2 = |P,K||P,O|$. If we had $K \in [P,U]$ we would have $|P,K| < |P,U|$ and so

$$|P,H|^2 = |P,O||P,K| < |P,O||P,U| = |P,T_1|^2.$$

From this it would follow that

$$|O,H|^2 = |O,P|^2 - |P,H|^2 > |O,P|^2 - |P,T_1|^2 = a^2,$$

and make H exterior to the circle. But H is the foot of the perpendicular from O to l, and by 7.1.1 this would cause l to have no point in common with the circle. This cannot occur and so we must have $K \in [O,U]$. By a similar argument it then follows that H is interior to the circle C and so l meets C in two points R and S.

By 7.2.1(iv) every point of the circle C is in the closed half-plane $\mathcal{H}_1$ with edge PT_1 and which contains O. By 2.2.3 $\mathcal{H}_1$ contains $U \in [P,O$ and then it also contains $T_2 \in [T_1,U$. Similarly every point of C is also in the closed half-plane with edge PT_2 and which contains T_1. It follows that every point of C lies in the interior region $\mathcal{IR}(|T_1PT_2)$. Now every point of $[P,R$ is in this interior region and so Q is. It follows that $Q \in [T_1,T_2]$ and so by 7.1.1 is interior to the circle; we thus must have $Q \in [R,S]$ by 7.1.1 again.

We let $x = |P,R|$, $y = |P,S|$, $z = |P,Q|$, and without loss of generality assume $|P,R| < |P,S|$ so that $x < y$. As P is outside the circle it is outside the segment $[R,S]$; as Q is on the segment $[R,S]$, it follows that $0 < x < z < y$. Then in turn

$$xy = \frac{1}{2}(x+y)z, \quad \frac{2}{z} = \frac{1}{x} + \frac{1}{y}, \quad \frac{1}{x} - \frac{1}{z} = \frac{1}{z} - \frac{1}{y},$$

$$\frac{z-x}{xz} = \frac{y-z}{zy}, \quad \frac{x}{z-x} = \frac{y}{y-z}, \quad \frac{|P,R|}{|R,Q|} = \frac{|P,S|}{|S,Q|}.$$

In the above we have assumed that l is not the line OP. When it is we have a simple case; l cuts the circle in points R_1, S_1 such that $[R_1,S_1]$ is a diameter. Then taking $|P,R_1| < |P,S_1|$, with the notation of 7.4.1 we have that

$$\frac{|S_1,P|}{|P,R_1|} = \frac{b+a}{b-a}, \quad \frac{|S_1,U|}{|U,R_1|} = \frac{a+a^2/b}{a-a^2/b},$$

and these are equal.

7.5   ANGLES STANDING ON ARCS OF CIRCLES

7.5.1

Let P,Q,R,S be points of a circle $C(O;k)$ such that R and S are on the same side of the line PQ. Then $|\angle PRQ|^\circ = |\angle PSQ|^\circ$.
   (i) When $O \in PQ$, $|\angle PRQ|^\circ = 90$;

(ii) when $O \notin PQ$ and R is on the same side of PQ as O is, then $|\angle PRQ|° = \frac{1}{2}|\angle POQ|°$;

(iii) when $O \notin PQ$ and R is on the opposite side of PQ from O, then $|\angle PRQ|°$ is equal to half of the degree-measure of the reflex-angle with support $|POQ$.

Proof. Now $R \notin PQ$ as by 7.1.1 a line cannot meet the circle in more than two points; for this reason also S cannot be on a side of the triangle $[P,Q,R]$. Moreover, neither can S be in $[P,Q,R]$ but not on a side, as then by the cross-bar theorem we would have $S \in [P,V]$ for some point V in $[Q,R]$ but not at an end-point. Then V is interior to the circle and P is on the circle, so by 7.1.1 every point of $[P,V]$, other than P, is interior to the circle; this would make S interior to the circle whereas it is on it.

Thus as $S \notin [P,Q,R]$ we must have at least one of the following

(a) S is on the opposite side of QR from P, (b) S is on the opposite side of RP from Q,

(c) S is on the opposite side of PQ from R,

and of course (c) is ruled out by assumption.

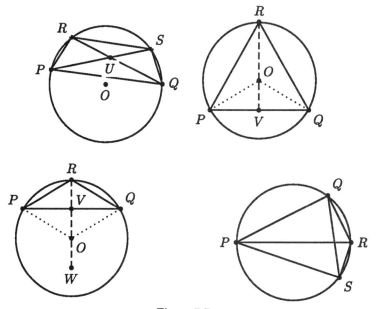

Figure 7.7.

We suppose that (a) holds as in the first figure; the case of (b) is treated similarly. Then there is a point $U \in [P,S] \cap QR$. As $U \in [P,S]$, U must be an interior point for the circle and hence we must have $U \in [Q,R]$. By 7.4.2

$$\frac{|U,P|}{|U,Q|} = \frac{|U,R|}{|U,S|},$$

and we also have $|\angle PUR|° = |\angle QUS|°$ as these are opposite angles. By 5.3.2, the triangles $[U,P,R], [U,Q,S]$ are similar. In particular $|\angle PRU|° = |\angle QSU|°$. The first diagram in Fig. 7.7 deals with this general case.

In (i) when $O \in PQ$, that we have a right-angle comes from 7.2.1 and there is a diagram for this in Fig. 7.2.

When $O \notin PQ$ we let V be the mid-point of $\{P, Q\}$ and for the second case (ii), as in the second diagram in Fig. 7.7, we take R to be point in which $[V, O$ meets the circle. Then by 5.2.2, Corollary, and 4.1.1(i) $|\angle VOP|^\circ = 2|\angle VRP|^\circ$, $|\angle VOQ|^\circ = 2|\angle VRQ|^\circ$. But $[R, V \subset \mathcal{IR}(|PRQ)$ so that $|\angle VRP|^\circ + |\angle VRQ|^\circ = |\angle PRQ|^\circ$. Moreover $[O, V \subset \mathcal{IR}(|POQ)$ so that $|\angle VOP|^\circ + |\angle VOQ|^\circ = |\angle POQ|^\circ$. By addition we then have that $|\angle POQ|^\circ = 2|\angle PRQ|^\circ$.

For the third case (iii), as in the third diagram in Fig. 7.7, we take R to be point in which $[O, V$ meets the circle and $W \ne O$ a point such that $O \in [V, W]$. Then by 5.2.2, Corollary, and 4.1.1(i) $|\angle WOP|^\circ = 2|\angle WRP|^\circ$, $|\angle WOQ|^\circ = 2|\angle WRQ|^\circ$. But $[R, W \subset \mathcal{IR}(|PRQ)$ so that $|\angle WRP|^\circ + |\angle WRQ|^\circ = |\angle PRQ|^\circ$. Moreover $[O, V \subset \mathcal{IR}(|POQ)$ so that by 3.7.1 $|\angle WOP|^\circ + |\angle WOQ|^\circ$ is equal to the degree-measure of the reflex- angle with support $|\angle POQ$. By addition we then have that the degree-measure of this reflex-angle is equal to $2|\angle PRQ|^\circ$.

Definition. If the vertices of a convex quadrilateral all lie on some circle, then the quadrilateral is said to be **cyclic**.

COROLLARY. *Let $[P, Q, R, S]$ be a convex cyclic quadrilateral. Then the sum of the degree-measures of a pair of opposite angles is 180.*

Proof. Using the fourth diagram in Fig. 7.7, we first we note that

$$|\angle RPQ|^\circ + |\angle PQR|^\circ + |\angle QRP|^\circ = 180.$$

Next as $[S, Q \subset \mathcal{IR}(|PSR)$, we have $|\angle PSR|^\circ = |\angle PSQ|^\circ + |\angle QSR|^\circ$. But $|\angle PSQ|^\circ = |\angle QRP|^\circ$, $|\angle QSR|^\circ = |\angle RPQ|^\circ$. Hence

$$|\angle PSR|^\circ + |\angle PQR|^\circ = |\angle RPQ|^\circ + |\angle PQR|^\circ + |\angle QRP|^\circ = 180.$$

7.5.2 Minor and major arcs of a circle

Definition. Let $P_1, P_2 \in \mathcal{C}(O; k)$ be distinct points such that $O \notin P_1 P_2$. Let $\mathcal{H}_5, \mathcal{H}_6$ be the closed half-planes with edge $P_1 P_2$, with $O \in \mathcal{H}_5$. Then $\mathcal{C}(O; k) \cap \mathcal{H}_5$, $\mathcal{C}(O; k) \cap \mathcal{H}_6$, are called, respectively, the **major** and **minor arcs** of $\mathcal{C}(O; k)$ with end-points P_1 and P_2.

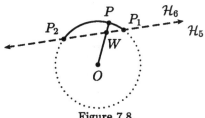

Figure 7.8.

The point $P \in \mathcal{C}(O; k)$ is in the minor arc with end-points P_1, P_2 if and only if $[O, P] \cap [P_1, P_2] \ne \emptyset$.

Proof. Let P be in the minor arc. Then $O \in \mathcal{H}_5, P \in \mathcal{H}_6$ so $[O, P]$ meets $P_1 P_2$ in some point W. As $W \in [O, P]$ we have $|O, W| \le k$ so by 7.1.1 $W \in [P_1, P_2]$.

Conversely suppose that $W \in [P_1, P_2]$ so that $|O, W| \le k$. Choose $P \in [O, W$ so that $|O, P| = k$. Then as $|O, W| \le |O, P|$ we have $W \in [O, P]$ so that as $O \in \mathcal{H}_5$ we have $P \in \mathcal{H}_6$.

7.6   SENSED DISTANCES

7.6.1   Sensed distance

Definition If l is a line, $\leq_l$ is a natural order on l and Z_1, $Z_2 \in l$, then we define $\overline{Z_1 Z_2}_{\leq_l}$ by

$$\overline{Z_1 Z_2}_{\leq_l} = \begin{cases} |Z_1, Z_2|, & \text{if } Z_1 \leq_l Z_2, \\ -|Z_1, Z_2|, & \text{if } Z_2 \leq_l Z_1, \end{cases}$$

and call this the **sensed distance** from Z_1 to Z_2. In knowing this rather than just the distance from Z_1 to Z_2 we have extra information which can be turned to good account. It can have negative as well as positive and zero values and it is related to the distance as $\overline{Z_1 Z_2}_{\leq_l} = \pm|Z_1, Z_2|$ or equivalently $|\overline{Z_1 Z_2}_{\leq_l}| = |Z_1, Z_2|$.

We note immediately the properties:

$$\overline{Z_1 Z_1}_{\leq_l} = 0, \tag{7.6.1}$$

$$\overline{Z_2 Z_1}_{\leq_l} = -\overline{Z_1 Z_2}_{\leq_l}, \tag{7.6.2}$$

in all cases. We can add sensed distances on a line and have the striking property that

$$\overline{Z_1 Z_2}_{\leq_l} + \overline{Z_2 Z_3}_{\leq_l} = \overline{Z_1 Z_3}_{\leq_l}, \tag{7.6.3}$$

for all Z_1, Z_2, $Z_3 \in l$. This is easily seen to hold by (7.6.1) when any two of the three points coincide, as e.g. when $Z_1 = Z_2$ it amounts to $0 + \overline{Z_1 Z_3}_{\leq_l} = \overline{Z_1 Z_3}_{\leq_l}$. Suppose then that Z_1, Z_2, Z_3 are all distinct and suppose first that $Z_1 \leq_l Z_2$. Then by 2.1.3 we have one of the cases

(a) $Z_3 \leq_l Z_1 \leq_l Z_2$,   (b) $Z_1 \leq_l Z_3 \leq_l Z_2$,   (c) $Z_1 \leq_l Z_2 \leq_l Z_3$.

In (a) we have

$$\overline{Z_1 Z_2}_{\leq_l} = |Z_1, Z_2|, \quad \overline{Z_2 Z_3}_{\leq_l} = -|Z_2, Z_3|, \quad \overline{Z_1 Z_3}_{\leq_l} = -|Z_1, Z_3|,$$

and as $Z_1 \in [Z_3, Z_2]$, $|Z_3, Z_1| + |Z_1, Z_2| = |Z_3, Z_2|$, which is $-\overline{Z_1 Z_3}_{\leq_l} + \overline{Z_1 Z_2}_{\leq_l} = -\overline{Z_2 Z_3}_{\leq_l}$. In (b) we have

$$\overline{Z_1 Z_2}_{\leq_l} = |Z_1, Z_2|, \quad \overline{Z_2 Z_3}_{\leq_l} = -|Z_2, Z_3|, \quad \overline{Z_1 Z_3}_{\leq_l} = |Z_1, Z_3|,$$

and as $Z_3 \in [Z_1, Z_2]$, $|Z_1, Z_3| + |Z_3, Z_2| = |Z_1, Z_2|$, which is $\overline{Z_1 Z_3}_{\leq_l} - \overline{Z_2 Z_3}_{\leq_l} = \overline{Z_1 Z_2}_{\leq_l}$. In (c) we have

$$\overline{Z_1 Z_2}_{\leq_l} = |Z_1, Z_2|, \quad \overline{Z_2 Z_3}_{\leq_l} = |Z_2, Z_3|, \quad \overline{Z_1 Z_3}_{\leq_l} = |Z_1, Z_3|,$$

and as $Z_2 \in [Z_1, Z_3]$, $|Z_1, Z_2| + |Z_2, Z_3| = |Z_1, Z_3|$, which is $\overline{Z_1 Z_2}_{\leq_l} + \overline{Z_2 Z_3}_{\leq_l} = \overline{Z_1 Z_3}_{\leq_l}$.

Next suppose that $Z_2 \leq_l Z_1$. Then on interchanging Z_1 and Z_2 in the cases just proved we have $\overline{Z_2 Z_1}_{\leq_l} + \overline{Z_1 Z_3}_{\leq_l} = \overline{Z_2 Z_3}_{\leq_l}$, for all $Z_3 \in l$ and by (7.6.2) this gives $-\overline{Z_1 Z_2}_{\leq_l} + \overline{Z_1 Z_3}_{\leq_l} = \overline{Z_2 Z_3}_{\leq_l}$. This completes the proof of (7.6.3) which shows that addition of sensed distances on a line is much simpler than addition of distances.

We next relate sensed distances to the parametric equations of l noted in 6.4.1, Corollary. Suppose that $W_0 \equiv (u_0, v_0)$, $W_1 \equiv (u_1, v_1)$ are distinct points on l and that $W_0 \leq_l W_1$. Then for points $Z_1 \equiv (x_1, y_1)$, $Z_2 \equiv (x_2, y_2)$ on l we have

$$x_1 = u_0 + s_1(u_1 - u_0), \; y_1 = v_0 + s_1(v_1 - v_0),$$
$$x_2 = u_0 + s_2(u_1 - u_0), \; y_2 = v_0 + s_2(v_1 - v_0),$$

and we recall that $Z_1 \leq_l Z_2$ if and only if $s_1 \leq s_2$. Moreover, by the distance formula

$$|Z_1, Z_2| = |s_2 - s_1| \, |W_0, W_1|.$$

From these we conclude that

$$\overline{Z_1 Z_2}_{\leq_l} = (s_2 - s_1)|W_0, W_1|. \tag{7.6.4}$$

In particular the simplest case of parametric representation in relation to sensed distances is when we additionally take $|W_0, W_1| = 1$ as we then have $\overline{Z_1 Z_2}_{\leq_l} = s_2 - s_1$.

When we consider the reciprocal natural order on l we note that

$$\overline{Z_1 Z_2}_{\geq_l} = -\overline{Z_1 Z_2}_{\leq_l},$$

so that changing to the reciprocal natural order multiplies the value by -1. As well as adding sensed distances on one line we can multiply or divide them. Now for Z_1, Z_2, Z_3, Z_4 in l,

$$\overline{Z_3 Z_4}_{\geq_l} \, \overline{Z_1 Z_2}_{\geq_l} = -\overline{Z_3 Z_4}_{\leq_l} (-1)\overline{Z_1 Z_2}_{\leq_l} = \overline{Z_3 Z_4}_{\leq_l} \, \overline{Z_1 Z_2}_{\leq_l},$$

so this **sensed product** is independent of which natural order is taken. Similarly, when $Z_1 \neq Z_2$, we can take a ratio of sensed distances

$$\frac{\overline{Z_3 Z_4}_{\leq_l}}{\overline{Z_1 Z_2}_{\leq_l}} = \frac{-\overline{Z_3 Z_4}_{\geq_l}}{-\overline{Z_1 Z_2}_{\geq_l}} = \frac{\overline{Z_3 Z_4}_{\geq_l}}{\overline{Z_1 Z_2}_{\geq_l}} = \frac{s_4 - s_3}{s_2 - s_1},$$

and see that this **sensed ratio** is independent of whichever of $\leq_l$, $\geq_l$ is used. When the line l is understood, we can relax our notation to $\overline{Z_3 Z_4} \, \overline{Z_1 Z_2}$ and $\frac{\overline{Z_3 Z_4}}{\overline{Z_1 Z_2}}$ for these products and ratios.

If for $Z_1, Z_2, Z \in l$ we take the parametric equations

$$x = x_1 + t(x_2 - x_1), \; y = y_1 + t(y_2 - y_1), \; (t \in \mathbf{R}),$$

then we have that

$$x = u_0 + [s_1 + t(s_2 - s_1)](u_1 - u_0), \; y = v_0 + [s_1 + t(s_2 - s_1)](v_1 - v_0),$$

and by (7.6.4) we have that

$$\overline{Z_1 Z}_{\leq_l} = t(s_2 - s_1)|W_0, W_1|, \; \overline{Z Z_2}_{\leq_l} = (1-t)(s_2 - s_1)|W_0, W_1|,$$

and so

$$\frac{\overline{Z_1 Z}}{\overline{Z Z_2}} = \frac{t}{1-t}. \tag{7.6.5}$$

Our main utilisation of these concepts is through sensed ratios; for example (Z_1, Z_2, Z_3, Z_4) is a harmonic range when $\overline{Z_1 Z_3}/\overline{Z_3 Z_2} = -\overline{Z_1 Z_4}/\overline{Z_4 Z_2}$. It is convenient to defer the details until Chapter 11. However we make one use of sensed products in the next subsection.

7.6.2 Sensed products and a circle

The conclusion of 7.4.2 can be strengthened to replace $|P,R||P,S|$ by $\overline{PR}\,\overline{PS}$. In fact the initial analytic proof gives this but it also easily follows from the stated result as $\overline{PR}\,\overline{PS} = -|P,R||P,S|$ when P is interior to the circle while $\overline{PR}\,\overline{PS} = |P,R||P,S|$ when P is exterior to the circle. We now look to a converse type of result.

Suppose that Z_1, Z_2 and Z_3 are fixed non-collinear points. For a variable point W let Z_1W meet Z_2Z_3 at W' and

$$\overline{W'W}\,\overline{W'Z_1} = \overline{W'Z_2}\,\overline{W'Z_3}.$$

Then W lies on the circle which passes through Z_1, Z_2 and Z_3.

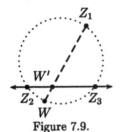

Figure 7.9.

Proof. Without loss of generality we may take our frame of reference so that $Z_1 \equiv (0, y_1)$, $Z_2 \equiv (x_2, 0)$, $Z_3 \equiv (x_3, 0)$, and we take $W \equiv (u, v)$, $W' \equiv (u', 0)$.

Then it is easily found that $u' = y_1 u/(y_1 - v)$, and so, first of all,

$$\overline{W'Z_2}\,\overline{W'Z_3} = \left(x_2 - \frac{y_1 u}{y_1 - v} \right) \left(x_3 - \frac{y_1 u}{y_1 - v} \right).$$

The line $W'W$ has parametric equations $x = u' + s(u - u')$, $y = 0 + s(v - 0)$, with $s = 0$ giving W' and $s = 1$ giving W. Thus $\overline{W'W} = |W', W|$. The point Z_1 has parameter given by $y_1 = sv$ and so $s = y_1/v$; then $\overline{W'Z_1} = \frac{y_1}{v}|W', W|$. It follows that

$$\overline{W'W}\,\overline{W'Z_1} = \frac{y_1}{v}|W', W|^2 = \frac{y_1}{v}\left[\left(u - \frac{y_1 u}{y_1 - v} \right)^2 + v^2 \right]$$

$$= y_1 v\left[\left(\frac{u}{y_1 - v} \right)^2 + 1 \right].$$

On equating the two expressions we have

$$\left(x_2 - \frac{y_1 u}{y_1 - v} \right) \left(x_3 - \frac{y_1 u}{y_1 - v} \right) = y_1 v\left[\left(\frac{u}{y_1 - v} \right)^2 + 1 \right],$$

which we re-write as

$$\frac{y_1 u^2 (y_1 - v)}{(y_1 - v)^2} = y_1 v - x_2 x_3 + \frac{y_1 (x_2 + x_3)u}{y_1 - v}.$$

On multiplying across by $y_1 - v$ we obtain

$$y_1 (u^2 + v^2) - y_1 (x_2 + x_3)u - (y_1^2 + x_2 x_3)v + y_1 x_2 x_3 = 0,$$

and this is the equation of a circle.

7.6.3   Radical axis and coaxal circles

In 7.4.2 our proof showed that if a line through the point Z meets the circle $C(Z_1, k_1)$ at the points R and S then $\overline{ZR}\,\overline{ZS} = |Z_1, Z|^2 - k_1^2$ depends only on the circle and the point Z. We call this expression the **power** of the point Z with respect to this circle. To cater for degenerate cases, when $k_1 = 0$ we also call $|Z_1, Z|^2$ the power of Z with respect to the point Z_1.

We let C_1 denote either $C(Z_1, k_1)$ or Z_1 and similarly consider C_2 which is either $C(Z_2, k_2)$ or Z_2. We ask *for what points Z its powers with respect to C_1 and C_2 are equal*. This occurs when $|Z_1, Z|^2 - k_1^2 = |Z_2, Z|^2 - k_2^2$ which simplifies to

$$2(x_2 - x_1)x + 2(y_2 - y_1)y + x_1^2 + y_1^2 - k_1^2 + x_2^2 + y_2^2 - k_2^2 = 0.$$

If $Z_1 \neq Z_2$ this is the equation of a line which is called the **radical axis** of C_1 and C_2. It is always perpendicular to the line $Z_1 Z_2$ and it passes through any points which C_1 and C_2 have in common.

More generally we also ask *for what points Z its powers with respect to C_1 and C_2 have a constant ratio*. For a real number λ which is not equal to 1 we consider when

$$|Z_1, Z|^2 - k_1^2 = \lambda \left[|Z_2, Z|^2 - k_2^2 \right]. \qquad (7.6.6)$$

When $\lambda = 0$ this yields C_1 and by considering $\mu \left[|Z_1, Z|^2 - k_1^2 \right] = |Z_2, Z|^2 - k_2^2$ as well, we also include C_2.

Now (7.6.6) expands to

$$x^2 + y^2 - 2\frac{x_1 - \lambda x_2}{1 - \lambda}x - 2\frac{y_1 - \lambda y_2}{1 - \lambda}y + \frac{x_1^2 + y_1^2 - k_1^2 - \lambda(x_2^2 + y_2^2 - k_2^2)}{1 - \lambda} = 0,$$

and on completing the squares in both x and y it becomes

$$\left[x - \frac{x_1 - \lambda x_2}{1 - \lambda} \right]^2 + \left[y - \frac{y_1 - \lambda y_2}{1 - \lambda} \right]^2$$
$$= \frac{1}{(1 - \lambda)^2} \left\{ k_1^2 + \left[(x_1 - x_2)^2 + (y_1 - y_2)^2 - k_1^2 - k_2^2 \right] \lambda + k_2^2 \lambda^2 \right\}.$$

This quadratic expression in λ is positive when $|\lambda|$ is large, so it has either a positive minimum, or its minimum is 0 attained at λ_1, say, or it has a negative minimum and so has the value 0 at λ_2 and λ_3, say, where $\lambda_2 < \lambda_3$. In the first of these cases (7.6.6) always represents a circle and in the second case it represents a circle for all $\lambda \neq \lambda_1$ and a point for $\lambda = \lambda_1$. In the third case it represents a circle when either $\lambda < \lambda_2$ or $\lambda > \lambda_3$, it represents a point when either $\lambda = \lambda_2$ or $\lambda = \lambda_3$, and it represents an empty locus when $\lambda_2 < \lambda < \lambda_3$. Thus it is the equation of a circle, a point or an empty locus.

Suppose that we consider two of these loci, corresponding to the values λ_4 and λ_5 of λ. They will then have equations

$$x^2 + y^2 - 2\frac{x_1 - \lambda_4 x_2}{1 - \lambda_4}x - 2\frac{y_1 - \lambda_4 y_2}{1 - \lambda_4}y + \frac{x_1^2 + y_1^2 - k_1^2 - \lambda_4(x_2^2 + y_2^2 - k_2^2)}{1 - \lambda_4} = 0,$$

$$x^2 + y^2 - 2\frac{x_1 - \lambda_5 x_2}{1 - \lambda_5}x - 2\frac{y_1 - \lambda_5 y_2}{1 - \lambda_5}y + \frac{x_1^2 + y_1^2 - k_1^2 - \lambda_5(x_2^2 + y_2^2 - k_2^2)}{1 - \lambda_5} = 0.$$

On subtracting the second of these from the first, and simplifying, we find that their radical axis is the line with equation

$$\frac{\lambda_4 - \lambda_5}{(1-\lambda_4)(1-\lambda_5)} \left[2(x_2-x_1)x + 2(y_2-y_1)y + x_1^2 + y_1^2 - k_1^2 + x_2^2 + y_2^2 - k_2^2\right] = 0.$$

As we can cancel the initial fraction we see that these loci have the same radical axis as did the original pair C_1 and C_2. For this reason all the loci considered are said to be **coaxal**.

Exercises

7.1 Prove that a circle cannot have more than one centre. [Hint. If O and O_1 are both centres, consider the intersection of OO_1 with the circle.]

7.2 Give an alternative proof of 7.2.1(iv) by showing that if $(x-k)^2 + y^2 = k^2$, where $k > 0$, then $x \geq 0$.

7.3 Prove that if the point X is interior to the circle $C(O;k), l$ is a line containing X, and $M = \pi_l(X)$, then M is also an interior point of this circle. Deduce that l is a secant line. Show too that if Y is also interior to this circle, then every point of the segment $[X,Y]$ is also interior.

7.4 Show that if A, B, C are non-collinear points, there is a unique circle to which the side-lines BC, CA, AB are all tangents.

7.5 Let $Z_1 \equiv (x_1,0), Z_2 \equiv (x_2,0)$ and $Z_3 \equiv (x_3,0)$ be distinct fixed collinear points and Z_3 not the mid-point of Z_1 and Z_2. For $W \notin Z_1 Z_2$ let l be the mid-line of $|\overline{Z_2 WZ_1}|$. Find the locus of W such that either l, or the line through W perpendicular to l, passes through Z_3.

7.6 Show that the locus of mid-points of chords of a circle on parallel lines is a diameter.

7.7 Show that if two tangents to a circle are parallel, then their points of contact are at the end-points of a diameter.

7.8 Show that if each of the side-lines of a rectangle is a tangent to a given circle, then it must be a square.

7.9 Consider the circle $C(O;a)$ and point $Z_1 \equiv (x_1,0)$ where $x_1 > a > 0$, so that Z_1 is an exterior point which lies on the diametral line AB, where $A \equiv (a,0), B \equiv (-a,0)$. Show that for all points $Z \equiv (x,y)$ on the circle,

$$|Z_1, A| \leq |Z_1, Z| \leq |Z_1, B|.$$

7.10 For $0 < a < b$, suppose that $A \equiv (0,a)$, $B \equiv (0,b)$. Show that the circles $C(A;a)$ and $C(B;b)$ both have the axis OI as a tangent at the point O, and that $A \in [O,B]$. Verify that every point of $C(A;a)$, other than O, is an interior point for $C(B;b)$. [Hint. Consider the equations of $C(A;a)$ and $C(B;b)$.]

7.11 Use Ex.6.3 to establish the equation of the mid-line l in 7.3.1 when P_1 and P_2 are not diametrically opposite.

8

Translations; axial symmetries; isometries

COMMENT. In this chapter we introduce translations and develop them and axial symmetries. These will be useful in later chapters. It is more convenient to frame our proofs for isometries generally.

8.1 TRANSLATIONS AND AXIAL SYMMETRIES

8.1.1

Definition. Given points $Z_1, Z_2 \in \Pi$, we define a **translation** t_{Z_1,Z_2} to be a function $t_{Z_1,Z_2} : \Pi \to \Pi$ such that, for all $Z \in \Pi$, $t_{Z_1,Z_2}(Z) = W$ where $\mathrm{mp}(Z_1, W) = \mathrm{mp}(Z_2, Z)$.

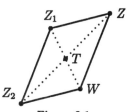

Figure 8.1.

Translations have the following properties:-

(i) *If $Z_1 \equiv (x_1, y_1)$ $Z_2 \equiv (x_2, y_2)$ $Z \equiv (x, y)$, $W \equiv (u, v)$, then $t_{Z_1,Z_2}(Z) = W$ if and only if*

$$u = x + x_2 - x_1, \ v = y + y_2 - y_1.$$

(ii) *In all cases $|t_{Z_1,Z_2}(Z_3), t_{Z_1,Z_2}(Z_4)| = |Z_3, Z_4|$, so that each translation preserves all distances.*

(iii) *For each $W \in \Pi$ the equation $t_{Z_1,Z_2}(Z) = W$ has a solution in Z, so that each translation is an onto function.*

(iv) *Each translation t_{Z_1,Z_2} has an inverse function $t_{Z_1,Z_2}^{-1} = t_{Z_2,Z_1}$.*

(v) *The translation t_{Z_1,Z_1} is the identity function on Π.*

(vi) *If $Z_1 \neq Z_2$, $Z \notin Z_1 Z_2$ and $W = t_{Z_1,Z_2}(Z)$, then $[Z_1, Z_2, W, Z]$ is a parallelogram.*

Proof.

(i) For $\text{mp}(Z_1, W) \equiv \left(\frac{1}{2}(x_1 + u), \frac{1}{2}(y_1 + v)\right)$, $\text{mp}(Z_2, Z) \equiv \left(\frac{1}{2}(x_2 + x), \frac{1}{2}(y_2 + y)\right)$ and these are equal if and only if $u = x + x_2 - x_1$, $v = y + y_2 - y_1$.

(ii) For if $W_3 = t_{Z_1, Z_2}(Z_3)$, $W_4 = t_{Z_1, Z_2}(Z_4)$, then

$$u_4 - u_3 = x_4 + x_2 - x_1 - (x_3 + x_2 - x_1) = x_4 - x_3,$$
$$v_4 - v_3 = y_4 + y_2 - y_1 - (y_3 + y_2 - y_1) = y_4 - y_3.$$

It follows that $|Z_3, Z_4| = |W_3, W_4|$.

(iii) By (i) the equation $t_{Z_1, Z_2}(Z) = W$ has the solution given by $x = u + x_1 - x_2$, $y = v + y_1 - y_2$.

(iv) By (ii) and (iii) the equation $t_{Z_1, Z_2}(Z) = W$ has a unique solution and this is denoted by $Z = t_{Z_1, Z_2}^{-1}(W)$. The correspondence from Π to Π given by $W \to Z$ is the inverse of t_{Z_1, Z_2} and is a function. As by the proof of (iii)

$$x = u + x_1 - x_2, \ y = v + y_1 - y_2,$$

by (i) this inverse function is t_{Z_2, Z_1}.

(v) For if $Z_1 = Z_2$, in (i) we have $u = x + x_1 - x_1 = x$, $v = y + y_1 - y_1 = y$.

(vi) We denote by T the common mid-point $\text{mp}(Z_1, W) = \text{mp}(Z_2, Z)$. First we note that $W \neq Z$, as $\text{mp}(Z_1, Z) = \text{mp}(Z_2, Z)$ would imply $Z_1 = Z_2$. As $Z \notin Z_1 Z_2$, we have $T \notin Z_1 Z_2$ and hence $W \notin Z_1 Z_2$. It follows that $T \notin ZW$ as otherwise we would have $Z_1 \in ZW$, $Z_2 \in ZW$ and so $Z \in Z_1 Z_2$. The triangles $[Z_1, T, Z_2]$, $[W, T, Z]$ are congruent in the correspondence $(Z_1, T, Z_2) \to (W, T, Z)$ by the SAS-principle. Hence the alternate angles $\angle W Z_1 Z_2$, $\angle Z_1 W Z$ have equal degree-measures and so $Z_1 Z_2 \parallel W Z$. Similarly $Z_1 Z \parallel Z_2 W$.

Axial symmetries have the following properties:-

(i) *In all cases* $|s_l(Z_3), s_l(Z_4)| = |Z_3, Z_4|$, *so that each axial symmetry preserves all distances.*

(ii) *Each axial symmetry* s_l *has an inverse function* $s_l^{-1} = s_l$.

Proof.

We note that by 6.6.1,

$$s_l(Z_3) \equiv \left(\frac{1}{a^2 + b^2}[(b^2 - a^2)x_3 - 2aby_3 - 2ac], \frac{1}{a^2 + b^2}[-2abx_3 - (b^2 - a^2)y_3 - 2bc]\right),$$

$$s_l(Z_4) \equiv \left(\frac{1}{a^2 + b^2}[(b^2 - a^2)x_4 - 2aby_4 - 2ac], \frac{1}{a^2 + b^2}[-2abx_4 - (b^2 - a^2)y_4 - 2bc]\right),$$

and thus

$$
\begin{aligned}
|s_l(Z_3), s_l(Z_4)|^2 &= \frac{1}{(a^2 + b^2)^2} \left\{ [(b^2 - a^2)(x_3 - x_4) - 2ab(y_3 - y_4)]^2 \right. \\
&\quad \left. + [-2ab(x_3 - x_4) - (b^2 - a^2)(y_3 - y_4)]^2 \right\} \\
&= \frac{1}{(a^2 + b^2)^2} \left\{ [(b^2 - a^2)^2 + 4a^2b^2][(x_3 - x_4)^2 + (y_3 - y_4)^2] \right. \\
&\quad \left. + [-4ab + 4ab](b^2 - a^2)(x_3 - x_4)(y_3 - y_4) \right\} \\
&= (x_3 - x_4)^2 + (y_3 - y_4)^2.
\end{aligned}
$$

(ii) For if m is the line through Z which is perpendicular to l and $W = s_l(Z)$, then $W \in m$ and so $\pi_l(W) = \pi_l(Z)$. Then $\pi_l(W) = \text{mp}(W, Z)$ so $Z = s_l(W)$. This shows that the function s_l is its own inverse.

8.2   ISOMETRIES

8.2.1

Definition. A function $f : \Pi \to \Pi$ which satisfies $|Z_1, Z_2| = |f(Z_1), f(Z_2)|$ for all points $Z_1, Z_2 \in \Pi$, is called an **isometry** of Π.

Each translation and each axial symmetry is an isometry.
Proof. This follows from 8.1.1.

Each isometry f has the following properties:-

(i) *The function $f : \Pi \to \Pi$ is one-one.*

(ii) *For all $Z_1, Z_2 \in \Pi$, $f([Z_1, Z_2]) = [f(Z_1), f(Z_2)]$, so that each segment is mapped onto a segment, with the end-points corresponding.*

(iii) *For all distinct points $Z_1, Z_2 \in \Pi$, $f([Z_1, Z_2) = [f(Z_1), f(Z_2)$, so that each half-line is mapped onto a half-line, with the initial points corresponding.*

(iv) *For all distinct points $Z_1, Z_2 \in \Pi$, $f(Z_1 Z_2) = f(Z_1)f(Z_2)$, so that each line is mapped onto a line. If $f(Z) \in f(Z_1)f(Z_2)$ then $Z \in Z_1 Z_2$.*

(v) *If Z_1, Z_2, Z_3 are noncollinear points, then $f([Z_1, Z_2, Z_3]) \equiv [f(Z_1), f(Z_2), f(Z_3)]$.*

(vi) *Let $Z_3 \notin l$ and $\mathcal{H}_1, \mathcal{H}_2$ be the closed half-planes with common edge l, with $Z_3 \in \mathcal{H}_1$. Let $\mathcal{H}_3, \mathcal{H}_4$ be the closed half-planes with common edge $f(l)$, with $f(Z_3) \in \mathcal{H}_3$. Then $f(\mathcal{H}_1) \subset \mathcal{H}_3$, $f(\mathcal{H}_2) \subset \mathcal{H}_4$.*

(vii) *The function $f : \Pi \to \Pi$ is onto.*

(viii) *In (vi), $f(\mathcal{H}_1) = \mathcal{H}_3$, $f(\mathcal{H}_2) = \mathcal{H}_4$.*

(ix) *If l and m are intersecting lines, then $f(l)$ and $f(m)$ are intersecting lines. If l and m are parallel lines, then $f(l)$ and $f(m)$ are parallel lines.*

(x) *If the points Z_1, Z_2 and Z_3 are non-collinear, then*

$$|\angle Z_2 Z_1 Z_3|^\circ = |\angle f(Z_2) f(Z_1) f(Z_3)|^\circ.$$

(xi) *If l and m are perpendicular lines, then $f(l) \perp f(m)$.*

(xii) *If a point Z has Cartesian coordinates (x, y) relative to the frame of reference $\mathcal{F} = ([O, I$, $[O, J$ $)$, then $f(Z)$ has Cartesian coordinates (x, y) relative to the frame of reference $([f(O), f(I)$, $[f(O), f(J)$ $)$.*

Proof.
(i) If $Z_1 \neq Z_2$ then $|Z_1, Z_2| > 0$ so that $|f(Z_1), f(Z_2)| > 0$, and so $f(Z_1) \neq f(Z_2)$.
(ii) If $Z_1 = Z_2$ the result is trivial, so suppose that $Z_1 \neq Z_2$. Then for all $Z \in [Z_1, Z_2]$, we have $|Z_1, Z| + |Z, Z_2| = |Z_1, Z_2|$ and so $|f(Z_1), f(Z)| + |f(Z), f(Z_2)| = |f(Z_1), f(Z_2)|$. It follows by 3.1.2 and 4.3.1 that $f(Z) \in [f(Z_1), f(Z_2)]$ and so $f([Z_1, Z_2]) \subset [f(Z_1), f(Z_2)]$.

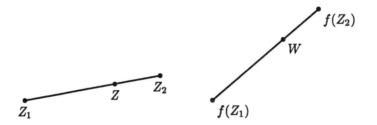

Figure 8.2.

Next let $W \in [f(Z_1), f(Z_2)]$. Then $|f(Z_1), W| \le |f(Z_1), f(Z_2)| = |Z_1, Z_2|$. Choose the point $Z \in [Z_1, Z_2$ so that $|Z_1, Z| = |f(Z_1), W|$; as $|Z_1, Z| \le |Z_1, Z_2|$ then $Z \in [Z_1, Z_2]$. Moreover $|f(Z_1), f(Z)| = |Z_1, Z| = |f(Z_1), W|$. Then $f(Z)$ and W are both in $[f(Z_1), f(Z_2)]$ and at the same distance from $f(Z_1)$ so $f(Z) = W$. Thus W is a value of f at some point of $[Z_1, Z_2]$. Hence $f([Z_1, Z_2]) = [f(Z_1), f(Z_2)]$.

(iii) By (i) $f(Z_1) \ne f(Z_2)$. Suppose that $Z \in [Z_1, Z_2$. Then either $Z \in [Z_1, Z_2]$ or $Z_2 \in [Z_1, Z]$. It follows from part (ii) of the present theorem, that then either $f(Z) \in [f(Z_1), f(Z_2)]$ or $f(Z_2) \in [f(Z_1), f(Z)]$. Thus $f(Z) \in [f(Z_1), f(Z_2)]$ and so $f([Z_1, Z_2) \subset [f(Z_1), f(Z_2)$.

If $W \in [f(Z_1), f(Z_2$) choose $Z \in [Z_1, Z_2$ so that $|Z_1, Z| = |f(Z_1), W|$. Then by the last paragraph $f(Z) \in [f(Z_1), f(Z_2)$ and as $|f(Z_1), f(Z)| = |f(Z_1), W|$, we have $f(Z) = W$. Thus W is a value of f at some point of $[Z_1, Z_2$. Hence $f([Z_1, Z_2) = [f(Z_1), f(Z_2)$.

(iv) Take $Z_3 \ne Z_1$ so that $Z_1 \in [Z_2, Z_3]$. Then $Z_1 Z_2 = [Z_1, Z_2 \cup [Z_1, Z_3$. Hence

$$
\begin{aligned}
f(Z_1 Z_2) &= f([Z_1, Z_2) \cup f([Z_1, Z_3) \\
&= [f(Z_1), f(Z_2) \cup [f(Z_1), f(Z_3) \\
&= f(Z_1) f(Z_2), \text{ as } f(Z_1) \in [f(Z_2), f(Z_3)].
\end{aligned}
$$

If $f(Z) \in f(Z_1) f(Z_2)$, then by the foregoing there is a point $Z_4 \in Z_1 Z_2$ such that $f(Z_4) = f(Z)$ and then as f is one-one $Z = Z_4 \in Z_1 Z_2$.

(v) For

$$
|Z_2, Z_3| = |f(Z_2), f(Z_3)|, \ |Z_3, Z_1| = |f(Z_3), f(Z_1)|, |Z_1, Z_2| = |f(Z_1), f(Z_2)|,
$$

so by the SSS-principle, these triangles are congruent in the correspondence

$$
(Z_1, Z_2, Z_3) \to (f(Z_1), f(Z_2), f(Z_3)) .
$$

(vi) Suppose that $f(\mathcal{H}_1)$ is not a subset of $\mathcal{H}_3$. Then there is some $Z_4 \in \mathcal{H}_1$ such that $f(Z_4) \in \mathcal{H}_4$, $f(Z_4) \notin f(l)$. Then $f(Z_3)$ and $f(Z_4)$ are on opposite sides of $f(l)$, so there is a point W on both $f(l)$ and $[f(Z_3), f(Z_4)]$. By (ii) there is a point $Z \in [Z_3, Z_4]$ such that $f(Z) = W$, and then by (i) and (iv) $Z \in l$. But this implies that $Z_4 \notin \mathcal{H}_1$ and so gives a contradiction. Hence $f(\mathcal{H}_1) \subset \mathcal{H}_3$ and by a similar argument $f(\mathcal{H}_2) \subset \mathcal{H}_4$.

(vii) Take distinct points Z_1, Z_2 in Π. If $W \in f(Z_1) f(Z_2)$, then by (iv) $f(Z) = W$ for some $Z \in Z_1 Z_2$ and so W is a value of f.

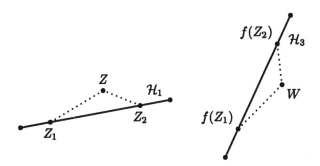

Figure 8.3.

Suppose then that $W \notin f(Z_1)f(Z_2)$ and let $\mathcal{H}_3$ be the closed half-plane with edge $f(Z_1)f(Z_2)$ which contains W. Let $\mathcal{H}_1$ be the closed half-plane with edge $Z_1 Z_2$ such that by (vi) $f(\mathcal{H}_1) \subset \mathcal{H}_3$. Take a point $Z \in \mathcal{H}_1$ such that $|\angle Z_2 Z_1 Z|^\circ = |\angle f(Z_2)f(Z_1) W|^\circ$ and $|Z_1, Z| = |f(Z_1), W|$. Then by the SAS-principle $[Z_1, Z_2, Z] \equiv [f(Z_1), f(Z_2), W]$, and so by (v) $[f(Z_1), f(Z_2), f(Z)] \equiv [f(Z_1), f(Z_2), W]$. In particular $|\angle f(Z_2)f(Z_1)f(Z)|^\circ = |\angle f(Z_2)f(Z_1) W|^\circ$. As $f(Z) \in \mathcal{H}_3$, $W \in \mathcal{H}_3$ we then have $f(Z) \in [f(Z_1), W$. But by the congruence we also have $|f(Z_1), f(Z)| = |f(Z_1), W|$. It follows that $f(Z) = W$ and so W is a value of f.

(viii) Let $W \in \mathcal{H}_3$. Then by (vii) $W = f(Z)$ for some $Z \in \Pi$. If $W \in f(l)$ then by (iv) $Z \in l \subset \mathcal{H}_1$. If $W \notin f(l)$ then $W \notin \mathcal{H}_4$ and by (vi) we cannot have $Z \in \mathcal{H}_2$ as that would imply $W \in \mathcal{H}_4$. Thus again $Z \in \mathcal{H}_1$. In both cases W is a value $f(Z)$ for some $Z \in \mathcal{H}_1$.

(ix) By part (iv) $f(l), f(m)$ are lines. If Z belongs to both l and m, then $f(Z)$ belongs to both $f(l)$ and $f(m)$ so these have a point in common.

On the other hand, if $l \parallel m$ suppose first that $l = m$. Then $f(l) = f(m)$ and so $f(l) \parallel f(m)$. Next suppose that $l \neq m$; then $l \cap m = \emptyset$. We now must have $f(l) \cap f(m) = \emptyset$, as if W were on both $f(l)$ and $f(m)$, by (iv) we would have $W = f(Z)$ for some $Z \in l$, $W = f(Z_0)$ for some $Z_0 \in m$. But by (i) $Z = Z_0$ so we would have Z on both l and m.

(x) By (v) the triangles $[Z_1, Z_2, Z_3], [f(Z_1), f(Z_2), f(Z_3)]$ are congruent, and so corresponding angles have equal degree-measures.

(xi) If l and m are perpendicular, let Z_1 be their point of intersection, and let Z_2, Z_3 be other points on l and m respectively. Then as in part (x), $\angle Z_2 Z_1 Z_3$ is a right-angle and so its image is also a right-angle.

(xii) For any line l and any point Z we recall that $\pi_l(Z)$ denotes the foot of the perpendicular from Z to l. For any point $Z \in \Pi$, let $U = \pi_{OI}(Z)$ and $V = \pi_{OJ}(Z)$. Let $O' = f(O)$, $I' = f(I)$, $J' = f(J)$. Then $O' \neq I'$, $O' \neq J'$ and $O'I' \perp O'J'$ so that we can take $([O', I', [O', J')$ as a frame of reference. Let $\mathcal{H}_1$, $\mathcal{H}_2$ be the half-planes with edge OI, with $J \in \mathcal{H}_1$, and $\mathcal{H}_3$, $\mathcal{H}_4$ the half-planes with edge OJ, with $I \in \mathcal{H}_3$. Similarly let $\mathcal{H}_1'$, $\mathcal{H}_2'$ be the half-planes with edge $O'I'$, with $J' \in \mathcal{H}_1'$, and $\mathcal{H}_3'$, $\mathcal{H}_4'$ the half-planes with edge $O'J'$, with $I' \in \mathcal{H}_3'$.

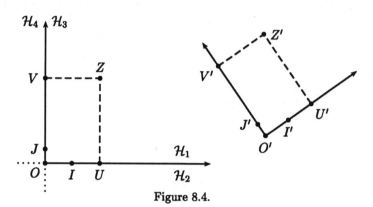

Figure 8.4.

If (x,y) are the Cartesian coordinates of Z relative to $([O,I\,,\,[O,J\,)$, then

$$x = \begin{cases} |O,U|, & \text{if } Z \in \mathcal{H}_3, \\ -|O,U|, & \text{if } Z \in \mathcal{H}_4. \end{cases}$$

But if $Z' = f(Z)$, $U' = f(U)$ we have $U' \in O'I'$, and if $Z \notin OI$ we have $ZU \perp OI$ and hence $Z'U' \perp O'I'$. It follows that $U' = \pi_{O'I'}(Z')$. Moreover $f(\mathcal{H}_3) = \mathcal{H}'_3$, $f(\mathcal{H}_4) = \mathcal{H}'_4$. Then if (x',y') are the Cartesian coordinates of Z' relative to $([O',I'\,,\,[O',J'\,)$, when $Z \in \mathcal{H}_3$ we have $Z' \in \mathcal{H}'_3$ and so

$$x' = |O',\pi_{O'I'}(Z')| = |O',U'| = |O,U| = x.$$

Similarly when $Z \in \mathcal{H}_4$ we have $Z' \in \mathcal{H}'_4$ and so

$$x' = -|O',\pi_{O'I'}(Z')| = -|O',U'| = -|O,U| = x.$$

Thus $x' = x$ in all cases, and by a similar argument $y' = y$.

8.2.2

If $l = \text{ml}(|\underline{BAC})$, then $s_l([A,B\,) = [A,C$ and $s_l([A,C\,) = [A,B\,.$

   Proof. We prove $s_l([A,B\,) = [A,C$ as the other then follows. As $A \in l$ we have $s_l(A) = A$ and so by 8.2.1(iii) $s_l([A,B\,) = [A,D$ for some point D.

   Suppose first that A,B,C are collinear. When $C \in [A,B$ we have $l = AB$, and so $s_l(P) = P$ for all $P \in [A,B\,$. As $[A,B\, = [A,C$ the conclusion is then immediate. On the other hand when $A \in [B,C]$ so that $|\underline{BAC}$ is straight, l is the perpendicular to AB at A. Then if $A = \text{mp}(B,D)$ we have $s_l([A,B\,) = [A,D\,$, and $[A,D\, = [A,C$ as $D \in [A,C\,.$

   Finally suppose that A,B,C are non-collinear. Now take $D \in [A,C$ so that $|A,D| = |A,B|$. If $M = \text{mp}(B,D)$ by 4.1.1(iv) we have that $l = AM$ and as $s_l(B) = D$ then $s_l([A,B\,) = [A,D\, = [A,C\,.$

8.3   TRANSLATION OF A FRAME OF REFERENCE

NOTATION. By using 8.2.1(iii), (vi) and (xi), we showed in 8.2.1(xii) that for any frame of reference $\mathcal{F} = ([O,I\,,\,[O,J\,)$ and any isometry f, $\mathcal{F}' =$

$([f(O), f(I)$, $[f(O), f(J)$) is also a frame of reference, and that Cartesian coordinates of Z relative to $\mathcal{F}$ are also Cartesian coordinates of $f(Z)$ relative to $\mathcal{F}'$. We denote $\mathcal{F}'$ by $f(\mathcal{F})$.

For any frame of reference $\mathcal{F} = ([O, I$, $[O, J$), *let* $Z_0 \equiv_{\mathcal{F}} (x_0, y_0)$ *and* $\mathcal{F}' = t_{O, Z_0}(\mathcal{F})$. *Then if* $Z \equiv_{\mathcal{F}} (x, y)$ *we have* $Z \equiv_{\mathcal{F}'} (x - x_0, y - y_0)$.

Proof. By 8.2.1(xii), $t_{O, Z_0}(Z)$ has coordinates (x, y) relative to $\mathcal{F}'$, and by 8.1.1(i) it also has coordinate $(x + x_0, y + y_0)$ relative to $\mathcal{F}$. Thus for all $(x, y) \in \mathbf{R} \times \mathbf{R}$ the point with coordinates $(x + x_0, y + y_0)$ relative to $\mathcal{F}$ has coordinates (x, y) relative to $\mathcal{F}'$. On replacing (x, y) by $(x - x_0, y - y_0)$, we conclude that the point with coordinates (x, y) relative to $\mathcal{F}$ has coordinates $(x - x_0, y - y_0)$ relative to $\mathcal{F}'$.

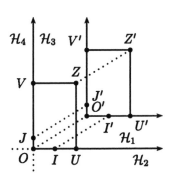

Figure 8.5.

Exercises

8.1 If $\mathcal{T}$ is the set of all translations of Π, show that $(\mathcal{T}, \circ)$ is a commutative group.

8.2 If $\mathcal{I}$ is the set of all isometries of Π, show that $(\mathcal{I}, \circ)$ is a group.

8.3 Given any half-lines $[A, B$, $[C, D$ show that there is an isometry f which maps $[A, B$ onto $[C, D$.

8.4 Show that each of the following concepts is an isometric invariant:- *interior region of an angle-support, triangle, dividing a pair of points in a given ratio, mid-point, centroid, circumcentre, orthocentre, mid-line, incentre, parallelogram, rectangle, square, area of a triangle, circle, tangent to a circle.*

8.5 For any line l, $s_l[\mathcal{C}(O; k)] = \mathcal{C}(s_l(O); k)$ so that, in particular, if $O \in l$ then $s_l[\mathcal{C}(O; k)] = \mathcal{C}(O; k)$.

9

Trigonometry; sine and cosine; addition formulae

COMMENT. In this chapter we go on to deal fully with reflex-angles as well as with wedge and straight ones, we define the cosine and sine of an angle and we deal with addition of angles. As a vitally convenient aid to identifying the two angles with a given support $|\underline{BAC}$, we start by introducing the notion of the indicator of an angle.

9.1 INDICATOR OF AN ANGLE

9.1.1

Definition. If α is an angle with support $|\underline{BAC}$, we call the other angle with support $|\underline{BAC}$ the **co-supported** angle for α, and denote it by $\mathrm{co-sp}\,\alpha$.

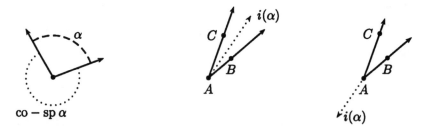

Figure 9.1. Co-supported angle.   Figure 9.2. Angle indicators.

Definition. Referring to 2.3.3, for each angle support $|\underline{BAC}$ let $l = \mathrm{ml}(|\underline{BAC})$ as in 3.6 and 4.1.1. When $A \notin [B, C]$, we call $l \cap \mathcal{IR}(|\underline{BAC})$ and $l \cap \mathcal{ER}(|\underline{BAC})$ the **indicators** of the wedge-angle $(|\underline{BAC}, \mathcal{IR}(|\underline{BAC}))$ and of the reflex-angle $(|\underline{BAC}, \mathcal{ER}(|\underline{BAC}))$, respectively. When $A \in [B, C]$ we call $l \cap \mathcal{H}_1$, $l \cap \mathcal{H}_2$ the indicators of the straight-angles $(|\underline{BAC}, \mathcal{H}_1)$, $(|\underline{BAC}, \mathcal{H}_2)$, respectively. In each case an indicator is a half-line of l with initial point the vertex A. We denote the indicator of an angle α by $i(\alpha)$.

COMMENT. The first use we make of the concept of indicator is in defining the cosine and sine of any angle.

9.2 COSINE AND SINE OF AN ANGLE

9.2.1

Definition. Starting with a support $\lfloor BAC$, let $\mathcal{H}_1$ be a closed half-plane with edge AB in which C lies. Let α be an angle with support $\lfloor BAC$ such that the indicator $i(\alpha)$ lies in $\mathcal{H}_1$. Then we define $\cos\alpha$ and $\sin\alpha$ as follows. Take any point $P \neq A$ on $[A, C$, let $Q \in [A, B$ be such that $|A, Q| = |A, P|$ and $R \in \mathcal{H}_1$ be such that $|A, R| = |A, P|$ and $AR \perp AB$.

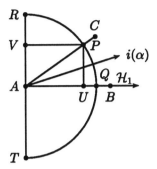

Figure 9.3. Cosine and Sine.

Let U, V be the feet of the perpendiculars from P to $AB = AQ$ and AR respectively. Then we define

$$\cos\alpha = \frac{|A, P| - |Q, U|}{|A, P|}, \quad \sin\alpha = \frac{|A, P| - |R, V|}{|A, P|}.$$

It follows from this definition that if $\mathcal{H}_2$ is the other half-plane with edge AB and if we take $T \in \mathcal{H}_2$ so that $|A, T| = |A, P|$ and $AT \perp AB$, then

$$\cos(\text{co} - \text{sp}\,\alpha) = \frac{|A, P| - |Q, U|}{|A, P|}, \quad \sin(\text{co} - \text{sp}\,\alpha) = \frac{|A, P| - |T, V|}{|A, P|}.$$

COMMENT. Two comments on this definition are in order. First we note that when A, B, C are collinear, $\mathcal{H}_1$ and $\mathcal{H}_2$ are not uniquely determined above but are interchangeable with each other, so that the angles α and co-spα are not uniquely determined. Our second comment is that to show that $\cos\alpha$, $\sin\alpha$ are well-defined it is first necessary to use the ratio results for triangles to show that the values of $\cos\alpha$ and $\sin\alpha$ do not depend on the particular point $P \in [A, C$ taken, and then to show that if the arms $[A, B$ and $[A, C$ are interchanged the outcome is unchanged.

To help us in our study of angles, it is convenient to fit a frame of reference to the situation in the definition. We take $O = A$, $I = B$ and $J \neq O$ a point in $\mathcal{H}_1$ so that $OI \perp OJ$. We let $\mathcal{H}_3$, $\mathcal{H}_4$ be the closed half-planes with edge OJ, with $I \in \mathcal{H}_3$.

With $k = |A, P| = |O, P|$, let Q be the point on $[O, I = [A, B$ such that $|O, Q| = k$, and let R be the point on $[O, J$ such that $|O, R| = k$. Choose S, T so that $O = mp(Q, S)$, $O = mp(R, T)$.

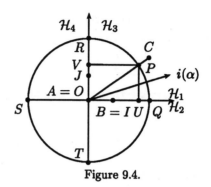

Figure 9.4.

The cosine and sine of an angle are well-defined.

Proof.

(i) When A, B and C are collinear there are two cases to be considered. One case is when $A \in [B, C]$ so that $|\underline{BAC}$ is straight. Then each of α, $co - sp \, \alpha$ is a straight-angle and as $P = S$, we have $U = S$, $V = A$ and so

$$\cos \alpha = \cos(co - sp \, \alpha) = -1, \ \sin \alpha = \sin(co - sp \, \alpha) = 0.$$

A second case is when $C \in [A, B$ so that one of α, $co - sp \, \alpha$ is a null-angle with indicator $[A, B$ and the other is a full-angle with indicator $[A, B_1$ where A is between B and B_1. Both of the indicators are in $\mathcal{H}_1$ and $\mathcal{H}_2$, but as $P = Q$ we have $U = Q$, $V = A$ and so

$$\cos \alpha = \cos(co - sp \, \alpha) = 1, \ \sin \alpha = \sin(co - sp \, \alpha) = 0.$$

Thus in neither case does the ambiguity affect the outcome.

(ii) We now use the ratio results for triangles to show that the values of $\cos \alpha$ and $\sin \alpha$ do not depend on the particular point $P \in [A, C$ chosen. Take $k_1 > 0$ and let P_1, Q_1, R_1 be the points in $[O, P$, $[O, Q$, $[O, R$, respectively, each at a distance k_1 from O. Let $U_1 = \tau_{OI}(P_1)$, $V_1 = \tau_{OJ}(P_1)$.

Suppose first that $P \notin OI$, $P \notin OJ$. As $PU \parallel P_1U_1$, by 5.3.1

$$\frac{|O, U|}{|O, U_1|} = \frac{|O, P|}{|O, P_1|},$$

and so

$$\frac{|O, U|}{k} = \frac{|O, U_1|}{k_1}.$$

Now if $P \in \mathcal{H}_3$ so that $U \in [Q, O]$ and so $|O, U| = k - |Q, U|$, by 2.2.3(iv) $P_1 \in \mathcal{H}_3$ and similarly $|O, U_1| = k_1 - |Q_1, U_1|$. On inserting this we get that

$$\frac{k - |Q, U|}{k} = \frac{k_1 - |Q_1, U_1|}{k_1}.$$

When instead $P \in \mathcal{H}_4$, we have $O \in [Q, U]$ so $|O, U| = |Q, U| - k$ and similarly $|O, U_1| = |Q_1, U_1| - k_1$. On inserting these we obtain

$$\frac{k - |Q, U|}{k} = \frac{k_1 - |Q_1, U_1|}{k_1}$$

again.

When $P \in OI$ we have either $P = Q$ or $P = S$. When $P = Q$, we have $P_1 = Q_1$ and the formula checks out. It checks out similarly in the cases when P is R, S or T.

By a similar proof we find that

$$\frac{k - |R, V|}{k} = \frac{k_1 - |R_1, V_1|}{k_1}.$$

Thus it makes no difference to the values of $\cos \alpha$ and $\sin \alpha$ if P is replaced by P_1.

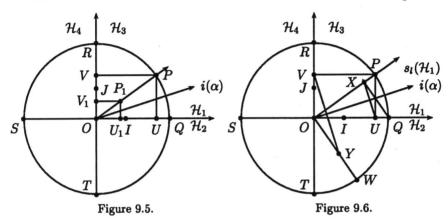

Figure 9.5.   Figure 9.6.

(iii) It remains to show that if the arms $[A, B$ and $[A, C$ are interchanged the outcome is unchanged. Let $l = ml(|QOP)$ so that $s_l(OQ) = OP$ and $s_l(\mathcal{H}_1)$ is a closed half-plane with edge OP. As $\overline{i(\alpha)} \subset \mathcal{H}_1$ we have $s_l(i(\alpha)) \subset s_l(\mathcal{H}_1)$; but as $i(\alpha) \subset l$, $s_l(i(\alpha)) = i(\alpha)$ and thus $i(\alpha) \subset s_l(\mathcal{H}_1)$. If $W = s_l(R)$ then $W \in s_l(\mathcal{H}_1)$ and as $OQ \perp OR$ we have $OP \perp OW$. Moreover $X = s_l(U) = \pi_{OP}(Q)$ and $Y = s_l(V) = \pi_{OW}(Q)$ satisfy $|P, X| = |Q, U|$, $|W, Y| = |R, V|$. Hence

$$\frac{k - |P, X|}{k} = \frac{k - |Q, U|}{k}, \quad \frac{k - |W, Y|}{k} = \frac{k - |R, V|}{k}.$$

This completes the proof.

9.2.2   Polar coordinates

For $Z \neq O$, let $k = |O, Z|$ and the angle α have support $|\underline{IOZ}$ and indicator $i(\alpha)$ in $\mathcal{H}_1$. Then if $Z \equiv_{\mathcal{F}} (x, y)$

$$x = k \cos \alpha, \; y = k \sin \alpha.$$

Proof. Let Q, R be the points where $\mathcal{C}(O; k)$ meets $[O, I$ and $[O, J$, respectively; then Q and R have Cartesian coordinates $(k, 0)$ and $(0, k)$, respectively. Let U, V be the feet of the perpendiculars from Z to the lines OI and OJ, so that these have Cartesian coordinates $(x, 0)$ and $(0, y)$ respectively. Now $O \equiv (0, 0)$ and $Z \equiv (x, y)$ so by the distance formula $(x - 0)^2 + (y - 0)^2 = k^2$. Thus $x^2 + y^2 = k^2$, so that $x^2 \leq k^2$ and as $k > 0$ we have $x \leq k$. Then by the distance formula

$$|Q, U| = \sqrt{(k - x)^2 + (0 - 0)^2} = k - x,$$

as $k - x \geq 0$, and similarly $|R, V| = k - y$. Hence

$$\cos \alpha = \frac{k - |Q, U|}{k} = \frac{k - (k - x)}{k} = \frac{x}{k}, \quad \sin \alpha = \frac{k - |R, V|}{k} = \frac{k - (k - y)}{k} = \frac{y}{k}.$$

Thus $x = k \cos \alpha$, $y = k \sin \alpha$.

We refer to k and α as **polar coordinates** of the point Z with respect to $\mathcal{F}$.

9.2.3

With the notation of 9.2.1, let α be an angle with support $|\underline{IOP} = |QOP$ and indicator $i(\alpha)$ in $\mathcal{H}_1$. Then we have the following properties:-

(i) *For all α, $\cos^2 \alpha + \sin^2 \alpha = 1$.*

(ii) *For $P \in \mathcal{Q}_1$, $\cos \alpha \geq 0$, $\sin \alpha \geq 0$; for $P \in \mathcal{Q}_2$, $\cos \alpha \leq 0$, $\sin \alpha \geq 0$; for $P \in \mathcal{Q}_3$, $\cos \alpha \leq 0$, $\sin \alpha \leq 0$; for $P \in \mathcal{Q}_4$, $\cos \alpha \geq 0$, $\sin \alpha \leq 0$.*

Proof.
(i) As in the proof in 9.2.1,

$$\cos \alpha = \pm \frac{|O, U|}{|O, P|}, \quad \sin \alpha = \pm \frac{|O, V|}{|O, P|}.$$

Now when O, U, P, V are not collinear they are the vertices of a rectangle and so $|O, V| = |U, P|$. Then by Pythagoras' theorem

$$|O, U|^2 + |U, P|^2 = |O, P|^2,$$

and the result follows. It can be verified directly when P is any of Q, R, S, T.

(ii) This follows directly from details in the proof in 9.2.1.

9.3 ANGLES IN STANDARD POSITION

9.3.1 Angles in standard position

COMMENT. The second use that we make of the concept of indicator of an angle is to identify angles with respect to a frame of reference.

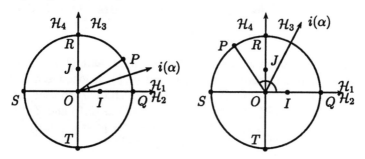

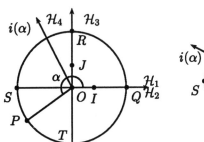
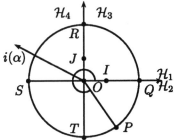

Figure 9.7.

Definition. We recall from 3.7
our extension of degree-measure
to reflex-angles. Let $|QOP$ be
a non-straight support, and let
$\mathcal{H}_1$, $\mathcal{H}_2$ be the closed-half planes
with edge OQ, with $P \in \mathcal{H}_2$. Let
α be the reflex angle with sup-
port $|QOP$, so that $i(\alpha) \subset \mathcal{H}_1$.
Let S be the point such that O
$= \mathrm{mp}(Q,S)$. Let β be the wedge
or straight angle with support
$|SOP$.

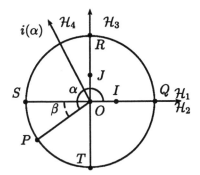

Figure 9.8. Measure of a reflex angle.

Then we defined the degree measure of α by

$$|\alpha|° = 180 + |\beta|°.$$

In particular if $P = Q$, then β is a straight angle, α is the full angle with support
$|POP = |QOQ$ and indicator $[O,S$, and $|\alpha|° = 360$.

Definition. Given a frame of reference $\mathcal{F} = ([O,I , [O,J)$, we denote by $\mathcal{A}^*(\mathcal{F})$
the set of angles α with arm $[O,I$ and with indicator $i(\alpha) \subset \mathcal{H}_1$.

If α and γ are different angles in $\mathcal{A}^*(\mathcal{F})$, then $|\alpha|° \neq |\gamma|°$.

Proof. This is evident if α and γ are both wedge or straight angles and hence, by
addition of 180, if they are both reflex or straight. If α is wedge or straight and γ is
reflex, then $|\alpha|° \leq 180$, $|\gamma|° > 180$.

NOTATION. Given any real number x such that $0 \leq x \leq 360$, we denote the angle
$\alpha \in \mathcal{A}^*(\mathcal{F})$ with $|\alpha|° = x$ by $x_\mathcal{F}$. Thus the null, straight and full angles in $\mathcal{A}^*(\mathcal{F})$ are
denoted by $0_\mathcal{F}$, $180_\mathcal{F}$ and $360_\mathcal{F}$, respectively.

9.3.2   Addition of angles

COMMENT. Given angles α, $\beta \in \mathcal{A}^*(\mathcal{F})$ we wish to define two closely related forms
of addition, the first suited to angle measure as to be dealt with in Chapter 12 and
the second suited to more general situations. As we make more use of the latter we
employ for it the common symbol $+$, and $\oplus$ for the former. As $\alpha \oplus \beta$ is to be an angle
we need to specify its support and its indicator; similarly for $\alpha + \beta$.

Definition. Let α, β be angles in $\mathcal{A}^*(\mathcal{F})$ with supports $|QOP_1|$, $|QOP_2|$, respectively. Let l be the midline of $|P_1OP_2|$ and let $P_3 = s_l(Q)$. Then $\alpha \oplus \beta$ is an angle with support $|QOP_3|$ for which $i(\alpha \oplus \beta) \subset \mathcal{H}_1$, so that $\alpha \oplus \beta \in \mathcal{A}^*(\mathcal{F})$. This identifies $\alpha \oplus \beta$ uniquely except when $P_3 = Q$; in this case both the null angle $0_{\mathcal{F}}$ and the full angle $360_{\mathcal{F}}$ have support $|QOQ|$ and we define $\alpha \oplus \beta$ to be this full angle $360_{\mathcal{F}}$ in every case except when α and β are both null; in the latter case we define the sum to be this null angle $0_{\mathcal{F}}$. We call $\alpha \oplus \beta$ the **sum** of the angles α and β.

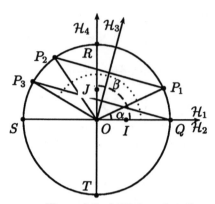

Figure 9.9. Addition of angles.

For all angles $\alpha, \beta \in \mathcal{A}^*(\mathcal{F})$,

(i) $\cos(\alpha \oplus \beta) = \cos\alpha\cos\beta - \sin\alpha\sin\beta$,

(ii) $\sin(\alpha \oplus \beta) = \sin\alpha\cos\beta + \cos\alpha\sin\beta$.

Proof. On using the notation of 7.3.1 and above, we have

$$a_1 = \cos\alpha, b_1 = \sin\alpha, a_2 = \cos\beta, b_2 = \sin\beta, a_3 = \cos(\alpha \oplus \beta), b_3 = \sin(\alpha \oplus \beta).$$

We note that in 7.3.1

$$(a_1 + a_2)^2 + (b_1 + b_2)^2 = 2(1 + a_1a_2 + b_1b_2),$$

as $a_1^2 + b_1^2 = a_2^2 + b_2^2 = 1$. Then, by 7.3.2, when P_1 and P_2 are not diametrically opposite,

$$\cos(\alpha \oplus \beta) - \cos\alpha\cos\beta + \sin\alpha\sin\beta$$
$$= \frac{(a_1 + a_2)^2 - (b_1 + b_2)^2 + 2(-a_1a_2 + b_1b_2)(1 + a_1a_2 + b_1b_2)}{2(1 + a_1a_2 + b_1b_2)},$$

and the numerator here is equal to

$$a_1^2 + a_2^2 - b_1^2 - b_2^2 - 2a_1^2a_2^2 + 2b_1^2b_2^2 = 2a_1^2 + 2a_2^2 - 2 - 2a_1^2a_2^2 + 2(1 - a_1^2)(1 - a_2^2) = 0.$$

Similarly

$$\sin(\alpha \oplus \beta) - \sin\alpha\cos\beta - \cos\alpha\sin\beta$$
$$= \frac{2(a_1 + a_2)(b_1 + b_2) - 2(a_1b_2 + a_2b_1)(1 + a_1a_2 + b_1b_2)}{2(1 + a_1a_2 + b_1b_2)},$$

and the numerator here is equal to twice

$$a_1b_1 + a_2b_2 - a_1b_1(a_2^2 + b_2^2) - a_2b_2(a_1^2 + b_1^2) = 0.$$

When P_1 and P_2 are diametrically opposite,

$$\cos(\alpha \oplus \beta) - \cos\alpha \cos\beta + \sin\alpha \sin\beta = b_1^2 - a_1^2 - a_1(-a_1) + b_1(-b_1) = 0,$$
$$\sin(\alpha \oplus \beta) - \sin\alpha \cos\beta - \cos\alpha \sin\beta = -2a_1b_1 - b_1(-a_1) - a_1(-b_1) = 0.$$

9.3.3   Modified addition of angles

COMMENT. In 9.3.2 we clearly exercised a choice in specifying what $\alpha \oplus \beta$ should be when $P_3 = Q$. The choice made there is what suits length of a circle and area of a disk which will be treated in Chapter 12, and that was the reason for the choice made. We now define modified addition $\alpha + \beta$ of angles, which is easier to use.

Definition. Let $\mathcal{A}(\mathcal{F}) = \mathcal{A}^*(\mathcal{F}) \setminus \{360_{\mathcal{F}}\}$, so that $\mathcal{A}(\mathcal{F})$ is the set of all non-full angles in $\mathcal{A}^*(\mathcal{F})$. We denote by $\angle_{\mathcal{F}} QOP = \angle_{\mathcal{F}} IOP$ the unique angle in $\mathcal{A}(\mathcal{F})$ with support $|QOP = |IOP$.

Definition. Let α, β be angles in $\mathcal{A}(\mathcal{F})$ with supports $|QOP_1$, $|QOP_2$. Let l be the midline of $|P_1 OP_2$ and let $P_3 = s_l(Q)$. Then $\alpha + \beta$ is the angle in $\mathcal{A}(\mathcal{F})$ with support $|QOP_3$. Note that when $P_3 = Q$ we have $\alpha + \beta = 0_{\mathcal{F}}$. We call $\alpha + \beta$ the **modified sum** of the angles α and β.

For all $\alpha, \beta \in \mathcal{A}(\mathcal{F})$,

$$\cos(\alpha + \beta) = \cos\alpha \cos\beta - \sin\alpha \sin\beta, \ \sin(\alpha + \beta) = \sin\alpha \cos\beta + \cos\alpha \sin\beta.$$

Proof. This follows immediately from 9.3.2 as $\cos 360_{\mathcal{F}} = \cos 0_{\mathcal{F}}$, $\sin 360_{\mathcal{F}} = \sin 0_{\mathcal{F}}$.

Modified addition + of angles has the following properties:-

(i) *For all* $\alpha, \beta \in \mathcal{A}(\mathcal{F})$, $\alpha + \beta$ *is uniquely defined and lies in* $\mathcal{A}(\mathcal{F})$.

(ii) *For all* $\alpha, \beta \in \mathcal{A}(\mathcal{F})$, $\alpha + \beta = \beta + \alpha$.

(iii) *For all* $\alpha, \beta, \gamma \in \mathcal{A}(\mathcal{F})$, $(\alpha + \beta) + \gamma = \alpha + (\beta + \gamma)$.

(iv) *For all* $\alpha \in \mathcal{A}(\mathcal{F})$, $\alpha + 0_{\mathcal{F}} = \alpha$.

(v) *Corresponding to each* $\alpha \in \mathcal{A}(\mathcal{F})$, *there is a* $\beta \in \mathcal{A}(\mathcal{F})$ *such that* $\alpha + \beta = 0_{\mathcal{F}}$.

Proof.
(i) This is evident from the definition.
(ii) This is evident as the roles of P_1 and P_2 are interchangeable in the definition.
(iii) We note that by the last result

$$\cos[(\alpha + \beta) + \gamma] = \cos(\alpha + \beta) \cos\gamma - \sin(\alpha + \beta) \sin\gamma$$

and then

$$\cos[(\alpha + \beta) + \gamma] = [\cos\alpha \cos\beta - \sin\alpha \sin\beta] \cos\gamma - [\sin\alpha \cos\beta + \cos\alpha \sin\beta] \sin\gamma,$$

while

$$\cos[\alpha + (\beta + \gamma)] = \cos\alpha \cos(\beta + \gamma) - \sin\alpha \sin(\beta + \gamma)$$
$$= \cos\alpha[\cos\beta \cos\gamma - \sin\beta \sin\gamma] - \sin\alpha[\sin\beta \cos\gamma + \cos\beta \sin\gamma],$$

and these are equal. Similarly

$$\sin[(\alpha + \beta) + \gamma] = \sin(\alpha + \beta)\cos\gamma + \cos(\alpha + \beta)\sin\gamma$$
$$= [\sin\alpha\cos\beta + \cos\alpha\sin\beta]\cos\gamma + [\cos\alpha\cos\beta - \sin\alpha\sin\beta]\sin\gamma,$$

while

$$\sin[\alpha + (\beta + \gamma)] = \sin\alpha\cos(\beta + \gamma) + \cos\alpha\sin(\beta + \gamma)$$
$$= \sin\alpha[\cos\beta\cos\gamma - \sin\alpha\sin\beta] + \cos\alpha[\sin\beta\cos\gamma + \cos\beta\sin\gamma],$$

and these are equal. Thus $(\alpha + \beta) + \gamma$ and $\alpha + (\beta + \gamma)$ are angles in $\mathcal{A}(\mathcal{F})$ with the same cosine and the same sine and so by 9.2.2 they are equal.

(iv) When $\beta = 0_{\mathcal{F}}$, in the definition we have $P_2 = Q$ and then l is the midline of $|QOP_1$ and so $P_3 = P_1$. Thus α and $\alpha + 0_{\mathcal{F}}$ are both in $\mathcal{A}(\mathcal{F})$ and they have the same support, so they must be equal.

(v) Given any angle $\alpha \in \mathcal{A}(\mathcal{F})$ with support $|QOP_1$, let $P_2 = s_{OI}(P_1)$ and β be the angle in $\mathcal{A}(\mathcal{F})$ with support $|QOP_2$. Then $l = OI$ is the midline of $|P_1OP_2$ and so in the definition $P_3 = s_l(Q) = \overline{Q}$. Thus $\alpha + \beta$ has support $|QOQ$ and so it is $0_{\mathcal{F}}$.

COMMENT. The properties just listed show that $(\mathcal{A}(\mathcal{F}), +)$ is a commutative group. Because of this the familiar properties of addition, subtraction and additive cancellation apply to it.

9.3.4   Subtraction of angles

Definition. For all $\alpha \in \mathcal{A}(\mathcal{F})$, we denote the angle β in 9.3.3(v) by $-\alpha$. The **difference** $\gamma - \alpha$ in $\mathcal{A}(\mathcal{F})$ is defined by specifying that $\gamma - \alpha = \gamma + (-\alpha)$. In this way we deal with **subtraction**.

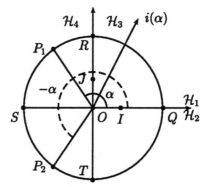

Figure 9.10.

For all $\alpha \in \mathcal{A}(\mathcal{F})$,

$$\cos(-\alpha) = \cos(\text{co} - \text{sp }\alpha) = \cos\alpha, \quad \sin(-\alpha) = \sin(\text{co} - \text{sp }\alpha) = -\sin\alpha.$$

Proof. With P_2 as in the proof of 9.3.3(v), we have

$$\cos(-\alpha) = \frac{k - |Q, U|}{k}, \quad \sin(-\alpha) = \frac{k - |R, V_1|}{k},$$

and $|R, V_1| = |T, V| = 2k - |R, V|$. We use this in conjunction with 9.2.1.

9.3.5 Integer multiples of an angle

Definition. For all $n \in \mathbf{N}$ and all $\alpha \in \mathcal{A}(\mathcal{F})$, $n\alpha$ is defined inductively by

$$1\alpha = \alpha,$$

$$(n+1)\alpha = n\alpha + \alpha, \quad \text{for all} \quad n \in \mathbf{N}.$$

We refer to $n\alpha$ as **integer multiples** of the angle α.

For all $\alpha \in \mathcal{A}(\mathcal{F})$,

(i) $\cos(2\alpha) = \cos^2 \alpha - \sin^2 \alpha = 2\cos^2 \alpha - 1 = 1 - 2\sin^2 \alpha$,

(ii) $\sin(2\alpha) = 2\cos\alpha \sin\alpha$.

Proof. These are immediate by 9.3.3 and 9.2.3.

9.3.6 Standard multiples of a right-angle

The angles $90_{\mathcal{F}}, 180_{\mathcal{F}}, 270_{\mathcal{F}}$ have the following properties:-

(i)

$$\cos 90_{\mathcal{F}} = 0, \; \sin 90_{\mathcal{F}} = 1, \; \cos 180_{\mathcal{F}} = -1,$$
$$\sin 180_{\mathcal{F}} = 0, \; \cos 270_{\mathcal{F}} = 0, \; \sin 270_{\mathcal{F}} = -1.$$

(ii) $2(90_{\mathcal{F}}) = 180_{\mathcal{F}}, \; 2(180_{\mathcal{F}}) = 0_{\mathcal{F}}$ *so that* $-180_{\mathcal{F}} = 180_{\mathcal{F}}$, *and* $90_{\mathcal{F}} + 270_{\mathcal{F}} = 0_{\mathcal{F}}$ *so that* $270_{\mathcal{F}} = -90_{\mathcal{F}}$.

(iii) *For all* $\alpha \in \mathcal{A}(\mathcal{F})$,

$$\cos(\alpha + 90_{\mathcal{F}}) = -\sin\alpha, \; \sin(\alpha + 90_{\mathcal{F}}) = \cos\alpha,$$
$$\cos(\alpha + 180_{\mathcal{F}}) = -\cos\alpha, \; \sin(\alpha + 180_{\mathcal{F}}) = -\sin\alpha,$$
$$\cos(\alpha + 270_{\mathcal{F}}) = \sin\alpha, \; \sin(\alpha + 270_{\mathcal{F}}) = -\cos\alpha.$$

Proof.
(i) These follow immediately from 9.2.1.
(ii) These follow from 9.2.1 and 9.3.4.
(iii) These follow immediately from 9.3.3 and (i) of the present theorem.

9.4 HALF ANGLES

9.4.1

Definition. Given any angle $\alpha \in \mathcal{A}^*(\mathcal{F})$ with support $|\underline{QOP}$, its indicator $i(\alpha)$ meets $\mathcal{C}(O; k)$ in a unique point P' which is in $\mathcal{H}_1$. Then the wedge or straight angle in $\mathcal{A}(\mathcal{F})$ with support $|\underline{QOP'}$ is denoted by $\frac{1}{2}\alpha$ and is called a **half-angle**.

Given any angle $\alpha \in \mathcal{A}(\mathcal{F})$, *the equation* $2\gamma = \alpha$ *has exactly two solutions in* $\mathcal{A}(\mathcal{F})$, *these being* $\frac{1}{2}\alpha$ *and* $\frac{1}{2}\alpha + 180_{\mathcal{F}}$.

Proof. In the definition we have $\frac{1}{2}\alpha = \angle_{\mathcal{F}}QOP'$ and take P'' so that $O = \mathrm{mp}(P', P'')$. Let $\beta = \angle_{\mathcal{F}}QOP''$ so that $\beta = \frac{1}{2}\alpha + 180_{\mathcal{F}}$. Then $s_{OP'}(Q) = P$ so that by 9.3.3, $2(\frac{1}{2}\alpha) = \alpha$, $2\beta = \alpha$. Now suppose that $\gamma, \delta \in \mathcal{A}(\mathcal{F})$ and $2\gamma = 2\delta = \alpha$. Then $\cos 2\delta = \cos 2\gamma$ so that $2\cos^2\delta - 1 = 2\cos^2\gamma - 1$, and hence $\cos\delta = \pm\cos\gamma$.

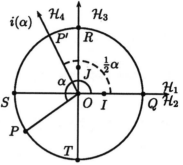

Figure 9.11.

Then also $\sin^2\delta = \sin^2\gamma$ so $\sin\delta = \pm\sin\gamma$. Moreover $\sin 2\delta = \sin 2\gamma$ so $2\sin\delta\cos\delta = 2\sin\gamma\cos\gamma$.

We first suppose that $\alpha \neq 180_{\mathcal{F}}$ so that $\cos\alpha \neq -1$ and so $\cos\gamma \neq 0$. Then if $\cos\delta = \cos\gamma$ we must have $\sin\delta = \sin\gamma$, and so $\delta = \gamma$. Alternatively we must have $\cos\delta = -\cos\gamma$, $\sin\delta = -\sin\gamma$ and so $\delta = \gamma + 180_{\mathcal{F}}$.

If $\alpha = 180_{\mathcal{F}}$ then $\cos\delta = 0$, so that $\sin\delta = \pm 1$ and so δ is either $90_{\mathcal{F}}$ or $270_{\mathcal{F}}$.

COMMENT. Our definition of a half-angle is the standard one for the angles we deal with, but it would not suit angles which we do not consider, for example ones with degree-magnitude greater than 360. The latter are difficult to give an account of geometrically. For us $\frac{1}{2}\alpha + \frac{1}{2}\beta$ and $\frac{1}{2}(\alpha + \beta)$ need not be equal; we shall deal with such matters in 12.1.1. Because of this, there is a danger of error if half-angles are used incautiously.

For any angles α, $\beta \in \mathcal{A}(\mathcal{F})$, *if* $\gamma = \frac{1}{2}\alpha + \frac{1}{2}\beta$ *and* $\delta = \frac{1}{2}\alpha - \frac{1}{2}\beta$, *then* $\gamma + \delta = \alpha$ *and* $\gamma - \delta = \beta$.

Proof. As we are dealing with a commutative group, we have

$$
\begin{aligned}
\gamma + \delta &= \left[\tfrac{1}{2}\alpha + \tfrac{1}{2}\beta\right] + \left[\tfrac{1}{2}\alpha + \left(-\tfrac{1}{2}\beta\right)\right] \\
&= \left[\tfrac{1}{2}\alpha + \tfrac{1}{2}\alpha\right] + \left[\tfrac{1}{2}\beta + \left(-\tfrac{1}{2}\beta\right)\right] \\
&= \alpha + 0_{\mathcal{F}} = \alpha.
\end{aligned}
$$

Similarly

$$
\begin{aligned}
\gamma - \delta &= \left[\tfrac{1}{2}\alpha + \tfrac{1}{2}\beta\right] - \left[\tfrac{1}{2}\alpha + \left(-\tfrac{1}{2}\beta\right)\right] \\
&= \left[\tfrac{1}{2}\alpha + \tfrac{1}{2}\beta\right] + \left[\left(-\tfrac{1}{2}\alpha\right) + \tfrac{1}{2}\beta\right] \\
&= \beta.
\end{aligned}
$$

9.5   THE COSINE AND SINE RULES

9.5.1   The cosine rule

NOTATION. Let A, B, C be non-collinear points. Then for the triangle $[A, B, C]$, we denote by a the length of the side which is opposite the vertex A, by b the length of the side opposite B, and by c the length of the side opposite C, so that

$$a = |B, C|, \ b = |C, A|, \ c = |A, B|.$$

We also use the notation

$$\alpha = \angle BAC, \ \beta = \angle CBA, \ \gamma = \angle ACB.$$

Let A, B, C be non-collinear points, let $D = \pi_{BC}(A)$ and write $x = |B, D|$. Then with the notation above, $2ax = a^2 + c^2 - b^2$ when $D \in [B, C]$ or $C \in [B, D]$, while $2ax = b^2 - a^2 - c^2$ when $B \in [D, C]$.

Proof. When $D \in [B, C]$ we have $|D, C| = a - x$, and when $C \in [B, D]$, $|D, C| = x - a$. In both of these cases, by Pythagoras' theorem used twice we have

$$|A, D|^2 = |A, B|^2 - |B, D|^2 = c^2 - x^2, \ |A, D|^2 = |A, C|^2 - |D, C|^2 = b^2 - (a - x)^2.$$

On equating these we have $c^2 - x^2 = b^2 - a^2 + 2ax - x^2$, giving $2ax = c^2 + a^2 - b^2$.

When $B \in [D, C]$ we have $|D, C| = a + x$, so by the formulae for $|A, D|^2$ above we have $c^2 - x = b^2 - (a + x)^2$. This simplifies to $2ax = b^2 - a^2 - c^2$.

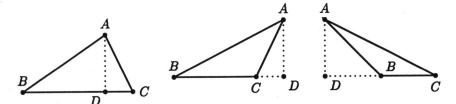

Figure 9.12.

THE COSINE RULE. *In each triangle $[A, B, C]$,*

$$\cos \alpha = \frac{b^2 + c^2 - a^2}{2bc}, \ \cos \beta = \frac{c^2 + a^2 - b^2}{2ca}, \ \cos \gamma = \frac{a^2 + b^2 - c^2}{2ab}.$$

Proof. On returning to the last proof, we note that when $D \in [B, C$ we have

$$\cos \alpha = \frac{|B, D|}{|B, A|} = \frac{x}{c},$$

while

$$x = \frac{c^2 + a^2 - b^2}{2a},$$

and so

$$\cos \alpha = \frac{c^2 + a^2 - b^2}{2ca}.$$

Similarly, when $B \in [D, C]$ we have

$$\cos \alpha = -\frac{|B, D|}{|B, A|} = -\frac{x}{c},$$

while

$$x = -\frac{c^2 + a^2 - b^2}{2a},$$

and this gives the same conclusion.

9.5.2 The sine rule

THE SINE RULE *In each triangle* $[A, B, C]$,

$$\frac{\sin \alpha}{a} = \frac{\sin \beta}{b} = \frac{\sin \gamma}{c}.$$

Proof. By the cosine rule

$$\frac{\cos^2 \alpha}{a^2} = \frac{1 - \sin^2 \alpha}{a^2} = \frac{(b^2 + c^2 - a^2)^2}{4a^2b^2c^2},$$

so that

$$\frac{\sin^2 \alpha}{a^2} = \frac{4b^2c^2 - (b^2 + c^2 - a^2)^2}{4a^2b^2c^2}$$

$$= \frac{2(b^2c^2 + c^2a^2 + a^2b^2) - (a^4 + b^4 + c^4)}{4a^2b^2c^2}.$$

As the right-hand side here is symmetrical in a, b and c we must have

$$\frac{\sin^2 \alpha}{a^2} = \frac{\sin^2 \beta}{b^2} = \frac{\sin^2 \gamma}{c^2}.$$

As the sines of wedge-angles are all positive, we may take square roots here and the result follows.

9.5.3

In a triangle $[A, B, C]$, *let the mid-line of* $\underline{|BAC}$ *meet* $[B, C]$ *at* D *and let* $d_1 = |A, D|$. *Then*

$$d_1 = \frac{2bc}{b+c} \cos \frac{1}{2}\alpha.$$

Proof. By 5.5

$$\frac{|B, D|}{|D, C|} = \frac{c}{b},$$

so that

$$|B, D| = \frac{c}{b+c}a.$$

On applying the sine rule to the triangle $[A, B, D]$ we have that

$$\frac{d_1}{\sin \beta} = \frac{ca}{b+c} \frac{1}{\sin \frac{1}{2}\alpha},$$

and so

$$d_1 = \frac{ca}{b+c} \frac{\sin \beta}{\sin \frac{1}{2}\alpha} = \frac{ca}{b+c} \frac{\sin \beta}{b} \frac{b}{\sin \frac{1}{2}\alpha} = \frac{ca}{b+c} \frac{\sin \alpha}{a} \frac{b}{\sin \frac{1}{2}\alpha} = \frac{2bc}{b+c} \cos \frac{1}{2}\alpha.$$

9.5.4 The Steiner-Lehmus theorem, 1842

Suppose that we are given a triangle $[A, B, C]$, that the mid-line
of $|CBA$ meets CA at E, that the
mid-line of $|ACB$ meets AB at
F, and that $|B, E| = |C, F|$. We
then wish to show that the triangle is isosceles. This is known
as the STEINER-LEHMUS THEO-
REM.

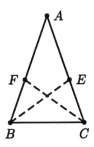

Figure 9.13. Steiner-Lehmus theorem.

Proof. By the last result we have

$$d_2 = \frac{2ca}{c+a} \cos \frac{1}{2}\beta, \ d_3 = \frac{2ab}{a+b} \cos \frac{1}{2}\gamma.$$

Then

$$
\begin{aligned}
d_2^2 - d_3^2 &= 4a^2 \left[\frac{c^2}{(c+a)^2} \cos^2 \frac{1}{2}\beta - \frac{b^2}{(a+b)^2} \cos^2 \frac{1}{2}\gamma \right] \\
&= 2a^2 \left[\frac{c^2}{(c+a)^2} (1 + \cos\beta) - \frac{b^2}{(a+b)^2} (1 + \cos\gamma) \right] \\
&= 2a^2 \left[\frac{c^2}{(c+a)^2} \left(1 + \frac{c^2 + a^2 - b^2}{2ca} \right) - \frac{b^2}{(a+b)^2} \left(1 + \frac{a^2 + b^2 - c^2}{2ab} \right) \right] \\
&= a \left[c - b + \frac{bc^2}{(a+b)^2} - \frac{b^2 c}{(c+a)^2} \right] \\
&= a \left[c - b + \frac{bc}{(a+b)^2(c+a)^2} [c(c+a)^2 - b(a+b)^2] \right] \\
&= a(c-b) \left[1 + \frac{bc}{(a+b)^2(c+a)^2} [a^2 + b^2 + c^2 + 2ab + bc + 2ca] \right]
\end{aligned}
$$

Then $b < c$ implies that $d_2 > d_3$.

9.6 COSINE AND SINE OF ANGLES EQUAL IN MAGNITUDE

9.6.1

*If angles α, β are such that $|\alpha|° = |\beta|°$, then $\cos\alpha = \cos\beta$ and $\sin\alpha = \sin\beta$. Con-
versely if $\cos\alpha = \cos\beta$ and $\sin\alpha = \sin\beta$, then $|\alpha|° = |\beta|°$ unless one of them is null
and the other is full.*

Proof. Let $\mathcal{F} = ([O, I, [O, J)$ and α have support $|QOP$ and indicator in $\mathcal{H}_1$,
where $|O, P| = |O, Q| = k$. Let $\mathcal{F}' = ([O', I', [O', J')$ and β have support $|Q'O'P'$
and indicator in $\mathcal{H}'_1$, where $|O', P'| = |O', Q'| = k$.

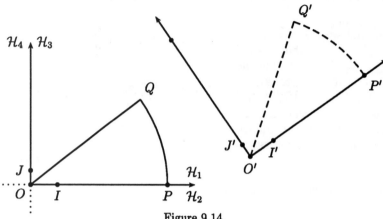

Figure 9.14.

We suppose first that $|\alpha|^\circ = |\beta|^\circ \leq 90$ so that $P \in \mathcal{Q}_1$, $P' \in \mathcal{Q}'_1$. Then $U \in [O, Q]$, $V \in [O, R]$, $U' \in [O', Q']$, $V' \in [O', R']$. The triangles $[O, U, P]$, $[O', U', P']$ are congruent by the ASA-principle, so $|O, U| = |O', U'|$, $|O, V| = |O', V'|$. Then $|Q, U| = |Q', U'|$, $|R, V| = |R', V'|$. Hence $\cos\alpha = \cos\beta$, $\sin\alpha = \sin\beta$.

Similar arguments work in the case of the other three quadrants of $\mathcal{F}$.

Conversely, let $\cos\alpha = \cos\beta$, $\sin\alpha = \sin\beta$. Suppose first that $\cos\alpha \geq 0$, $\sin\alpha \geq 0$. Then $P \in \mathcal{Q}_1$, $P' \in \mathcal{Q}'_1$. But $|Q, U| = |Q', U'|$, $|R, V| = |R', V'|$ and so $|O, U| = |O', U'|$, $|U, P| = |U', P'|$. By the SSS-principle, the triangles $[O, U, P]$, $[O', U', P']$ are congruent so $|\alpha|^\circ = |\beta|^\circ$, unless we have a degeneration from a triangle and one angle is null and the other is full.

A similar argument works for the other three quadrants of $\mathcal{F}$.

Exercises

9.1 Prove that for all angles $\alpha \in \mathcal{A}^*(\mathcal{F})$,

$$-1 \leq \cos\alpha \leq 1, \quad -1 \leq \sin\alpha \leq 1.$$

9.2 Let $\mathcal{C}_1$ be the circle with centre O and radius of length k. Let $Z_1 \equiv (k\cos\theta, k\sin\theta)$, $Z_2 \equiv (-k, 0)$, $Z_3 \equiv (k, 0)$, so that Z_1 is a point on this circle, and $[Z_2, Z_3]$ is a diameter. Let $\mathcal{C}_2$ be the circle with $[Z_1, Z_3]$ as diameter. Find the coordinates of the second point in which $\mathcal{C}_2$ meets the line $Z_2 Z_3$. How does this relate to 4.3.3?

9.3 If D is the mid-point of the side $[B, C]$ of the triangle $[A, B, C]$ and $d_1 = |A, D|$, prove that

$$4d_1^2 = b^2 + c^2 + 2bc\cos\alpha.$$

Deduce that $2d_1 > a$ if and only if α is an acute angle.

9.4 Prove the identities

$$\cos\alpha + \cos\beta = 2\cos(\tfrac{1}{2}\alpha + \tfrac{1}{2}\beta)\cos(\tfrac{1}{2}\alpha - \tfrac{1}{2}\beta),$$
$$\cos\alpha - \cos\beta = -2\sin(\tfrac{1}{2}\alpha + \tfrac{1}{2}\beta)\sin(\tfrac{1}{2}\alpha - \tfrac{1}{2}\beta),$$

and find similar results for $\sin \alpha + \sin \beta$ and $\sin \alpha - \sin \beta$.

9.5 Show that

$$\sin 270_\mathcal{F} + \sin 210_\mathcal{F} = -\frac{3}{2},$$

and yet

$$2\sin[\tfrac{1}{2}(270_\mathcal{F} + 270_\mathcal{F})]\cos[\tfrac{1}{2}(270_\mathcal{F} - 270_\mathcal{F})] = \frac{3}{2}.$$

10

Complex coordinates; sensed angles; rotations; applications to circles; angles between lines

COMMENT. In this chapter we utilise complex coordinates, develop sensed angles and rotations, complete our formulae for axial symmetries and identify isometries in terms of translations, rotations and axial symmetries. We go on to establish more results on circles and consider a variant on the angles we have been dealing with.

10.1 COMPLEX COORDINATES

10.1.1

We now introduce the field of complex numbers $(\mathbf{C}, +, .)$ as an aid. This has an added convenience when doing coordinate geometry. We recall that any $z \in \mathbf{C}$ can be written uniquely in the form $z = x + \imath y$, where $x, y \in \mathbf{R}$ and $\imath^2 = -1$. We use the notations $|z| = \sqrt{x^2 + y^2}$, $\bar{z} = x - \imath y$ for the modulus or absolute value, and complex conjugate, respectively, of z. As well as having the familiar properties for addition, subtraction, multiplication and division (except division by 0), these have the further properties:

$$\bar{\bar{z}} = z, \ \overline{z_1 z_2} = \overline{z_1} \ \overline{z_2}, \ \forall z, z_1, z_2 \in \mathbf{C}; \ \bar{z} = z \text{ iff } z \in \mathbf{R};$$

$$|z_1 z_2| = |z_1||z_2|, \ |\bar{z}| = |z|, \ \forall z, z_1, z_2 \in \mathbf{C}; \ |z| = z \text{ iff } z \in \mathbf{R} \text{ and } z \geq 0;$$

$$z\bar{z} = |z|^2, \ \forall z \in \mathbf{C}; \ \frac{1}{z} = \frac{\bar{z}}{|z|^2} \ \forall z \neq 0.$$

Definition. Let $\mathcal{F} = ([O, I \ , [O, J \ )$ be a frame of reference for Π and for any point $Z \in \Pi$ we recall the Cartesian coordinates (x, y) of Z relative to $\mathcal{F}$, $Z \equiv_{\mathcal{F}} (x, y)$. If $z = x + \imath y$, we also write $Z \sim_{\mathcal{F}} z$, and call z a **Cartesian complex coordinate** of the point Z relative to $\mathcal{F}$. When $\mathcal{F}$ can be understood, we can relax our notation and denote this by $Z \sim z$.

Complex coordinates have the following properties:-

(i) $|z_2 - z_1| = |Z_1, Z_2|$ *for all* Z_1, Z_2.

(ii) *If* $Z_1 \neq Z_2$, *then* $Z \in Z_1 Z_2$ *if and only if* $z - z_1 = t(z_2 - z_1)$ *for some* $t \in \mathbf{R}$.

(iii) *If* $Z_1 \neq Z_2$, *then* $Z \in [Z_1, Z_2$ *if and only if* $z - z_1 = t(z_2 - z_1)$ *for some* $t \geq 0$.

(iv) *If* $Z_1 \neq Z_2$, *then* $Z \in [Z_1, Z_2]$ *if and only if* $z - z_1 = t(z_2 - z_1)$ *for some* t *such that* $0 \leq t \leq 1$.

(v) *For* $Z_1 \neq Z_2$ *and* $Z_3 \neq Z_4$, $Z_1 Z_2 \parallel Z_3 Z_4$ *if and only if* $z_4 - z_3 = t(z_2 - z_1)$ *for some* $t \in \mathbf{R} \setminus \{0\}$.

(vi) *For* $Z_1 \neq Z_2$ *and* $Z_3 \neq Z_4$, $Z_1 Z_2 \perp Z_3 Z_4$ *if and only if* $z_4 - z_3 = tı(z_2 - z_1)$ *for some* $t \in \mathbf{R} \setminus \{0\}$.

Proof.
(i) For $|z_2 - z_1|^2 = |x_2 - x_1 + ı(y_2 - y_1)|^2 = (x_2 - x_1)^2 + (y_2 - y_1)^2 = |Z_1, Z_2|^2$.
(ii) For $z - z_1 = t(z_2 - z_1)$ if and only if $x - x_1 + ı(y - y_1) = t[x_2 - x_1 + ı(y_2 - y_1)]$. If this happens for some $t \in \mathbf{R}$, then $x - x_1 = t(x_2 - x_1)$, $y - y_1 = t(y_2 - y_1)$. By 6.4.1, Corollary (i), this implies that $Z \in Z_1 Z_2$.
Conversely if $Z \in Z_1 Z_2$, by the same reference there is such a $t \in \mathbf{R}$ and it follows that $z - z_1 = t(z_2 - z_1)$.
(iii) and (iv). In (ii) we have $Z \in [Z_1, Z_2$ when $t \geq 0$ by 6.4.1, Corollary, and similarly $Z \in [Z_1, Z_2]$ when $0 \leq t \leq 1$.
(v) By 6.5.1, Corollary (ii), $Z_1 Z_2$ and $Z_3 Z_4$ are parallel only if

$$-(y_2 - y_1)(x_4 - x_3) + (y_4 - y_3)(x_2 - x_1) = 0. \qquad (10.1.1)$$

We note that as $Z_1 \neq Z_2$ we must have either $x_1 \neq x_2$ or $y_1 \neq y_2$.
Suppose first that $z_4 - z_3 = t(z_2 - z_1)$ for some $t \in \mathbf{R}$. Then

$$x_4 - x_3 + ı(y_4 - y_3) = t(x_2 - x_1) + ıt(y_2 - y_1),$$

and so as t is real,

$$x_4 - x_3 = t(x_2 - x_1), \ y_4 - y_3 = t(y_2 - y_1).$$

Then

$$-(y_2 - y_1)(x_4 - x_3) + (y_4 - y_3)(x_2 - x_1)$$
$$= -(y_2 - y_1)t(x_2 - x_1) + t(y_2 - y_1)(x_2 - x_1) = 0,$$

so that (10.1.1) holds and hence the lines are parallel.
Conversely suppose that the lines are parallel so that (10.1.1) holds. When $x_1 \neq x_2$, we let

$$t = \frac{x_4 - x_3}{x_2 - x_1},$$

so that $x_4 - x_3 = t(x_2 - x_1)$. On inserting this in (10.1.1), we have

$$-t(y_2 - y_1)(x_2 - x_1) + (y_4 - y_3)(x_2 - x_1) = 0,$$

from which we have $y_4 - y_3 = t(y_2 - y_1)$.

When $x_2 - x_1 = 0$, by (10.1.1) we must have $x_4 - x_3 = 0$. We now let

$$t = \frac{y_4 - y_3}{y_2 - y_1},$$

so that $y_4 - y_3 = t(y_2 - y_1)$. For this t we also have, trivially, $x_4 - x_3 = t(x_2 - x_1)$.

Thus in both cases $x_4 - x_3 = t(x_2 - x_1)$, $y_4 - y_3 = t(y_2 - y_1)$, and so on combining these

$$x_4 - x_3 + \imath(y_4 - y_3) = t(x_2 - x_1) + \imath t(y_2 - y_1).$$

Thus $z_4 - z_3 = t(z_2 - z_1)$.

(vi) By 6.5.1, Corollary (i), these lines are perpendicular if and only if

$$(y_2 - y_1)(y_4 - y_3) + (x_2 - x_1)(x_4 - x_3) = 0. \tag{10.1.2}$$

Suppose first that $z_4 - z_3 = t\imath(z_2 - z_1)$ for some $t \in \mathbf{R}$. Then

$$x_4 - x_3 + \imath(y_4 - y_3) = \imath t(x_2 - x_1) - t(y_2 - y_1),$$

and so as t is real, $x_4 - x_3 = -t(y_2 - y_1)$, $y_4 - y_3 = t(x_2 - x_1)$. Then

$$(y_2 - y_1)(y_4 - y_3) + (x_2 - x_1)(x_4 - x_3)$$
$$= (y_2 - y_1)t(x_2 - x_1) - t(x_2 - x_1)(y_2 - y_1) = 0,$$

so that (10.1.2) holds, and hence the lines are perpendicular.

Conversely suppose that the lines are perpendicular so that (10.1.2) holds. When $x_1 \neq x_2$, we let

$$t = \frac{y_4 - y_3}{x_2 - x_1}$$

so that $y_4 - y_3 = t(x_2 - x_1)$. On inserting this in (10.1.2), we have

$$(y_2 - y_1)t(x_2 - x_1) + (x_2 - x_1)(x_2 - x_1) = 0,$$

from which we have $x_4 - x_3 = -t(y_2 - y_1)$.

When $x_2 - x_1 = 0$, by (10.1.2) we must have $y_4 - y_3 = 0$. We now let

$$t = -\frac{x_4 - x_3}{y_2 - y_1},$$

so that $x_4 - x_3 = -t(y_2 - y_1)$. For this t we also have, trivially, $y_4 - y_3 = t(x_2 - x_1)$.

Thus in both cases $x_4 - x_3 = -t(y_2 - y_1)$, $y_4 - y_3 = t(x_2 - x_1)$, and so on combining these

$$x_4 - x_3 + \imath(y_4 - y_3) = -t(y_2 - y_1) + \imath t(x_2 - x_1) = \imath t[x_2 - x_1 + \imath(y_2 - y_1)].$$

Thus $z_4 - z_3 = t\imath(z_2 - z_1)$.

10.2 COMPLEX-VALUED DISTANCE

10.2.1 Complex-valued distance

The material in 7.6 is long established; we can generalise those concepts of sensed distances and sensed ratios as follows.

Definition Let $\mathcal{F}$ be a frame of reference for Π. If $Z_1 \sim_{\mathcal{F}} z_1$, $Z_2 \sim_{\mathcal{F}} z_2$ we define $\overline{Z_1 Z_2}_{\mathcal{F}} = z_2 - z_1$, and call this a **complex-valued distance** from Z_1 to Z_2. We then consider also $\frac{\overline{Z_3 Z_4}_{\mathcal{F}}}{\overline{Z_1 Z_2}_{\mathcal{F}}}$, a ratio of complex-valued distances or **complex ratio** when $Z_1 \neq Z_2$.

We show that this latter reduces to the sensed ratio

$$\frac{\overline{Z_3 Z_4}_{\leq_l}}{\overline{Z_1 Z_2}_{\leq_l}}$$

when Z_1, Z_2, Z_3, Z_4 are points of a line l. As in 7.6.1 we suppose that l is the line $W_0 W_1$ where $W_0 \equiv (u_0, v_0)$ and $W_1 \equiv (u_1, v_1)$, and has parametric equations

$$x = u_0 + s(u_1 - u_0), \quad y = v_0 + s(v_1 - v_0).$$

By 10.1.1(ii) l then has complex parametric equation $z = w_0 + s(w_1 - w_0)$. If Z_1, Z_2, Z_3, Z_4 have parameters s_1, s_2, s_3, s_4, respectively, then

$$z_2 - z_1 = (z_2 - w_0) - (z_1 - w_0) = (s_2 - s_1)(w_1 - w_0), \quad z_4 - z_3 = (s_4 - s_3)(w_1 - w_0),$$

and so

$$\frac{\overline{Z_3 Z_4}_{\mathcal{F}}}{\overline{Z_1 Z_2}_{\mathcal{F}}} = \frac{s_4 - s_3}{s_2 - s_1}.$$

By 7.6.1 this is equal to the sensed ratio. This shows that for four collinear points a ratio of complex-valued distances reduces to the corresponding ratio of sensed distances.

COMMENT. We could make considerable use of this concept in our notation for the remainder of this chapter but in fact we use it sparingly.

10.2.2 A complex-valued trigonometric function

For $Z_0 \sim_{\mathcal{F}} z_0$ and $\mathcal{F}' = t_{O, z_0}(\mathcal{F})$, let $I_0 = t_{O, z_0}(I)$; we recall from 8.3 that $Z \sim_{\mathcal{F}'} z - z_0$. Then if $Z \neq Z_0$, $Z \sim_{\mathcal{F}} z$ and $\theta = \angle_{\mathcal{F}'} I_0 Z_0 Z$, by 9.2.2 we have

$$x - x_0 = r \cos \theta, \quad y - y_0 = r \sin \theta,$$

where $r = |Z_0, Z| = |z - z_0|$. It follows that $z - z_0 = r(\cos \theta + i \sin \theta)$.

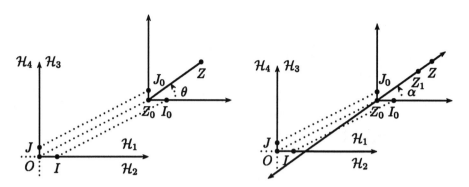

Figure 10.1.

If $Z_1 \neq Z_0$, $Z_1 \sim_{\mathcal{F}} z_1$ and $\alpha = \angle_{\mathcal{F}'} I_0 Z_0 Z_1$, then by this

$$x_1 - x_0 = k \cos \alpha, \quad y_1 - y_0 = k \sin \alpha,$$

where $k = |Z_0, Z_1|$. On inserting this in 6.3.1 Corollary, we see that

$$Z_0 Z_1 = \{ Z \equiv (x, y) : (x - x_0) \sin \alpha - (y - y_0) \cos \alpha = 0 \}.$$

When $Z_0 Z_1$ is not parallel to OJ we have that $\cos \alpha \neq 0$ and this equation of the line $Z_0 Z_1$ can be re-written as $y - y_0 = \tan \alpha (x - x_0)$ where $\tan \alpha = \sin \alpha / \cos \alpha$. We call $\tan \alpha$ the **slope** of this line.

Notation. For any angle θ we write cis $\theta = \cos \theta + \imath \sin \theta$.

The complex-valued function cis *has the properties:-*

(i) *For all* $\theta, \phi \in \mathcal{A}(\mathcal{F})$, cis $(\theta + \phi) = $ cis θ.cis ϕ.

(ii) cis $0_{\mathcal{F}} = 1$.

(iii *For all* $\theta \in \mathcal{A}(\mathcal{F})$, $\frac{1}{\text{cis } \theta} = $ cis $(-\theta)$.

(iii) *For all* $\theta \in \mathcal{A}(\mathcal{F})$, $\overline{\text{cis } \theta} = $ cis $(-\theta)$, *where* $\bar{z}$ *denotes the complex conjugate of* z.

(iv) *For all* θ, $|\text{cis } \theta| = 1$.

Proof.

(i) For

$$\begin{aligned} \text{cis } \theta.\text{cis } \phi &= (\cos \theta + \imath \sin \theta)(\cos \phi + \imath \sin \phi) \\ &= \cos \theta \cos \phi - \sin \theta \sin \phi + \imath[\sin \theta \cos \phi + \cos \theta \sin \phi] \\ &= \cos(\theta + \phi) + \imath \sin(\theta + \phi) = \text{cis } (\theta + \phi). \end{aligned}$$

(ii) For cis $0_{\mathcal{F}} = \cos 0_{\mathcal{F}} + \imath \sin 0_{\mathcal{F}} = 1 + \imath 0 = 1$.

(iii) For by (i) and (ii) of the present theorem,

$$\text{cis } \theta.\text{cis } (-\theta) = \text{cis } (\theta - \theta) = \text{cis } 0_{\mathcal{F}} = 1.$$

(iv) For the complex conjugate of $\cos \theta + \imath \sin \theta$ is $\cos \theta - \imath \sin \theta = \cos(-\theta) + \imath \sin(-\theta)$.

(v) For $|\text{cis } \theta|^2 = \cos^2 \theta + \sin^2 \theta = 1$.

10.3 ROTATIONS AND AXIAL SYMMETRIES

10.3.1 Rotations

Definition. Let $Z_0 \sim_{\mathcal{F}} z_0$, t be the translation t_{O,z_0}, $\mathcal{F}' = t(\mathcal{F})$ and $I_0 = t(I)$. Let $\alpha \in \mathcal{A}(\mathcal{F}')$. The function $r_{\alpha;z_0} : \Pi \to \Pi$ defined by

$$Z \sim_{\mathcal{F}} z, \ Z' \sim_{\mathcal{F}} z', \ r_{\alpha;z_0}(Z) = Z' \text{ if } z' - z_0 = (z - z_0)\text{cis } \alpha,$$

is called **rotation** about the point Z_0 through the angle α.

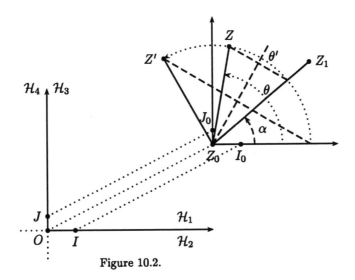

Figure 10.2.

If $r_{\alpha;z_0}(Z) = Z'$ we have the following properties:-

(i) *In all cases $|Z_0, Z'| = |Z_0, Z|$, and hence in particular $r_{\alpha;z_0}(Z_0) = Z_0$.*

(ii) *If $Z \neq Z_0$, $\theta = \angle_{\mathcal{F}'} I_0 Z_0 Z$ and $\theta' = \angle_{\mathcal{F}'} I_0 Z_0 Z'$, then $\theta' = \theta + \alpha$.*

(iii) *If $Z_0 \sim_{\mathcal{F}} z_0$, $Z \sim_{\mathcal{F}} z$, $Z' \sim_{\mathcal{F}} z'$, then $r_{\alpha;z_0}$ has the real coordinates form*

$$\begin{aligned} x' - x_0 &= \cos\alpha.(x - x_0) - \sin\alpha.(y - y_0), \\ y' - y_0 &= \sin\alpha.(x - x_0) + \cos\alpha.(y - y_0), \end{aligned}$$

which has the matrix form

$$\begin{pmatrix} x' - x_0 \\ y' - y_0 \end{pmatrix} = \begin{pmatrix} \cos\alpha & -\sin\alpha \\ \sin\alpha & \cos\alpha \end{pmatrix} \begin{pmatrix} x - x_0 \\ y - y_0 \end{pmatrix}.$$

Proof.
(i) For $|z' - z_0| = |(z - z_0)\text{cis } \alpha| = |z - z_0||\text{cis } \alpha| = |z - z_0|$.
(ii) For by 10.2.2

$$z - z_0 = |z - z_0|\text{cis } \theta, \ z' - z_0 = |z' - z_0|\text{cis } \theta',$$

and so $|z' - z_0|$cis $\theta' = |z - z_0|$cis θcis α. Hence cis $\theta' = $ cis θcis $\alpha = $ cis $(\theta + \alpha)$ and so $\theta' = \theta + \alpha$ by 9.2.2.

(iii) Now $x' - x_0 + \imath(y' - y_0) = (\cos\alpha + \imath\sin\alpha)[x - x_0 + \imath(y - y_0)]$ and so

$$x' - x_0 = \cos\alpha.(x - x_0) - \sin\alpha.(y - y_0),$$
$$y' - y_0 = \sin\alpha.(x - x_0) + \cos\alpha.(y - y_0).$$

COMMENT. The rotation $r_{\alpha;z_0}$ is characterised by (i) and (ii), as the steps can be traced backwards. Why a frame of reference $\mathcal{F}'$ is prominent in this characterisation stems from the need to identify the angles α, θ, θ'.

10.3.2   Formula for an axial symmetry

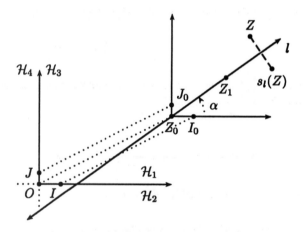

Figure 10.3.

The form of equation of a line noted in 10.2.2 can be used in the formula in 6.6.1(iii) for an axial symmetry. However, for practice with complex-valued coordinates we deduce the result independently.

Let l be the line $Z_0 Z_1$, $Z_0 \sim_{\mathcal{F}} z$, $\mathcal{F}' = t_{O,z_0}(\mathcal{F})$, $I_0 = t_{O,z_0}(I)$ and $\alpha = \angle_{\mathcal{F}'} I_0 Z_0 Z_1$. Then $s_l(Z) = Z'$ where

$$Z \sim_{\mathcal{F}} z, \quad Z' \sim_{\mathcal{F}} z', \quad z' - z_0 = (\bar{z} - \bar{z}_0)\text{cis } 2\alpha,$$

so that s_l has the real coordinates form

$$x' - x_0 = \cos 2\alpha.(x - x_0) + \sin 2\alpha.(y - y_0),$$
$$y' - y_0 = \sin 2\alpha.(x - x_0) - \cos 2\alpha.(y - y_0),$$

and so has the matrix form

$$\begin{pmatrix} x' - x_0 \\ y' - y_0 \end{pmatrix} = \begin{pmatrix} \cos 2\alpha & \sin 2\alpha \\ \sin 2\alpha & -\cos 2\alpha \end{pmatrix} \begin{pmatrix} x - x_0 \\ y - y_0 \end{pmatrix}.$$

Proof. To find a formula for $Z' = s_l(Z)$ we first show that if $W = \pi_l(Z)$ and $W \sim_{\mathcal{F}} w$ then

$$w - z_0 = \Re\left[\frac{z - z_0}{z_1 - z_0}\right](z_1 - z_0), \quad z - w = \imath\Im\left[\frac{z - z_0}{z_1 - z_0}(z_1 - z_0)\right].$$

To start on this we note that

$$z - z_0 = \frac{z - z_0}{z_1 - z_0}(z_1 - z_0) = \left[\Re\frac{z - z_0}{z_1 - z_0}\right](z_1 - z_0) + \imath\left[\Im\frac{z - z_0}{z_1 - z_0}\right](z_1 - z_0).$$

If we now define w by

$$w - z_0 = \left[\Re\frac{z - z_0}{z_1 - z_0}\right](z_1 - z_0)$$

then $W \in Z_0 Z_1$ as $w - z_0$ is a real multiple of $z_1 - z_0$. But then

$$z - w = \imath\left[\Im\frac{z - z_0}{z_1 - z_0}\right](z_1 - z_0),$$

so W is on a line through Z which is perpendicular to $Z_0 Z_1$. Thus W is the foot of the perpendicular from Z to $Z_0 Z_1$.

From this, as $z' + z = 2w$, we have $z' - z = z' - w - (z - w) = -2(z - w)$ so

$$z' - z = -2\imath\left[\Im\frac{z - z_0}{z_1 - z_0}\right](z_1 - z_0).$$

As $z_1 - z_0 = k\operatorname{cis}\alpha$ for some $k > 0$, we then have

$$\begin{aligned}
z' - z &= -2\imath\left[\Im\frac{z - z_0}{k\operatorname{cis}\alpha}\right]k\operatorname{cis}\alpha = -2\imath\{\Im[(z - z_0)\operatorname{cis}(-\alpha)]\}\operatorname{cis}\alpha \\
&= -[(z - z_0)\operatorname{cis}(-\alpha) - (\bar{z} - \bar{z}_0)\operatorname{cis}\alpha]\operatorname{cis}\alpha = -(z - z_0) + (\bar{z} - \bar{z}_0)\operatorname{cis}2\alpha
\end{aligned}$$

and so $z' - z_0 = (\bar{z} - \bar{z}_0)\operatorname{cis}2\alpha$. Hence $x' - x_0 + \imath(y' - y_0) = [x - x_0 - \imath(y - y_0)](\cos 2\alpha + \imath\sin 2\alpha)$, so that

$$\begin{aligned}
x' - x_0 &= \cos 2\alpha.(x - x_0) + \sin 2\alpha.(y - y_0), \\
y' - y_0 &= \sin 2\alpha.(x - x_0) - \cos 2\alpha.(y - y_0).
\end{aligned}$$

We can express this in matrix form as stated.

We denote s_l by $s_{\alpha;z_0}$ as well.

10.4   SENSED ANGLES

10.4.1

Definition. For $\mathcal{F}' = t_{O,z_0}(\mathcal{F})$, let $I_0 = t_{O,z_0}(I)$. Then if $Z_1 \neq Z_0$, $Z_2 \neq Z_0$, we let $\theta_1 = \angle_{\mathcal{F}'} I_0 Z_0 Z_1$ and $\theta_2 = \angle_{\mathcal{F}'} I_0 Z_0 Z_2$. We define the **sensed-angle** $\angle_{\mathcal{F}} Z_1 Z_0 Z_2$ to be $\theta_2 - \theta_1$.

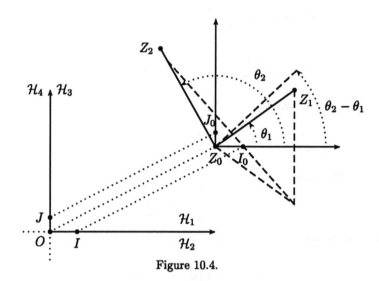

Figure 10.4.

Sensed angles have the following properties.   Throughout $Z_0 \sim_{\mathcal{F}} z_0$, $Z_1 \sim_{\mathcal{F}}$ z_1, $Z_2 \sim_{\mathcal{F}} z_2$.

(i) *If the points Z_1 and Z_2 are both distinct from Z_0, and ϕ is the sensed-angle $\angle_{\mathcal{F}} Z_1 Z_0 Z_2$, then*

$$\frac{\overline{Z_0 Z_2}_{\mathcal{F}}}{\overline{Z_0 Z_1}_{\mathcal{F}}} = \frac{z_2 - z_0}{z_1 - z_0} = \frac{|Z_0, Z_2|}{|Z_0, Z_1|} \operatorname{cis} \phi.$$

(ii) *The sensed-angle $\angle_{\mathcal{F}} Z_1 Z_0 Z_2$ is wedge or reflex according as*

$$\Im \frac{z_2 - z_0}{z_1 - z_0}$$

is positive or negative, and this occurs according as

$$\tfrac{1}{2}[(y_2 - y_0)(x_1 - x_0) - (x_2 - x_0)(y_1 - y_0)]$$

is positive or negative.

(iii) *If the points Z_1 and Z_2 are both distinct from Z_0, then*

$$\angle_{\mathcal{F}} Z_1 Z_0 Z_2 = -\angle_{\mathcal{F}} Z_2 Z_0 Z_1.$$

(iv) *If Z_1, Z_2, Z_3 are all distinct from Z_0, then*

$$\angle_{\mathcal{F}} Z_1 Z_0 Z_2 + \angle_{\mathcal{F}} Z_2 Z_0 Z_3 = \angle_{\mathcal{F}} Z_1 Z_0 Z_3.$$

(v) *If $\phi = \angle_{\mathcal{F}} Z_1 Z_0 Z_2$, then $r_{\phi; z_0}([Z_0, Z_1) = [Z_0, Z_2 .$*

(vi) *In 10.3.1(ii), $\angle_{\mathcal{F}} Z Z_0 Z' = \alpha$.*

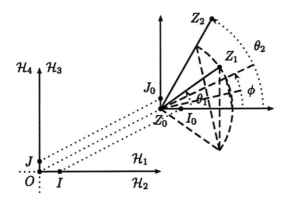

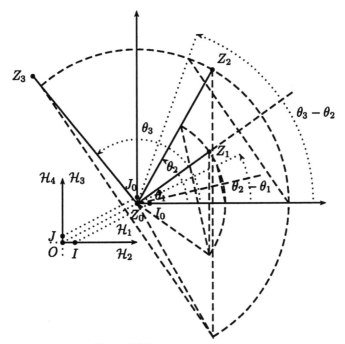

Figure 10.5.

Proof.

(i) For $z_1 - z_0 = |z_1 - z_0| \operatorname{cis} \theta_1$, $z_2 - z_0 = |z_2 - z_0| \operatorname{cis} \theta_2$ and so

$$\frac{z_2 - z_0}{z_1 - z_0} = \frac{|z_2 - z_0| \operatorname{cis} \theta_2}{|z_1 - z_0| \operatorname{cis} \theta_1} = \frac{|z_2 - z_0|}{|z_1 - z_0|} \operatorname{cis} (\theta_2 - \theta_1).$$

(ii) From (i)

$$\Im \frac{z_2 - z_0}{z_1 - z_0} = \frac{|z_2 - z_0|}{|z_1 - z_0|} \sin(\theta_2 - \theta_1),$$

and this is positive or negative according as $\theta_2 - \theta_1$ is wedge or reflex. Moreover

$$|z_1 - z_0|^2 \Im \frac{z_2 - z_0}{z_1 - z_0} = \Im[(z_2 - z_0)(\bar{z}_1 - \bar{z}_0)] = (y_2 - y_0)(x_1 - x_0) - (x_2 - x_0)(y_1 - y_0).$$

(iii) For the first is $\theta_2 - \theta_1$ and the second is $\theta_1 - \theta_2$.

(iv) For if $\theta_1 = \angle_{\mathcal{F}'} I_0 Z_0 Z_1$, $\theta_2 = \angle_{\mathcal{F}'} I_0 Z_0 Z_2$, $\theta_3 = \angle_{\mathcal{F}'} I_0 Z_0 Z_3$, then

$$\theta_2 - \theta_1 + (\theta_3 - \theta_2) = \theta_3 - \theta_1.$$

(v) For if $Z \in [Z_0, Z_1$, then $z = z_0 + t(z_1 - z_0) = z_0 + t|z_1 - z_0|\mathrm{cis}\,\theta_1$ for some $t \geq 0$. Hence $r_{\theta_2 - \theta_1; z_0}(Z) = Z'$ where

$$z' - z_0 = (z - z_0)\mathrm{cis}\,(\theta_2 - \theta_1) = t|z_1 - z_0|\mathrm{cis}\,\theta_1\mathrm{cis}\,(\theta_2 - \theta_1) = t|z_1 - z_0|\mathrm{cis}\,\theta_2.$$

Thus $Z' \in [Z_0, Z_2$.

(vi) For $\theta' - \theta = \alpha$.

If the points Z_1 and Z_2 are both distinct from Z_0 and $\phi = \angle_{\mathcal{F}} Z_1 Z_0 Z_2$, then

$$|Z_1, Z_2|^2 = |Z_0, Z_1|^2 + |Z_0, Z_2|^2 - 2|Z_0, Z_1||Z_0, Z_2|\cos\phi.$$

Proof. For by (i) in the last result,

$$z_2 - z_0 = \frac{|Z_0, Z_2|}{|Z_0, Z_1|}(\cos\phi + \imath\sin\phi)(z_1 - z_0),$$

so that

$$z_2 - z_1 = \left[\frac{|Z_0, Z_2|}{|Z_0, Z_1|}(\cos\phi + \imath\sin\phi) - 1\right](z_1 - z_0).$$

Then

$$|Z_1, Z_2|^2 = \left\{\left[\frac{|Z_0, Z_2|}{|Z_0, Z_1|}\cos\phi - 1\right]^2 + \left[\frac{|Z_0, Z_2|}{|Z_0, Z_1|}\sin\phi\right]^2\right\}|Z_0, Z_1|^2,$$

and the result follows on expanding the right-hand side here.

For a non-collinear triple (Z_0, Z_1, Z_2), let α be the wedge-angle $\angle Z_1 Z_0 Z_2$ and ϕ be the sensed angle $\angle_{\mathcal{F}} Z_1 Z_0 Z_2$. Then $\cos\phi = \cos\alpha$ so that $|\phi|° = |\alpha|°$ when ϕ is wedge, and $|\phi|° = 360 - |\alpha|°$ when ϕ is reflex.

Proof. By the last result,

$$|Z_1, Z_2|^2 = |Z_0, Z_1|^2 + |Z_0, Z_2|^2 - 2|Z_0, Z_1||Z_0, Z_2|\cos\phi,$$

while by the cosine rule for a triangle in 9.5.1

$$|Z_1, Z_2|^2 = |Z_0, Z_1|^2 + |Z_0, Z_2|^2 - 2|Z_0, Z_1||Z_0, Z_2|\cos\alpha.$$

Hence $\cos\phi = \cos\alpha$ so that $\sin^2\phi = \sin^2\alpha$ and hence $\sin\phi = \pm\sin\alpha$. The result follows from 9.3.4 and 9.6.

10.5   SENSED-AREA

10.5.1

For points $Z_0 \equiv_{\mathcal{F}} (x_0, y_0)$, $Z_1 \equiv_{\mathcal{F}} (x_1, y_1)$, $Z_2 \equiv_{\mathcal{F}} (x_2, y_2)$ such that $Z_1 \neq Z_0$, $Z_2 \neq Z_0$, and $\theta = \angle_{\mathcal{F}} Z_1 Z_0 Z_2$ we have

(i)
$$\tfrac{1}{2}|Z_0, Z_1||Z_0, Z_2|\sin\theta = \tfrac{1}{2}[(x_1 - x_0)(y_2 - y_0) - (x_2 - x_0)(y_1 - y_0)],$$

(ii)
$$\tfrac{1}{2}|Z_0, Z_1||Z_0, Z_2|\cos\theta = \tfrac{1}{2}[(x_1 - x_0)(x_2 - x_0) + (y_1 - y_0)(y_2 - y_0)].$$

Proof. By 8.3 and 9.2.2, if $k_1 = |Z_0, Z_1|$, $k_2 = |Z_0, Z_2|$, then

$$
\begin{aligned}
x_1 - x_0 &= k_1 \cos\theta_1, \; y_1 - y_0 = k_1 \sin\theta_1, \\
x_2 - x_0 &= k_2 \cos\theta_2, \; y_2 - y_0 = k_2 \sin\theta_2.
\end{aligned}
$$

Then by 9.3.3 and 9.3.4,

$$
\begin{aligned}
k_1 k_2 \sin(\theta_2 - \theta_1) &= k_2 \sin\theta_2 k_1 \cos\theta_1 - k_2 \cos\theta_2 k_1 \sin\theta_1 \\
&= (y_2 - y_0)(x_1 - x_0) - (x_2 - x_0)(y_1 - y_0).
\end{aligned}
$$

Similarly

$$
\begin{aligned}
k_1 k_2 \cos(\theta_2 - \theta_1) &= k_2 \cos\theta_2 k_1 \cos\theta_1 + k_2 \sin\theta_2 k_1 \sin\theta_1 \\
&= (x_2 - x_0)(x_1 - x_0) + (y_2 - y_0)(y_1 - y_0).
\end{aligned}
$$

10.5.2   Sensed-area of a triangle

For an ordered triple of points (Z_1, Z_2, Z_3) of points and a frame of reference $\mathcal{F}$, if $Z_1 \equiv_{\mathcal{F}} (x_1, y_1)$, $Z_2 \equiv_{\mathcal{F}} (x_2, y_2)$ and $Z_3 \equiv_{\mathcal{F}} (x_3, y_3)$, we recall from 6.6.2 and 10.5.1(i) $\delta_{\mathcal{F}}(Z_1, Z_2, Z_3)$ defined by the formula

$$
\begin{aligned}
\delta_{\mathcal{F}}(Z_1, Z_2, Z_3) &= \tfrac{1}{2}[x_1(y_2 - y_3) - y_1(x_2 - x_3) + x_2 y_3 - x_3 y_2] \\
&= \tfrac{1}{2}[(x_2 - x_1)(y_3 - y_1) - (x_3 - x_1)(y_2 - y_1)] \\
&= \frac{1}{2} \det \begin{pmatrix} x_1 & y_1 & 1 \\ x_2 & y_2 & 1 \\ x_3 & y_3 & 1 \end{pmatrix}.
\end{aligned}
$$

By 6.6.2, when Z_1, Z_2, Z_3 are non-collinear $|\delta_{\mathcal{F}}(Z_1, Z_2, Z_3)|$ is equal to the area of the triangle $[Z_1, Z_2, Z_3]$. In this case we refer to $\delta_{\mathcal{F}}(Z_1, Z_2, Z_3)$ as the **sensed-area** of the triangle $[Z_1, Z_2, Z_3]$, with the order of vertices (Z_1, Z_2, Z_3). This was first introduced by Möbius in 1827.

Note that

$$
\begin{aligned}
\delta_{\mathcal{F}}(Z_1, Z_2, Z_3) &= \delta_{\mathcal{F}}(Z_2, Z_3, Z_1) = \delta_{\mathcal{F}}(Z_3, Z_1, Z_2) \\
&= -\delta_{\mathcal{F}}(Z_1, Z_3, Z_2) = -\delta_{\mathcal{F}}(Z_2, Z_1, Z_3) = -\delta_{\mathcal{F}}(Z_3, Z_2, Z_1),
\end{aligned}
$$

so that its value is unchanged if Z_1, Z_2, Z_3 are permuted cyclically, and its value is multiplied by -1 if the order of these is changed.

We note that 10.4.1(ii) can be restated as that the sensed-angle $\angle_{\mathcal{F}} Z_1 Z_0 Z_2$ is wedge or reflex according as

$$\Im \frac{z_2 - z_0}{z_1 - z_0} = \Im \frac{\overline{Z_0 Z_2}_{\mathcal{F}}}{\overline{Z_0 Z_1}_{\mathcal{F}}}$$

is positive or negative, and this occurs according as $\delta_{\mathcal{F}}(Z_0, Z_1, Z_2)$ is positive or negative.

10.5.3   A basic feature of sensed-area

A basic feature of sensed-area is given by the follow-ing. Let the points $Z_3 \equiv (x_3, y_3), Z_4 \equiv (x_4, y_4), Z_5 \equiv (x_5, y_5)$ be such that

$$x_3 = (1 - s)x_4 + sx_5, \quad y_3 = (1 - s)y_4 + sy_5,$$

for some $s \in \mathbf{R}$. Then for all Z_1, Z_2,

$$\delta_{\mathcal{F}}(Z_1, Z_2, Z_3) = (1 - s)\delta_{\mathcal{F}}(Z_1, Z_2, Z_4) + s\delta_{\mathcal{F}}(Z_1, Z_2, Z_5).$$

For

$$\delta_{\mathcal{F}}(Z_1, Z_2, Z_3) = \frac{1}{2} \det \begin{pmatrix} x_1 & y_1 & 1 \\ x_2 & y_2 & 1 \\ (1-s)x_4 + sx_5 & (1-s)y_4 + sy_5 & (1-s)+s \end{pmatrix}$$

$$= \frac{1}{2} \det \begin{pmatrix} x_1 & y_1 & 1 \\ x_2 & y_2 & 1 \\ (1-s)x_4 & (1-s)y_4 & 1-s \end{pmatrix} + \frac{1}{2} \det \begin{pmatrix} x_1 & y_1 & 1 \\ x_2 & y_2 & 1 \\ sx_5 & sy_5 & s \end{pmatrix}$$

$$= \frac{1}{2}(1-s) \det \begin{pmatrix} x_1 & y_1 & 1 \\ x_2 & y_2 & 1 \\ x_4 & y_4 & 1 \end{pmatrix} + \frac{1}{2}s \det \begin{pmatrix} x_1 & y_1 & 1 \\ x_2 & y_2 & 1 \\ x_5 & y_5 & 1 \end{pmatrix}$$

$$= (1-s)\delta_{\mathcal{F}}(Z_1, Z_2, Z_4) + s\delta_{\mathcal{F}}(Z_1, Z_2, Z_5).$$

10.5.4   An identity for sensed-area

An identity that we have for sensed-area is that for any points Z_1, Z_2, Z_3, Z_4,

$$\delta_{\mathcal{F}}(Z_4, Z_2, Z_3) + \delta_{\mathcal{F}}(Z_4, Z_3, Z_1) + \delta_{\mathcal{F}}(Z_4, Z_1, Z_2) = \delta_{\mathcal{F}}(Z_1, Z_2, Z_3).$$

For the left-hand side is equal to

$$\frac{1}{2}\det\begin{pmatrix} x_4 & y_4 & 1 \\ x_2 & y_2 & 1 \\ x_3 & y_3 & 1 \end{pmatrix} + \frac{1}{2}\det\begin{pmatrix} x_4 & y_4 & 1 \\ x_3 & y_3 & 1 \\ x_1 & y_1 & 1 \end{pmatrix} + \frac{1}{2}\det\begin{pmatrix} x_4 & y_4 & 1 \\ x_1 & y_1 & 1 \\ x_2 & y_2 & 1 \end{pmatrix}$$

$$= \frac{1}{2}\det\begin{pmatrix} x_4 & y_4 & 1 \\ x_2 & y_2 & 1 \\ x_3 & y_3 & 1 \end{pmatrix} - \frac{1}{2}\det\begin{pmatrix} x_4 & y_4 & 1 \\ x_1 & y_1 & 1 \\ x_3 & y_3 & 1 \end{pmatrix} + \frac{1}{2}\det\begin{pmatrix} x_4 & y_4 & 1 \\ x_1 & y_1 & 1 \\ x_2 & y_2 & 1 \end{pmatrix}$$

$$= \frac{1}{2}\det\begin{pmatrix} x_4 & y_4 & 1 \\ x_2-x_1 & y_2-y_1 & 0 \\ x_3 & y_3 & 1 \end{pmatrix} + \frac{1}{2}\det\begin{pmatrix} x_4 & y_4 & 1 \\ x_1 & y_1 & 1 \\ x_2 & y_2 & 1 \end{pmatrix}$$

$$= \frac{1}{2}\det\begin{pmatrix} x_4 & y_4 & 1 \\ x_2-x_1 & y_2-y_1 & 0 \\ x_3 & y_3 & 1 \end{pmatrix} + \frac{1}{2}\det\begin{pmatrix} x_4 & y_4 & 1 \\ x_1 & y_1 & 1 \\ x_2-x_1 & y_2-y_1 & 0 \end{pmatrix}$$

$$= \frac{1}{2}\det\begin{pmatrix} x_4 & y_4 & 1 \\ x_2-x_1 & y_2-y_1 & 0 \\ x_3 & y_3 & 1 \end{pmatrix} - \frac{1}{2}\det\begin{pmatrix} x_4 & y_4 & 1 \\ x_2-x_1 & y_2-y_1 & 0 \\ x_1 & y_1 & 1 \end{pmatrix}$$

$$= \frac{1}{2}\det\begin{pmatrix} x_4 & y_4 & 1 \\ x_2-x_1 & y_2-y_1 & 0 \\ x_3-x_1 & y_3-y_1 & 0 \end{pmatrix}$$

$$= \frac{1}{2}\det\begin{pmatrix} x_1 & y_1 & 1 \\ x_2-x_1 & y_2-y_1 & 0 \\ x_3-x_1 & y_3-y_1 & 0 \end{pmatrix} = \frac{1}{2}\det\begin{pmatrix} x_1 & y_1 & 1 \\ x_2 & y_2 & 1 \\ x_3 & y_3 & 1 \end{pmatrix},$$

and this is equal to the right-hand side. This was first proved by Möbius.

10.6   ISOMETRIES AS COMPOSITIONS

10.6.1

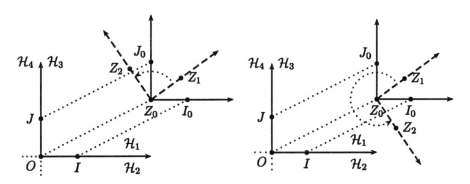

Figure 10.6.

Let $\mathcal{F} = ([O, I\ ,\ [O, J\ )$ and $\mathcal{F}_1 = ([Z_0, Z_1\ ,\ [Z_0, Z_2\ )$ be frames of reference. Let $t_{O, z_0}(I) = I_0$, $t_{O, z_0}(J) = J_0$, $\mathcal{F}' = ([Z_0, I_0\ ,\ [Z_0, J_0\ )$ and $\alpha = \angle_{\mathcal{F}'} I_0 Z_0 Z_1$. Then there is a unique isometry g such that

$$g([O, I\ ) = [Z_0, Z_1\ ,\quad g([O, J\ ) = [Z_0, Z_2\ .$$

When $\angle_{\mathcal{F}} Z_1 Z_0 Z_2$ is a wedge-angle and so a right-angle $90_{\mathcal{F}'}$,

$$g = r_{\alpha; z_0} \circ t_{O, z_0}.$$

When $\angle_{\mathcal{F}} Z_1 Z_0 Z_2$ is a reflex-angle $270_{\mathcal{F}'}$ and so its co-supported angle is a right-angle,

$$g = s_{\frac{1}{2}\alpha; z_0} \circ t_{O, z_0}.$$

Proof. Without loss of generality we take $|O, I| = |O, J| = |Z_0, Z_1| = |Z_0, Z_2| = 1$. Let $Z_0 \sim_{\mathcal{F}} z_0$, $Z_1 \sim_{\mathcal{F}} z_1$, $Z_2 \sim_{\mathcal{F}} z_2$ and note that $I \sim_{\mathcal{F}} 1$, $J \sim_{\mathcal{F}} \imath$, $z_1 - z_0 = \text{cis}\,\alpha$. As $Z_0 Z_1 \perp Z_0 Z_2$, by 10.1.1(vi) we have

$$z_2 - z_0 = \imath(z_1 - z_0) \text{ when } \angle_{\mathcal{F}} Z_1 Z_0 Z_2 \text{ is a wedge- angle,} \qquad (10.6.1)$$

and

$$z_2 - z_0 = -\imath(z_1 - z_0) \text{ when } \angle_{\mathcal{F}} Z_1 Z_0 Z_2 \text{ is a reflex- angle.} \qquad (10.6.2)$$

In case (10.6.1) we take the transformation $Z' = g(Z)$ where

$$z' = z_0 + z\,\text{cis}\,\alpha = z_0 + (z + z_0 - z_0)\text{cis}\,\alpha.$$

Then for $z = t \geq 0$, $z' = z_0 + t(z_1 - z_0)$ so $g([O, I\ ) = [Z_0, Z_1$. Similarly for $z = \imath t$ $(t \geq 0)$, $z' = z_0 + t(z_2 - z_0)$ so $g([O, J\ ) = [Z_0, Z_2$.

In case (10.6.2) we take the transformation $Z' = g(Z)$ where

$$z' = z_0 + \bar{z}\,\text{cis}\,\alpha = z_0 + (\overline{z + z_0 - z_0})\text{cis}\,\alpha.$$

Then for $z = t \geq 0$, $z' = z_0 + t(z_1 - z_0)$ so $g([O, I\ ) = [Z_0, Z_1$. Similarly for $z = \imath t$ $(t \geq 0)$, $z' = z_0 + t(z_2 - z_0)$ so $g([O, J\ ) = [Z_0, Z_2$.

This establishes the existence of g. As to uniqueness, suppose that f is also an isometry such that $f(\mathcal{F}) = \mathcal{F}_1$. Then by 8.2.1(xii), if $Z \sim_{\mathcal{F}} z$ we have $f(Z) \sim_{\mathcal{F}_1} z$, $g(Z) \sim_{\mathcal{F}_1} z$, and so $f(Z) = g(Z)$ for all $Z \in \Pi$.

COROLLARY. Let f be any isometry. Then f can be expressed in one or other of the forms

$$\text{(a) } f = r_{\alpha; z_0} \circ t_{O, z_0}, \quad \text{(b)} f = s_{\frac{1}{2}\alpha; z_0} \circ t_{O, z_0}.$$

Proof. In the theorem, take $Z_0 = f(O)$, $Z_1 = f(I)$, $Z_2 = f(J)$ and consequently f is equal to the function g as defined in the proof.

10.7 ORIENTATION OF A TRIPLE OF NON- COLLINEAR POINTS

10.7.1

Definition. We say that an ordered triple (Z_0, Z_1, Z_2) of non-collinear points is **positively** or **negatively oriented** with respect to $\mathcal{F}$ according as the sensed angle

$\angle_F Z_1 Z_0 Z_2$ is wedge or reflex. By 10.4.1(ii) this occurs according as $\delta_F(Z_0, Z_1, Z_2)$ is positive or negative.

Definition. Let $\mathcal{F} = ([O,I\,, [O,J\,)$ and $\mathcal{F}_1 = ([Z_0, Z_1\,, [Z_0, Z_2\,)$ be frames of reference. We say that $\mathcal{F}_1$ is **positively or negatively oriented** with respect to $\mathcal{F}$ according as (Z_0, Z_1, Z_2) is positively or negatively oriented with respect to $\mathcal{F}$.

The special isometries have the following effects on orientation:-

(i) *Each translation preserves the orientations with respect to $\mathcal{F}$ of all non-collinear triples.*

(ii) *Each rotation preserves the orientations with respect to $\mathcal{F}$ of all non-collinear triples.*

(iii) *Each axial symmetry reverses the orientations with respect to $\mathcal{F}$ of all non-collinear triples.*

Proof.
(i) Let $f = t_{Z_0, Z_1}$ and $Z_2 \sim_F z_2$, $Z_3 \sim_F z_3$, $Z_4 \sim_F z_4$. Then

$$z_3' = z_3 + (z_1 - z_0), \quad z_2' = z_2 + (z_1 - z_0),$$

so that $z_3' - z_2' = z_3 - z_2$, and similarly $z_4' - z_2' = z_4 - z_2$. Hence

$$\frac{z_4' - z_2'}{z_3' - z_2'} = \frac{z_4 - z_2}{z_3 - z_2},$$

and so by 10.4.1(ii) the result follows.
(ii) Let $f = r_{\alpha; Z_0}$. Then by 10.3.1

$$z_2' - z_0 = (z_2 - z_0)\text{cis } \alpha, \quad z_3' - z_0 = (z_3 - z_0)\text{cis } \alpha,$$

and so

$$z_3' - z_2' = (z_3 - z_2)\text{cis } \alpha, \quad z_4' - z_2' = (z_4 - z_2)\text{cis } \alpha.$$

Hence

$$\frac{z_4' - z_2'}{z_3' - z_2'} = \frac{z_4 - z_2}{z_3 - z_2},$$

and so by 10.4.1(ii) the result follows.
(iii) Let $f = s_{\alpha; Z_0}$. Then by 10.3.2

$$z_2' - z_0 = (\bar{z}_2 - \bar{z}_0)\text{cis } 2\alpha, \quad z_3' - z_0 = (\bar{z}_3 - \bar{z}_0)\text{cis } 2\alpha,$$

and so

$$z_3' - z_2' = (\bar{z}_3 - \bar{z}_2)\text{cis } 2\alpha, \quad z_4' - z_2' = (\bar{z}_4 - \bar{z}_2)\text{cis } 2\alpha.$$

Hence

$$\frac{z_4' - z_2'}{z_3' - z_2'} = \overline{\left(\frac{z_4 - z_2}{z_3 - z_2}\right)},$$

so that

$$\Im\frac{z_4' - z_2'}{z_3' - z_2'} = -\Im\frac{z_4 - z_2}{z_3 - z_2},$$

and so by 10.4.1(ii) the result follows.

Let $\mathcal{F}, \mathcal{F}_1$ be frames of reference and Z_3, Z_4, Z_5 non-collinear points. Let $\theta = \angle_{\mathcal{F}} Z_4 Z_3 Z_5$ and $\phi = \angle_{\mathcal{F}_1} Z_4 Z_3 Z_5$. Then $|\phi|°$ is equal to $|\theta|°$ or $360 - |\theta|°$, according as $\mathcal{F}_1$ is positively or negatively oriented with respect to $\mathcal{F}$.

Proof. We use the notation of 10.6.1. When $\mathcal{F}_1$ is positively oriented with respect to $\mathcal{F}$, we recall that for $f(Z) = Z'$ with $z' = z_0 + z\text{cis}\,\alpha$, we have $f(\mathcal{F}) = \mathcal{F}_1$. On solving this for z and then interchanging z and z', we see that

$$f^{-1}(Z) \sim_{\mathcal{F}} (z - z_0)\text{cis}\,(-\alpha).$$

Then by 8.2.1(xii), $Z = f(f^{-1}(Z)) \sim_{\mathcal{F}_1} (z - z_0)\text{cis}\,(-\alpha)$.

Letting $Z_j \sim_{\mathcal{F}} z_j$, $Z'_j \sim_{\mathcal{F}_1} z'_j$ we then have $z'_j = (z_j - z_0)\text{cis}\,(-\alpha)$. Thus

$$\frac{z'_5 - z'_3}{z'_4 - z'_3} = \frac{(z_5 - z_0)\text{cis}\,(-\alpha) - (z_3 - z_0)\text{cis}\,(-\alpha)}{(z_4 - z_0)\text{cis}\,(-\alpha) - (z_3 - z_0)\text{cis}\,(-\alpha)} = \frac{z_5 - z_3}{z_4 - z_3}.$$

But by 10.4.1(i),

$$\frac{z_5 - z_3}{z_4 - z_3} = \frac{|Z_3, Z_5|}{|Z_3, Z_4|}\text{cis}\,\theta, \qquad \frac{z'_5 - z'_3}{z'_4 - z'_3} = \frac{|Z_3, Z_5|}{|Z_3, Z_4|}\text{cis}\,\phi.$$

Thus $\text{cis}\,\phi = \text{cis}\,\theta$ and so $|\phi|° = |\theta|°$.

When $\mathcal{F}_1$ is negatively oriented with respect to $\mathcal{F}$, we take instead $f(Z) = Z'$ with $z' = z_0 + \bar{z}\text{cis}\,\alpha$. Now $f^{-1}(Z) \sim_{\mathcal{F}} (\bar{z} - \bar{z}_0)\text{cis}\,\alpha$ and so

$$\frac{z'_5 - z'_3}{z'_4 - z'_3} = \frac{(\bar{z}_5 - \bar{z}_0)\text{cis}\,(\alpha) - (\bar{z}_3 - \bar{z}_0)\text{cis}\,(\alpha)}{(\bar{z}_4 - \bar{z}_0)\text{cis}\,(\alpha) - (\bar{z}_3 - \bar{z}_0)\text{cis}\,(\alpha)} = \frac{\bar{z}_5 - \bar{z}_3}{\bar{z}_4 - \bar{z}_3}.$$

Thus $\text{cis}\,\phi = \overline{\text{cis}\,\theta} = \text{cis}\,(-\theta)$ and so $|\phi|° = |(-\theta)|° = 360 - |\theta|°$.

Let $\mathcal{F}$ and $\mathcal{F}_1$ be frames of reference. Then the ratios of complex-valued distances

$$\rho = \frac{\overline{Z_3 Z_4}_{\mathcal{F}}}{\overline{Z_1 Z_2}_{\mathcal{F}}}, \qquad \sigma = \frac{\overline{Z_3 Z_4}_{\mathcal{F}_1}}{\overline{Z_1 Z_2}_{\mathcal{F}_1}},$$

defined in 10.2.1, satisfy $\sigma = \rho$ when $\mathcal{F}_1$ is positively oriented with respect to $\mathcal{F}$, and $\sigma = \bar{\rho}$ when $\mathcal{F}_1$ is negatively oriented with respect to $\mathcal{F}$.

Proof. We use the notation of 10.6.1. In the case (10.6.1) $z' = z_0 + z\text{cis}\,\alpha$ so that

$$\sigma = \frac{z'_4 - z'_3}{z'_2 - z'_1} = \frac{(z_4 - z_3)\text{cis}\,\alpha}{(z_2 - z_1)\text{cis}\,\alpha} = \rho.$$

In the case (10.6.2) $z' = z_0 + \bar{z}\text{cis}\,\alpha$ so that

$$\sigma = \frac{z'_4 - z'_3}{z'_2 - z'_1} = \frac{(\bar{z}_4 - \bar{z}_3)\text{cis}\,\alpha}{(\bar{z}_2 - \bar{z}_1)\text{cis}\,\alpha} = \bar{\rho}.$$

10.8   SENSED ANGLES OF TRIANGLES, THE SINE RULE

10.8.1

Definition. For any non-full angle θ, we denote by $\theta_{\mathcal{F}}$ the angle in $\mathcal{A}(\mathcal{F})$ such that $|\theta_{\mathcal{F}}|^{\circ} = |\theta|^{\circ}$.

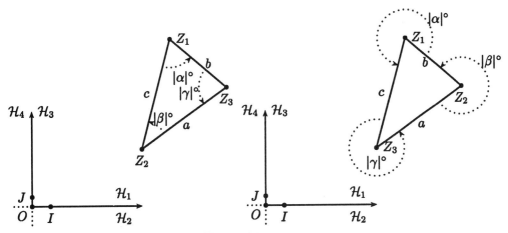

Figure 10.7.

NOTATION. For non-collinear points Z_1, Z_2, Z_3, we use as standard notation

$$|Z_2, Z_3| = a, \ |Z_3, Z_1| = b, \ |Z_1, Z_2| = c,$$

$$u = \frac{b}{c}, \ v = \frac{c}{a}, \ w = \frac{a}{b},$$

$$\alpha = \measuredangle_{\mathcal{F}} Z_2 Z_1 Z_3, \ \beta = \measuredangle_{\mathcal{F}} Z_3 Z_2 Z_1, \ \gamma = \measuredangle_{\mathcal{F}} Z_1 Z_3 Z_2.$$

Note that by comparison with 9.5.1 we are now using sensed-angles instead of wedge-angles.

For non-collinear points Z_1, Z_2, Z_3 if

$$\alpha = \measuredangle_{\mathcal{F}} Z_2 Z_1 Z_3, \ \beta = \measuredangle_{\mathcal{F}} Z_3 Z_2 Z_1, \ \gamma = \measuredangle_{\mathcal{F}} Z_1 Z_3 Z_2,$$

and $\theta = \alpha_{\mathcal{F}}, \ \phi = \beta_{\mathcal{F}}, \ \psi = \gamma_{\mathcal{F}}$, then $\theta + \phi + \psi = 180_{\mathcal{F}}$.

   Proof. For

$$\frac{z_3 - z_1}{z_2 - z_1} = \frac{b}{c} \text{cis} \, \theta, \ \frac{z_1 - z_2}{z_3 - z_2} = \frac{c}{a} \text{cis} \, \phi, \ \frac{z_2 - z_3}{z_1 - z_3} = \frac{a}{b} \text{cis} \, \psi.$$

On multiplying these together, we find that

$$-1 = \text{cis} \, \theta . \text{cis} \, \phi . \text{cis} \, \psi = \text{cis} \, (\theta + \phi + \psi).$$

As cis $180_{\mathcal{F}} = -1$ it follows that $\theta + \phi + \psi = 180_{\mathcal{F}}$.

With the above notation, the lengths of the sides and the sensed-angles of a triangle $[Z_1, Z_2, Z_3]$ have the properties:-

(i) In each case

$$v\text{cis}\,\beta = \frac{1}{1 - u\text{cis}\,\alpha},$$

and two pairs of similar identities obtained from these on advancing cyclically through (u, v, w) and (α, β, γ).

(ii) In each case

$$c = b\cos\alpha + a\cos\beta, \quad \frac{\sin\alpha}{a} = \frac{\sin\beta}{b},$$

and two pairs of similar identities obtained from these on advancing cyclically through (a, b, c) and (α, β, γ).

Proof.

(i) For $z_3 - z_1 = u\text{cis}\,\alpha.(z_2 - z_1)$ so that $z_3 - z_2 = (1 - u\text{cis}\,\alpha)(z_1 - z_2)$, while $z_1 - z_2 = v\text{cis}\,\beta.(z_3 - z_2)$, which give $(1 - u\text{cis}\,\alpha)v\text{cis}\,\beta = 1$.

(ii) From (i)

$$1 - u[\cos\alpha + \imath\sin\alpha] = \frac{1}{v}[\cos\beta - \imath\sin\beta],$$

so equating real parts gives $c = b\cos\alpha + a\cos\beta$, while equating imaginary parts gives $\sin\alpha/a = \sin\beta/b$.

This result re-derives the sine rule for a triangle.

If Z_1, Z_2, Z_3 are distinct points, then

$$\frac{\overline{Z_3 Z_1}_\mathcal{F}}{\overline{Z_3 Z_2}_\mathcal{F}} + + \frac{\overline{Z_2 Z_1}_\mathcal{F}}{\overline{Z_2 Z_3}_\mathcal{F}} = 1.$$

Proof. For

$$\frac{z_1 - z_3}{z_2 - z_3} + \frac{z_1 - z_2}{z_3 - z_2} = 1.$$

10.9   SOME RESULTS ON CIRCLES

10.9.1   A necessary condition to lie on a circle

In this section we provide some results on circles which are conveniently proved using complex coordinates.

Let Z_1, Z_2 be fixed distinct points, and Z a variable point, all on the circle $\mathcal{C}(Z_0; k)$. Let $\mathcal{F}' = t_{0,z_0}(\mathcal{F})$ and $\alpha = \angle_{\mathcal{F}'}I_0Z_0Z_1$, $\beta = \angle_{\mathcal{F}'}I_0Z_0Z_2$ and $\gamma = \frac{1}{2}(\beta - \alpha)$. As Z varies on the circle, in one of the open half-planes with edge Z_1Z_2 the sensed angle $\angle_{\mathcal{F}}Z_1ZZ_2$ is equal in measure to γ, while in the other open half-plane with edge Z_1Z_2 it is equal in measure to $\gamma + 180_{\mathcal{F}'}$. Note that $2\gamma = \angle_{\mathcal{F}}Z_1Z_0Z_2$.

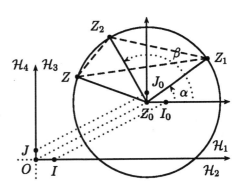

Figure 10.8.

Proof. Now $z_1 - z_0 = k\mathrm{cis}\,\alpha$, $z_2 - z_0 = k\mathrm{cis}\,\beta$ and if $\theta = \angle_{\mathcal{F}}I_0Z_0Z$, then $z - z_0 = k\mathrm{cis}\,\theta$. We write $\phi = \angle_{\mathcal{F}}Z_1ZZ_2$ so that

$$\frac{z_2 - z}{z_1 - z} = l\mathrm{cis}\,\phi, \text{ where } l = \frac{|Z, Z_2|}{|Z, Z_1|}.$$

Then

$$l\mathrm{cis}\,\phi = \frac{\mathrm{cis}\,\beta - \mathrm{cis}\,\theta}{\mathrm{cis}\,\alpha - \mathrm{cis}\,\theta},$$

while on taking complex conjugates here

$$l\mathrm{cis}\,(-\phi) = \frac{\mathrm{cis}\,(-\beta) - \mathrm{cis}\,(-\theta)}{\mathrm{cis}\,(-\alpha) - \mathrm{cis}\,(-\theta)} = \frac{\mathrm{cis}\,\alpha\,\mathrm{cis}\,\theta - \mathrm{cis}\,\beta}{\mathrm{cis}\,\beta\,\mathrm{cis}\,\theta - \mathrm{cis}\,\alpha}.$$

By division

$$\mathrm{cis}\,2\phi = \frac{\mathrm{cis}\,\beta}{\mathrm{cis}\,\alpha} = \mathrm{cis}\,(\beta - \alpha).$$

Thus $2(\mathrm{cis}\,\phi)^2 = (\mathrm{cis}\,\gamma)^2$ so that $\mathrm{cis}\,\phi = \pm\mathrm{cis}\,\gamma$. Thus either $\mathrm{cis}\,\phi = \mathrm{cis}\,\gamma$ or $\mathrm{cis}\,\phi = \mathrm{cis}\,(\gamma + 180_{\mathcal{F}'})$, and accordingly

$$\Im\frac{z_2 - z}{z_1 - z} = l\sin\gamma \text{ or } \Im\frac{z_2 - z}{z_1 - z} = l\sin(\gamma + 180_{\mathcal{F}'}).$$

As $\sin\gamma > 0$, the first of these occurs when Z is in the half-plane with edge Z_1Z_2 in which $\Im\frac{z_2-z}{z_1-z} > 0$, and the second when Z is in the half-plane with edge Z_1Z_2 in which $\Im\frac{z_2-z}{z_1-z} < 0$.

10.9.2   A sufficient condition to lie on a circle

Let Z_1, Z_2 be fixed distinct points and Z a variable point. As Z varies in one of the half-planes with edge $Z_1 Z_2$, for the sensed angle $\theta = \measuredangle_{\mathcal{F}} Z_1 Z Z_2$ let $|\theta|° = |\gamma|°$ where γ is a fixed non-null and non-straight angle in $A(\mathcal{F}')$, while as Z varies in the other half-plane with edge $Z_1 Z_2$, let $|\theta|° = |\gamma + 180_{\mathcal{F}'}|°$. Then Z lies on a circle which passes through Z_1 and Z_2.

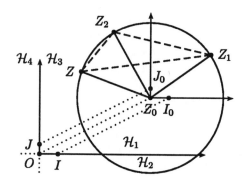

Figure 10.9.

Proof. We have

$$\frac{z_2 - z}{z_1 - z} = t\operatorname{cis}\gamma$$

for some $t \in \mathbf{R} \setminus \{0\}$. Then

$$z = \frac{z_2 - t z_1 \operatorname{cis}\gamma}{1 - t\operatorname{cis}\gamma}$$

so that with $\cot\gamma = \cos\gamma / \sin\gamma$,

$$
\begin{aligned}
&z - \frac{1}{2}(z_1 + z_2) - \frac{1}{2}\imath \cot\gamma.(z_2 - z_1) \\
&= \frac{z_2 - t z_1 \operatorname{cis}\gamma}{1 - t\operatorname{cis}\gamma} - \frac{1}{2}(z_1 + z_2) - \frac{1}{2}i \cot\gamma.(z_2 - z_1) \\
&= \frac{\frac{1}{2}(z_2 - z_1)[1 + t\operatorname{cis}\gamma - \imath \cot\gamma(1 - t\operatorname{cis}\gamma)]}{1 - t\operatorname{cis}\gamma} \\
&= \frac{\frac{1}{2}(z_2 - z_1)[\sin\gamma(1 + t\operatorname{cis}\gamma) - \imath \cos\gamma(1 - t\operatorname{cis}\gamma)]}{\sin\gamma(1 - t\operatorname{cis}\gamma)} \\
&= \frac{\frac{1}{2}(z_2 - z_1)[\sin\gamma + \imath(t - \cos\gamma)]}{\sin\gamma(1 - t\operatorname{cis}\gamma)}
\end{aligned}
$$

and this has absolute value

$$\frac{|z_2 - z_1|}{2|\sin\gamma|}.$$

This shows that Z lies on a circle, the centre and length of radius of which are evident.

10.9.3 Complex cross-ratio

Let Z_2, Z_3, Z_4 be non-collinear points and C the circle that contains them. Then $Z \notin Z_3 Z_4$ lies in C if and only if

$$\Im \frac{(z - z_3)(z_2 - z_4)}{(z - z_4)(z_2 - z_3)} = 0.$$

When this holds and Z and Z_2 are on the same side of $Z_3 Z_4$, then

$$\frac{(z - z_3)(z_2 - z_4)}{(z - z_4)(z_2 - z_3)} > 0.$$

Proof. The given condition is equivalent to

$$\frac{z - z_4}{z - z_3} = t \frac{z_2 - z_4}{z_2 - z_3}, \tag{10.9.1}$$

for some $t \neq 0$ in $\mathbf{R}$. Let $\mathcal{G}_1, \mathcal{G}_2$ be the open half-planes with common edge $Z_3 Z_4$, with $Z_2 \in \mathcal{G}_1$. Let $\theta = \measuredangle_{\mathcal{F}} Z_3 Z_2 Z_4$ and $\phi = \measuredangle_{\mathcal{F}} Z_3 Z Z_4$.

Suppose first that (10.9.1) holds. For $Z \in \mathcal{G}_1$,

$$\Im \frac{z - z_4}{z - z_3} \text{ and } \Im \frac{z_2 - z_4}{z_2 - z_3}$$

must have the same sign and so $t > 0$; it follows that $\phi = \theta$. For $Z \in \mathcal{G}_2$,

$$\Im \frac{z - z_4}{z - z_3} \text{ and } \Im \frac{z_2 - z_4}{z_2 - z_3}$$

must have opposite signs and so $t < 0$; it follows that $\phi = \theta + 180_{\mathcal{F}'}$. By 10.9.2 $Z \in C$ in both cases.

Conversely let $Z \in C$. Then by (10.9.1) for $Z \in \mathcal{G}_1$ we have $\phi = \theta$, while for $Z \in \mathcal{G}_2$ we have $\phi = \theta + 180_{\mathcal{F}'}$ and the result now follows.

The expression $\frac{(z-z_3)(z_2-z_4)}{(z-z_4)(z_2-z_3)}$ is called the **cross-ratio** of the ordered set of points (Z, Z_2, Z_3, Z_4).

10.9.4 Ptolemy's theorem, c.200A.D.

Let Z_2, Z_3, Z_4 be non-collinear points and C the circle that contains them. Let $Z \in C$ be such that Z and Z_3 are on opposite sides of $Z_2 Z_4$. Then $|Z, Z_4||Z_2, Z_3| + |Z, Z_2||Z_3, Z_4| = |Z, Z_3||Z_2, Z_4|$.

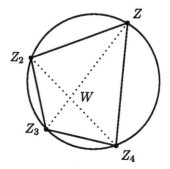

Figure 10.10.

Proof. By multiplying out, it can be checked that

$$(z - z_4)(z_2 - z_3) + (z - z_2)(z_3 - z_4) = (z - z_3)(z_2 - z_4).$$

This is an identity due to Euler and from it

$$\frac{(z - z_4)(z_2 - z_3)}{(z - z_3)(z_2 - z_4)} + \frac{(z - z_2)(z_4 - z_3)}{(z - z_3)(z_4 - z_2)} = 1. \qquad (10.9.2)$$

By 10.9.3 both fractions on the left are real-valued. As Z and Z_3 are on opposite sides of Z_2Z_4, there is a point W of $[Z, Z_3]$ on Z_2Z_4. Then W is an interior point of the circle, and so $W \in [Z_2, Z_4]$ as the only points of the line Z_2Z_4 which are interior to the circle are in this segment. It follows that Z_2 and Z_4 are on opposite sides of ZZ_3. Then $[Z, Z_2]$, $[Z_3, Z_4]$ are in different closed half-planes with common edge the line ZZ_3, so they have no points in common. It follows that Z and Z_2 are on the one side of Z_3Z_4 so the first of the fractions in (10.9.2) is positive, and so equal to its own absolute value. But $[Z, Z_4]$ and $[Z_3, Z_2]$ are in different closed half-planes with common edge ZZ_3 so they have no point in common. It follows that Z and Z_4 are on the one side of Z_2Z_3, so the second fraction in (10.9.2) is positive and so equal to its own absolute value. Hence

$$\frac{|(z - z_4)(z_2 - z_3)|}{|(z - z_3)(z_2 - z_4)|} + \frac{|(z - z_2)(z_4 - z_3)|}{|(z - z_3)(z_4 - z_2)|} = 1.$$

This is known as PTOLEMY'S THEOREM.

From the original identity (10.9.2) with Z_1 replacing Z we can deduce that for four distinct points Z_1, Z_2, Z_3, Z_4

$$\frac{\overline{Z_1 Z_4}_{\mathcal{F}}}{\overline{Z_1 Z_2}_{\mathcal{F}}} \frac{\overline{Z_2 Z_3}_{\mathcal{F}}}{\overline{Z_2 Z_4}_{\mathcal{F}}} + \frac{\overline{Z_1 Z_2}_{\mathcal{F}}}{\overline{Z_1 Z_3}_{\mathcal{F}}} \frac{\overline{Z_4 Z_3}_{\mathcal{F}}}{\overline{Z_4 Z_2}_{\mathcal{F}}} = 1.$$

This can be expanded as

$$\frac{|Z_1, Z_4|}{|Z_1, Z_3|} \text{cis } \alpha \frac{|Z_2, Z_3|}{|Z_2, Z_4|} \text{cis } \beta + \frac{|Z_1, Z_2|}{|Z_1, Z_3|} \text{cis } \gamma \frac{|Z_4, Z_3|}{|Z_2, Z_4|} \text{cis } \delta = 1,$$

where

$$\alpha = \measuredangle_{\mathcal{F}} Z_3 Z_1 Z_4, \ \beta = \measuredangle_{\mathcal{F}} Z_4 Z_2 Z_3, \ \gamma = \measuredangle_{\mathcal{F}} Z_3 Z_1 Z_2, \ \delta = \measuredangle_{\mathcal{F}} Z_4 Z_2 Z_3.$$

From this we have that

$$\frac{|Z_1, Z_4|}{|Z_1, Z_3|} \frac{|Z_2, Z_3|}{|Z_2, Z_4|} \text{cis } (\alpha_{\mathcal{F}} + \beta_{\mathcal{F}}) + \frac{|Z_1, Z_2|}{|Z_1, Z_3|} \frac{|Z_4, Z_3|}{|Z_2, Z_4|} \text{cis } (\gamma_{\mathcal{F}} + \delta_{\mathcal{F}}) = 1.$$

We get two relationships on equating the real parts in this and also equating the imaginary parts.

NOTE. For other applications of complex numbers to geometry, see Chapter 11 and Hahn [8].

10.10 ANGLES BETWEEN LINES

10.10.1 Motivation

Since $\cos(180_{\mathcal{F}} + \theta) = -\cos\theta$, $\sin(180_{\mathcal{F}} + \theta) = -\sin\theta$, we have that $\tan(180_{\mathcal{F}} + \theta) = \tan\theta$. Thus results that $\tan\theta$ is constant do not imply that θ is an angle of constant magnitude. To extract more information from such situations, we develop new material. This also deals with the rather abrupt transitions in results such as those in 10.9.1 and 10.9.2.

10.10.2 Duo-sectors

Let l_1, l_2 be lines intersecting at a point Z_1. When $l_1 \neq l_2$, let $Z_2, Z_3 \in l_1$ with Z_1 between Z_2 and Z_3, and let $Z_4, Z_5 \in l_2$ with Z_1 between Z_4 and Z_5. Then the union

$$\mathcal{IR}(|Z_2 Z_1 Z_4) \cup \mathcal{IR}(|Z_3 Z_1 Z_5)$$

we shall call a **duo-sector** with side-lines l_1 and l_2; we shall denote it by $\mathcal{D}_1$. Similarly

$$\mathcal{IR}(|Z_2 Z_1 Z_5) \cup \mathcal{IR}(|Z_3 Z_1 Z_4)$$

is also a duo-sector with side-lines l_1 and l_2, and we shall denote it by $\mathcal{D}_2$.

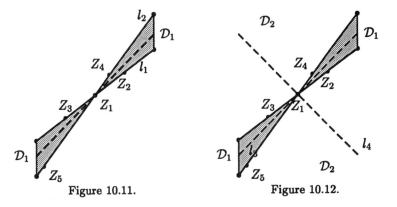

Figure 10.11.   Figure 10.12.

The mid-line l_3 of $|Z_2 Z_1 Z_4$ is also the mid-line of $|Z_3 Z_1 Z_5$ and it lies entirely in $\mathcal{D}_1$. The mid-line l_4 of $|\overline{Z_2 Z_1 Z_5}$ is also the mid-line of $|Z_3 Z_1 Z_4$ and it lies entirely in $\mathcal{D}_2$. We call $\{l_3, l_4\}$ the **bisectors** of the line pair $\{l_1, l_2\}$ and use l_3 to identify $\mathcal{D}_1$, l_4 to identify $\mathcal{D}_2$. When $\delta_{\mathcal{F}}(Z_1, Z_2, Z_4) > 0$ we note that

$$\mathcal{D}_1 = \{Z \in \Pi : \delta_{\mathcal{F}}(Z_1, Z_2, Z)\delta_{\mathcal{F}}(Z_1, Z_4, Z) \leq 0\},$$
$$\mathcal{D}_2 = \{Z \in \Pi : \delta_{\mathcal{F}}(Z_1, Z_2, Z)\delta_{\mathcal{F}}(Z_1, Z_4, Z) \geq 0\},$$

and get a similar characterisation when $\delta_{\mathcal{F}}(Z_1, Z_2, Z_4) < 0$.

   When $l_1 = l_2$, we take $\mathcal{D}_1 = l_1$; we could also take $\mathcal{D}_2 = \Pi$ but do not make any use of this.

10.10.3 Duo-angles

When l_1, l_2 are distinct lines, intersecting at Z_1, we call the pairs

$$(\{l_1, l_2\}, \mathcal{D}_1), \quad (\{l_1, l_2\}, \mathcal{D}_2),$$

duo-angles, with arms l_1, l_2; in this $\mathcal{D}_1$, $\mathcal{D}_2$ are the duo-sectors of 10.10.2. We denote these duo-angles by α_d, β_d, respectively. We call the bisector l_3 the **indicator** of α_d, and the bisector l_4 the indicator of β_d. We define the degree-magnitudes of these by

$$|\alpha_d|° = |\angle Z_2 Z_1 Z_4|° = |\angle Z_3 Z_1 Z_5|°, \quad |\beta_d|° = |\angle Z_2 Z_1 Z_5|° = |\angle Z_3 Z_1 Z_4|°.$$

If $l_1 \perp l_2$ we have that $|\alpha_d|° = |\beta_d|° = 90$, and we call these **right duo-angles**.

When $l_1 = l_2$ we take $\alpha_d = (\{l_1, l_2\}, l_1)$ to be a duo-angle with arms l_1, l_1, and call it a **null duo-angle**. Its indicator is l_1, and we define its degree-measure to be 0. We do not define a straight duo-angle. Thus the measure of a duo-angle γ_d always satisfies $0 \le |\gamma_d|° < 180$.

When $l_1 \ne l_2$ we define

$$\begin{aligned}
\sin \alpha_d &= \sin(\angle Z_2 Z_1 Z_4) = \sin(\angle Z_3 Z_1 Z_5), \\
\cos \alpha_d &= \cos(\angle Z_2 Z_1 Z_4) = \cos(\angle Z_3 Z_1 Z_5), \\
\sin \beta_d &= \sin(\angle Z_2 Z_1 Z_5) = \sin(\angle Z_3 Z_1 Z_4), \\
\cos \beta_d &= \cos(\angle Z_2 Z_1 Z_5) = \cos(\angle Z_3 Z_1 Z_4).
\end{aligned}$$

For a right duo-angle these have the values 1 and 0, respectively.

When l_1 and l_2 are not perpendicular, we can define as well $\tan \alpha_d = \frac{\sin \alpha_d}{\cos \alpha_d}$, $\tan \beta_d = \frac{\sin \beta_d}{\cos \beta_d}$.

If α_d is a null duo-angle we define $\sin \alpha_d = 0$, $\cos \alpha_d = 1$, $\tan \alpha_d = 0$.

10.10.4 Duo-angles in standard position

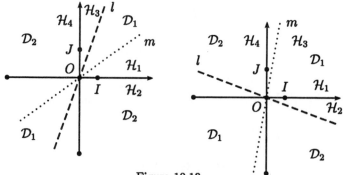

Figure 10.13.

We extend our frame of reference $\mathcal{F}$ by taking in connection with the line pair $\{OI, OJ\}$ a canonical pair of duo-sectors $\mathcal{D}_1$ and $\mathcal{D}_2$, with $\mathcal{D}_1$ the union of the first and third quadrants $\mathcal{Q}_1$ and $\mathcal{Q}_3$, and $\mathcal{D}_2$ the union of the second and fourth quadrants.

For any line l through the origin O, we consider the duo-angle α_d with side-lines OI and l, such that the indicator m of α_d lies in the duo-sector $\mathcal{D}_1$, that is the bisector of the line-pair $\{OI, l\}$ which lies in the duo-sector of α_d also lies in $\mathcal{D}_1$. We denote by $\mathcal{DA}(\mathcal{F})$ the set of such duo-angles, and we say that they are in standard position with respect to $\mathcal{F}$.

If $l \neq OI$ and $Z_4 \equiv (x_4, y_4)$ is a point other than O on l, then so is the point with coordinates $(-x_4, -y_4)$; thus, without loss of generality, we may assume that $y_4 > 0$ in identifying l as OZ_4. Then $Z_4 \in \mathcal{H}_1$ and

$$
\begin{aligned}
|\alpha_d|^\circ &= |\angle IOZ_4|^\circ, \\
\cos \alpha_d &= \cos(\angle IOZ_4) = \frac{x_4}{\sqrt{x_4^2 + y_4^2}}, \\
\sin \alpha_d &= \sin(\angle IOZ_4) = \frac{y_4}{\sqrt{x_4^2 + y_4^2}}.
\end{aligned}
$$

When α_d is not a right duo-angle, we have

$$
\tan \alpha_d = \frac{y_4}{x_4}.
$$

We identify $l = OI$ as OZ_4 where $Z_4 \equiv (x_4, 0)$ and $x_4 > 0$. Thus for the null duo-angle in standard position we have $\cos \alpha_d = 1$, $\sin \alpha_d = 0$, $\tan \alpha_d = 0$. We denote this null duo-angle by $0_{d\mathcal{F}}$ and the right duo-angle in standard position by $90_{d\mathcal{F}}$.

We now note that if α_d, $\beta_d \in \mathcal{DA}(\mathcal{F})$ and $\tan \alpha_d = \tan \beta_d$, then $\alpha_d = \beta_d$.

Proof. For this we let α_d, β_d have pairs of side-lines (OI, OZ_4), (OI, OZ_5), respectively, where $|0, Z_4| = |0, Z_5| = k$, and either $y_4 > 0$ or $x_4 > 0$, $y_4 = 0$, and similarly either $y_5 > 0$ or $x_5 > 0$, $y_5 = 0$. Then neither α_d nor β_d is $90_{d\mathcal{F}}$ and

$$
\frac{y_4}{x_4} = \frac{y_5}{x_5}, \quad x_4^2 + y_4^2 = x_5^2 + y_5^2 = k^2.
$$

If $y_4 = 0$ then $y_5 = 0$ and both duo-angles are null. Suppose then that $y_4 \neq 0$ so that $y_4 > 0$; it follows that $y_5 > 0$. Then

$$
k^2 = x_5^2 + y_5^2 = \frac{y_5^2}{y_4^2} x_4^2 + y_5^2 = \frac{y_5^2}{y_4^2}(x_4^2 + y_4^2) = \frac{y_5^2}{y_4^2} k^2.
$$

Hence $y_5^2 = y_4^2$, and so $y_5 = y_4$. It follows that $x_5 = x_4$.

10.10.5   Addition of duo-angles in standard position

To deal with addition of duo-
angles in standard position, let
$Q \equiv (k,0)$, $R \equiv (0,k)$ for some
$k > 0$, and α_d have side-lines
OQ and OZ_4, β_d have side-lines
OQ and OZ_5, where $|0, Z_4| =$
$|0, Z_5| = k$ and both have their
indicators in $\mathcal{D}_1$. Without loss of
generality, we may suppose that
either $y_4 > 0$ or $y_4 = 0$, $x_4 >$
0, and similarly with respect to
(x_5, y_5).

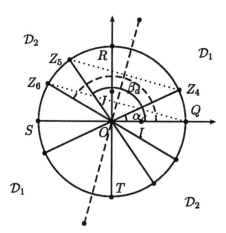

Figure 10.14. Addition of duo-angles.

Then the line through Q which is parallel to $Z_4 Z_5$ will meet the circle $C(O; k)$ in a
second point, which we denote by $Z_6 \equiv (x_6, y_6)$. The line through Q parallel to $Z_4 Z_5$
has parametric equations

$$x = k + t(x_5 - x_4), \; y = t(y_5 - y_4),$$

and so meets the circle again when $t \neq 0$ satisfies

$$[k + t(x_5 - x_4)]^2 + [t(y_5 - y_4)]^2 = k^2.$$

This yields

$$t = -\frac{2k(x_5 - x_4)}{(x_5 - x_4)^2 + (y_5 - y_4)^2},$$

and so we find for (x_6, y_6) that

$$x_6 = k\frac{(y_5 - y_4)^2 - (x_5 - x_4)^2}{(x_5 - x_4)^2 + (y_5 - y_4)^2}, \; y_6 = -2k\frac{(x_5 - x_4)(y_5 - y_4)}{(x_5 - x_4)^2 + (y_5 - y_4)^2}. \qquad (10.10.1)$$

We define the **sum** $\alpha_d + \beta_d = \gamma_d$, where γ_d has side-lines OQ and OZ_6 and has its
indicator in $\mathcal{D}_1$. When $Z_4 = Z_5$ we take QZ_6 as the line through Q which is parallel
to the tangent to the circle at Z_4. This is analogous to the modified sum of angles.
    It can be checked that

$$\frac{y_6}{k} - \frac{x_5 y_4 + x_4 y_5}{k^2} = 0, \; \frac{x_6}{k} - \frac{x_4 x_5 - y_4 y_5}{k^2} = 0. \qquad (10.10.2)$$

To see this we first note that $(y_5 - y_4)^2 + (x_5 - x_4)^2 = 2[k^2 - (x_4 x_5 + y_4 y_5)]$. Then
the numerator in

$$\frac{(x_5 - x_4)(y_5 - y_4)}{x_4 x_5 + y_4 y_5 - k^2} - \frac{x_5 y_4 + x_4 y_5}{k^2}$$

is equal to

$$k^2[(x_5 - x_4)(y_5 - y_4) + x_5 y_4 + x_4 y_5] - (x_5 y_4 + x_4 y_5)(x_4 x_5 + y_4 y_5)$$
$$= k^2(x_5 y_5 + x_4 y_4) - [x_4 y_4(x_5^2 + y_5^2) + x_5 y_5(x_4^2 + y_4^2)]$$
$$= (k^2 - k^2)(x_5 y_5 + x_4 y_4) = 0.$$

Similarly the numerator in

$$\frac{1}{2} \frac{(y_5 - y_4)^2 - (x_5 - x_4)^2}{k^2 - (x_4 x_5 + y_4 y_5)} - \frac{x_4 x_5 - y_4 y_5}{k^2}$$

equals

$$k^2[(y_5 - y_4)^2 - (x_5 - x_4)^2 - 2(x_4 x_5 - y_4 y_5)] + 2(x_4 x_5 + y_4 y_5)(x_4 x_5 - y_4 y_5)$$
$$= k^2[y_5^2 + y_4^2 - x_5^2 - x_4^2] + 2(x_4^2 x_5^2 - y_4^2 y_5^2)$$
$$= k^2[y_5^2 + y_4^2 - x_5^2 - x_4^2] + 2[x_4^2(k^2 - y_5^2) - y_4^2 y_5^2]$$
$$= k^2[y_5^2 + y_4^2 - x_5^2 - x_4^2] + 2[x_4^2 k^2 - y_5^2(x_4^2 + y_4^2)]$$
$$= k^2[y_5^2 + y_4^2 - x_5^2 - x_4^2 + 2x_4^2 - 2y_5^2] = 0.$$

To apply these we note that by 10.10.4

$$\sin \alpha_d = \frac{y_4}{k}, \quad \cos \alpha_d = \frac{x_4}{k}, \quad \sin \beta_d = \frac{y_5}{k}, \quad \cos \beta_d = \frac{x_5}{k}.$$

The sum $\gamma_d = \alpha_d + \beta_d$ has side-lines OQ and OZ_6, and we sub-divide into two major cases. First we suppose that $x_5 y_4 + x_4 y_5 > 0$ or equivalently $|\alpha_d|° + |\beta_d|° < 180$. Then $y_6 > 0$ and we have

$$\sin \gamma_d = \frac{y_6}{k}, \quad \cos \gamma_d = \frac{x_6}{k}.$$

It follows from (10.10.2) that

$$\sin(\alpha_d + \beta_d) = \sin \alpha_d \cos \beta_d + \cos \alpha_d \sin \beta_d,$$
$$\cos(\alpha_d + \beta_d) = \cos \alpha_d \cos \beta_d - \sin \alpha_d \sin \beta_d.$$

Secondly we suppose that $x_5 y_4 + x_4 y_5 < 0$ or equivalently $|\alpha_d|° + |\beta_d|° > 180$. Then $y_6 < 0$ so we have

$$\sin \gamma_d = -\frac{y_6}{k}, \quad \cos \gamma_d = -\frac{x_6}{k}.$$

It follows from (10.10.2) that

$$-\sin(\alpha_d + \beta_d) = \sin \alpha_d \cos \beta_d + \cos \alpha_d \sin \beta_d,$$
$$-\cos(\alpha_d + \beta_d) = \cos \alpha_d \cos \beta_d - \sin \alpha_d \sin \beta_d.$$

There is a further case when $x_5 y_4 + x_4 y_5 = 0$ and we obtain these formulae according as $x_4 x_5 - y_4 y_5$ is positive or negative, respectively. Thus the addition formulae for sine and cosine of duo-angles are more complicated than those of angles.

10.10.6   Addition formulae for tangents of duo-angles

(i) We first note that if $\alpha_d, \beta_d \in \mathcal{DA}(\mathcal{F})$ and $\alpha_d + \beta_d = 90_{d\mathcal{F}}$, neither duo-angle being null or right, then $\tan \alpha_d \tan \beta_d = 1$. For we have that $x_6 = 0$, so that by (10.10.2) $x_4 x_5 - y_4 y_5 = 0$ and thus

$$\frac{y_4}{x_4} \frac{y_5}{x_5} = 1.$$

(ii) Next we note that, as $\tan \alpha_d = \sin \alpha_d / \cos \alpha_d$, it follows from the above addition formulae for cosine and sine that

$$\tan(\alpha_d + \beta_d) = \frac{\tan \alpha_d + \tan \beta_d}{1 - \tan \alpha_d \tan \beta_d},$$

provided that 0 does not occur in a denominator, that is provided none of α_d, β_d, $\alpha_d + \beta_d$ is a right duo-angle; this can be done separately for the cases considered in 10.10.5. In fact this addition formula for the tangent function can be verified without subdivision into cases, as

$$\frac{y_6}{x_6} = -2\frac{(y_5 - y_4)(x_5 - x_4)}{(y_5 - y_4)^2 - (x_5 - x_4)^2},$$

and we wish to show that this is equal to

$$\frac{y_5/x_5 + y_4/x_4}{1 - y_4 y_5 / x_4 x_5} = \frac{x_4 y_5 + x_5 y_4}{x_4 x_5 - y_4 y_5}. \tag{10.10.3}$$

On subtracting the first of these expressions from the second, we obtain a quotient the numerator of which is equal to

$$(x_4 y_5 + x_5 y_4)[(y_5 - y_4)^2 - (x_5 - x_4)^2] + 2(x_4 x_5 - y_4 y_5)(y_5 - y_4)(x_5 - x_4)$$
$$= (x_4 y_5 + x_5 y_4)[y_5^2 + y_4^2 - x_5^2 - x_4^2 + 2(x_4 x_5 - y_4 y_5)]$$
$$+ 2(x_4 x_5 - y_4 y_5)[x_4 y_4 + x_5 y_5 - (x_4 y_5 + x_5 y_4)]$$
$$= (x_4 y_5 + x_5 y_4)[y_5^2 + y_4^2 - x_5^2 - x_4^2] + 2(x_4 x_5 - y_4 y_5)(x_4 y_4 + x_5 y_5)$$
$$= (x_4 y_5 + x_5 y_4)[y_5^2 + y_4^2 - x_5^2 - x_4^2] + 2(x_4^2 x_5 y_4 - y_4^2 x_4 y_5 + x_5^2 x_4 y_5 - y_5^2 x_5 y_4)$$
$$= (x_5^2 + y_5^2 - x_4^2 - y_4^2)(x_4 y_5 - x_5 y_4) = 0,$$

as $x_4^2 + y_4^2 = x_5^2 + y_5^2 = k^2$. This identity then implies the standard addition formula for the tangents of duo-angles.

(iii) We also wish to show that

$$\tan(\alpha_d + 90_{d\mathcal{F}}) = \frac{-1}{\tan \alpha_d},$$

when α_d is neither null nor right. For with $x_5 = 0$, $y_5 = k$ (10.10.1) gives

$$x_6 = k\frac{(k - y_4)^2 - x_4^2}{x_4^2 + (k - y_4)^2}, \quad y_6 = 2k\frac{-x_4(k - y_4)}{x_4^2 + (k - y_4)^2},$$

so that

$$\frac{y_6}{x_6} = \frac{2x_4(k - y_4)}{(k - y_4)^2 - x_4^2}.$$

Then

$$\frac{y_6}{x_6}\frac{y_4}{x_4} = 2\frac{y_4(k - y_4)}{k^2 - 2k y_4 + y_4^2 - x_4^2} = 2\frac{y_4(k - y_4)}{2y_4^2 - 2k y_4} = -1.$$

10.10.7 Associativity of addition of duo-angles

With the notation of 10.10.5, suppose that α_d, β_d and γ_d are duo-angles in $\mathcal{DA}(\mathcal{F})$, with pairs side- lines (OQ, OZ_4), (OQ, OZ_5), (OQ, OZ_6), respectively. We wish to consider the sums $(\alpha_d + \beta_d) + \gamma_d$ and $\alpha_d + (\beta_d + \gamma_d)$. We suppose that $\alpha_d + \beta_d$ has side-lines (OQ, OZ_7) and that $(\alpha_d + \beta_d) + \gamma_d$ has side-lines (OQ, OZ_9). Similarly we suppose that $\beta_d + \gamma_d$ has side-lines (OQ, OZ_8) and $\alpha_d + (\beta_d + \gamma_d)$ has side-lines (OQ, OZ_{10}). Then by (10.10.2) applied several times we have that

$$y_7 = \frac{x_5 y_4 + x_4 y_5}{k}, \ x_7 = \frac{x_4 x_5 - y_4 y_5}{k},$$

$$y_9 = \frac{x_6 y_7 + x_7 y_6}{k} = \frac{x_6 \frac{x_5 y_4 + x_4 y_5}{k} + \frac{x_4 x_5 - y_4 y_5}{k} y_6}{k}$$

$$= \frac{x_6 (x_5 y_4 + x_4 y_5) + (x_4 x_5 - y_4 y_5) y_6}{k^2},$$

$$x_9 = \frac{x_7 x_6 - y_7 y_6}{k} = \frac{\frac{x_4 x_5 - y_4 y_5}{k} x_6 - \frac{x_5 y_4 + x_4 y_5}{k} y_6}{k}$$

$$= \frac{(x_4 x_5 - y_4 y_5) x_6 - (x_5 y_4 + x_4 y_5) y_6}{k^2}.$$

Similarly

$$y_8 = \frac{x_6 y_5 + x_5 y_6}{k}, \ x_8 = \frac{x_5 x_6 - y_5 y_6}{k},$$

$$y_{10} = \frac{x_8 y_4 + x_4 y_8}{k} = \frac{\frac{x_5 x_6 - y_5 y_6}{k} y_4 + x_4 \frac{x_6 y_5 + x_5 y_6}{k}}{k}$$

$$= \frac{(x_5 x_6 - y_5 y_6) y_4 + x_4 (x_6 y_5 + x_5 y_6)}{k^2},$$

$$x_{10} = \frac{x_4 x_8 - y_4 y_8}{k} = \frac{x_4 \frac{x_5 x_6 - y_5 y_6}{k} - y_4 \frac{x_6 y_5 + x_5 y_6}{k}}{k}$$

$$= \frac{x_4 (x_5 x_6 - y_5 y_6) - y_4 (x_6 y_5 + x_5 y_6)}{k^2}.$$

From these we can see that $Z_9 = Z_{10}$ and so we have that

$$(\alpha_d + \beta_d) + \gamma_d = \alpha_d + (\beta_d + \gamma_d).$$

Thus addition of duo-angles is associative on $\mathcal{DA}(\mathcal{F})$.

10.10.8 Group properties of duo-angles; sensed duo-angles

We note the following properties of addition of duo-angles:-
   (i) *Given any duo-angles α_d, β_d in $\mathcal{DA}(\mathcal{F})$, the sum $\alpha_d + \beta_d$ is a unique object γ_d and it lies in $\mathcal{DA}(\mathcal{F})$.*
   (ii) *Addition of duo-angles is commutative, that is*

$$\alpha_d + \beta_d = \beta_d + \alpha_d,$$

for all α_d, $\beta_d \in \mathcal{DA}(\mathcal{F})$.

(iii) *Addition of duo-angles is associative on* $\mathcal{DA}(\mathcal{F})$.

(iv) *The null angle* $0_{d\mathcal{F}}$ *is a neutral element for* $+$ *on* $\mathcal{DA}(\mathcal{F})$.

(v) *Each* $\alpha_d \in \mathcal{DA}(\mathcal{F})$ *has an additive inverse in* $\mathcal{DA}(\mathcal{F})$.

Proof.

(i) This is evident from the definition.

(ii) This is evident as the definition is symmetrical in the roles of the two duo-angles.

(iii) This was established in 10.10.7.

(iv) For $\alpha_d + 0_{d\mathcal{F}} = \alpha_d$, for all $\alpha_d \in \mathcal{DA}(\mathcal{F})$.

(v) With the notation of 10.10.5 let $Z_5 = s_{OJ}(Z_4)$ so that $Z_5 \equiv (-x_4, y_4)$, and let δ_d be the duo-angle in $\mathcal{DA}(\mathcal{F})$ with arms OI, OZ_5. Then, straightforwardly, $\alpha_d + \delta_d = 0_{d\mathcal{F}}$. Thus this duo-angle δ_d is an additive inverse for α_d in $\mathcal{DA}(\mathcal{F})$. We denote it by $-\alpha_d$.

These properties show that we have a commutative group. We note that

$$\sin(-\alpha_d) = \frac{y_4}{k} = \sin\alpha_d, \quad \cos(-\alpha_d) = -\frac{x_4}{k} = -\cos\alpha_d,$$

$$\tan(-\alpha_d) = -\frac{y_4}{x_4} = -\tan\alpha_d.$$

If $\alpha = \angle_{\mathcal{F}}QOZ_4$ is a wedge-angle in $\mathcal{A}(\mathcal{F})$, with $Z_4 \equiv (x_4, y_4)$ and $y_4 > 0$, we recall that $-\alpha = \angle_{\mathcal{F}}QOZ_6$ where $Z_6 = s_{OI}(Z_4) \equiv (x_4, -y_4)$. If α_d is the duo-angle in $\mathcal{DA}(\mathcal{F})$ with side-lines (OQ, OZ_4) then $-\alpha_d$ is the duo-angle in $\mathcal{DA}(\mathcal{F})$ with side-lines (OQ, OZ_5) where $Z_5 = s_{OJ}(Z_4) \equiv (-x_4, y_4)$. This inverse angle and inverse duo-angle are linked in that $OZ_5 = OZ_6$ and so $|-\alpha|° = |-\alpha_d|° + 180$.

We define $\beta_d - \alpha_d = \beta_d + (-\alpha_d)$, and this is the duo-angle in standard position with side-lines OQ and OZ_7, where $Z_7 \equiv (x_7, y_7)$ is the point where the line through Q and parallel to $Z_5 s_{OJ}(Z_4)$ meets the circle $\mathcal{C}(O; k)$ again. We call $\beta_d - \alpha_d$ the sensed **duo-angle** with side-lines OZ_4, OZ_5 and denote it by $\blacktriangleleft_{\mathcal{F}}(OZ_4, OZ_5)$. If $\mathcal{F}'$ is any frame of reference obtained from $\mathcal{F}$ by translation, we also define

$$\blacktriangleleft_{\mathcal{F}'}(OZ_4, OZ_5) = \blacktriangleleft_{\mathcal{F}}(OZ_4, OZ_5).$$

Earlier names for this were a 'complete angle' and a 'cross'; see Forder [7] for applications and exercises, and Forder [6, pages 120–121, 151–154] for applications, the terminology used being 'cross'. Sensed duo-angles were also used by Johnson [9, pages 11–15] under the name of 'directed angles'.

We have

$$\tan(\beta_d - \alpha_d) = \frac{\tan\beta_d - \tan\alpha_d}{1 + \tan\alpha_d \tan\beta_d},$$

provided none of α_d, β_d, $\beta_d - \alpha_d$ is a right duo-angle. For a coordinate formula to utilise this we replace x_4 by $-x_4$ in (10.10.3) and translate to parallel axes through Z_1. Thus for $\gamma_d = \blacktriangleleft_{\mathcal{F}}(Z_1 Z_4, Z_1 Z_5)$ we have

$$\tan\gamma_d = \frac{\frac{y_5 - y_1}{x_5 - x_1} - \frac{y_4 - y_1}{x_4 - x_1}}{1 + \frac{y_5 - y_1}{x_5 - x_1}\frac{y_4 - y_1}{x_4 - x_1}}$$

when γ_d is not right, and

$$1 + \frac{y_5 - y_1}{x_5 - x_1}\frac{y_4 - y_1}{x_4 - x_1} = 0$$

when it is.

10.10.9 An application

For fixed points Z_4 and Z_5, consider the locus of points Z such that $\sphericalangle_\mathcal{F}(ZZ_4, ZZ_5)$ has constant magnitude. If it is a right duo-angle we will have

$$1 + \frac{y_5 - y}{x_5 - x}\frac{y_4 - y}{x_4 - x} = 0,$$

and so the points $Z \notin Z_4 Z_5$ lie on the circle on $[Z_4, Z_5]$ as diameter. Otherwise, we have that

$$\frac{\frac{y_5 - y}{x_5 - x} - \frac{y_4 - y}{x_4 - x}}{1 + \frac{y_5 - y}{x_5 - x}\frac{y_4 - y}{x_4 - x}} = 1 - \lambda,$$

for some $\lambda \neq 1$, and then the points $Z \notin Z_4 Z_5$ lie on a circle which passes through Z_4 and Z_5. In fact we obtain a set of coaxal circles through Z_4 and Z_5. This should be compared with 7.5.1 and 10.9.1.

10.11 A CASE OF PASCAL'S THEOREM, 1640

10.11.1

Let Z_1, W_1, Z_2, W_2 be distinct points on the circle $\mathcal{C}(O; k)$. Then $Z_1 W_2 \parallel W_1 Z_2$ if and only if $\sphericalangle_\mathcal{F} Z_2 O W_2 = \sphericalangle_\mathcal{F} Z_1 O W_1$.

Proof. We let $z_1 \sim k\operatorname{cis}\theta_1$, $z_2 \sim k\operatorname{cis}\theta_2$, $w_1 \sim k\operatorname{cis}\phi_1$, $w_2 \sim k\operatorname{cis}\phi_2$. Then $Z_1 W_2$ and $W_1 Z_2$ are parallel if and only if

$$\frac{k\operatorname{cis}\phi_2 - k\operatorname{cis}\theta_1}{k\operatorname{cis}\theta_2 - k\operatorname{cis}\phi_1} = t$$

for some $t \neq 0$ in $\mathbf{R}$. By 9.4.1 the left-hand side is equal to

$$\begin{aligned}
&\frac{\cos\phi_2 - \cos\theta_1 + \imath(\sin\phi_2 - \sin\theta_1)}{\cos\theta_2 - \cos\phi_1 + \imath(\sin\theta_2 - \sin\phi_1)}\\
&= \frac{-2\sin(\frac{1}{2}\phi_2 + \frac{1}{2}\theta_1)\sin(\frac{1}{2}\phi_2 - \frac{1}{2}\theta_1) + 2\imath\cos(\frac{1}{2}\phi_2 + \frac{1}{2}\theta_1)\cos(\frac{1}{2}\phi_2 - \frac{1}{2}\theta_1)}{-2\sin(\frac{1}{2}\theta_2 + \frac{1}{2}\phi_1)\sin(\frac{1}{2}\theta_2 - \frac{1}{2}\phi_1) + 2\imath\cos(\frac{1}{2}\theta_2 + \frac{1}{2}\phi_1)\cos(\frac{1}{2}\theta_2 - \frac{1}{2}\phi_1)}\\
&= \frac{\sin(\frac{1}{2}\phi_2 - \frac{1}{2}\theta_1)\operatorname{cis}(\frac{1}{2}\phi_2 + \frac{1}{2}\theta_1)}{\sin(\frac{1}{2}\theta_2 - \frac{1}{2}\phi_1)\operatorname{cis}(\frac{1}{2}\theta_2 + \frac{1}{2}\phi_1)}\\
&= \frac{\sin(\frac{1}{2}\phi_2 - \frac{1}{2}\theta_1)}{\sin(\frac{1}{2}\theta_2 - \frac{1}{2}\phi_1)}\operatorname{cis}(\frac{1}{2}\phi_2 + \frac{1}{2}\theta_1 - \frac{1}{2}\theta_2 - \frac{1}{2}\phi_1).
\end{aligned}$$

Thus

$$\operatorname{cis}(\tfrac{1}{2}\phi_2 + \tfrac{1}{2}\theta_1 - \tfrac{1}{2}\theta_2 - \tfrac{1}{2}\phi_1) = \frac{\sin(\frac{1}{2}\theta_2 - \frac{1}{2}\phi_1)}{\sin(\frac{1}{2}\phi_2 - \frac{1}{2}\theta_1)}t.$$

But the absolute value here is 1, so the right-hand side is ± 1. Thus we have either

$$\tfrac{1}{2}\phi_2 + \tfrac{1}{2}\theta_1 - \tfrac{1}{2}\theta_2 - \tfrac{1}{2}\phi_1 = 0_\mathcal{F},$$

or

$$\tfrac{1}{2}\phi_2 + \tfrac{1}{2}\theta_1 - \tfrac{1}{2}\theta_2 - \tfrac{1}{2}\phi_1 = 180_\mathcal{F}.$$

In each case, we have that $\phi_2 - \phi_1 = \theta_2 - \theta_1$ and so $\measuredangle_{\mathcal{F}} Z_2 O W_2 = \measuredangle_{\mathcal{F}} Z_1 O W_1$.

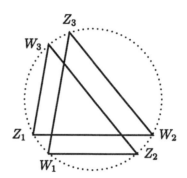

If (Z_1, W_1), (Z_2, W_2), (Z_3, W_3) *are distinct pairs of points all on a circle and such that $Z_1 W_2 \parallel W_1 Z_2$ and $Z_2 W_3 \parallel W_2 Z_3$ then $Z_1 W_3 \parallel W_1 Z_3$.*
Proof. This follows immediately from the last subsection. It is a case of what is known as PAS-CAL'S THEOREM .

Figure 10.15. A case of Pascal's theorem

COROLLARY. *If (Z_1, W_1), (Z_2, W_2), (Z_3, W_3), (Z_4, W_4) are four distinct pairs of points all on a circle and such that*

$$Z_1 W_2 \parallel W_1 Z_2, \quad Z_2 W_3 \parallel W_2 Z_3, \quad Z_3 W_4 \parallel W_3 Z_4,$$

then $Z_1 W_4 \parallel W_1 Z_4$.

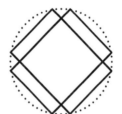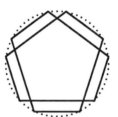

Figure 10.16. Very symmetrical cases.

Proof. For from the first two we deduce that $Z_1 W_3 \parallel W_1 Z_3$ and on combining this with the third relation we obtain the conclusion.

NOTE. Clearly this last result can be extended to any number of pairs of points on a circle.

10.11.2

Starting more generally than in the last subsection, for pairs of distinct points let $(Z_1, W_1) \sim (Z_2, W_2)$ if and only if $Z_1 W_2 \parallel W_1 Z_2$. Then clearly the relation $\sim$ is reflexive and symmetric. We ask when it is also transitive and thus an equivalence relation.

Now if $Z_1 \equiv (x_1, y_1)$, $Z_2 \equiv (x_2, y_2)$, $W_1 \equiv (u_1, v_1)$, $W_2 \equiv (u_2, v_2)$, we have $(Z_1, W_1) \sim (Z_2, W_2)$ if and only if

$$(v_2 - y_1)(x_2 - u_1) = (u_2 - x_1)(y_2 - v_1). \qquad (10.11.1)$$

Similarly we have $(Z_2, W_2) \sim (Z, W)$ if and only if

$$(v - y_2)(x - u_2) = (u - x_2)(y - v_2). \qquad (10.11.2)$$

We wish (10.11.1) and (10.11.2) to imply that

$$(v - y_1)(x - u_1) = (u - x_1)(y - v_1).\tag{10.11.3}$$

From (10.11.1) we have that

$$v_2 x_2 - u_2 y_2 = u_1 v_2 - u_2 v_1 + x_2 y_1 - x_1 y_2 + x_1 v_1 - y_1 u_1,$$

and from (10.11.2)

$$v_2 x_2 - u_2 y_2 = u v_2 - u_2 v + x_2 y - x y_2 + v x - u y,$$

so together these give

$$v x - u y = u_1 v_2 - u_2 v_1 + x_2 y_1 - x_1 y_2 + x_1 v_1 - y_1 u_1 - u v_2 + u_2 v - x_2 y + x y_2.$$

We need for (10.11.3) that

$$v x - u y = v u_1 - u v_1 + y_1 x - x_1 y - y_1 u_1 + x_1 v_1$$

and so our condition for transitivity is got by equating the two right- hand sides here. This turns out to be $\delta_{\mathcal{F}}(Z_1, Z_2, Z) = \delta_{\mathcal{F}}(W_1, W_2, W)$.

Now (10.11.2) and (10.11.3) simultaneously give a transformation under which $Z \to W$ as we see by writing them as

$$\frac{v - y_2}{u - x_2} = \frac{y - v_2}{x - u_2}, \quad \frac{v - y_1}{u - x_1} = \frac{y - v_1}{x - u_1}.\tag{10.11.4}$$

On solving for u and v in this we obtain

$$u = \frac{y_2 - y_1 + x_1 \frac{y - v_1}{x - u_1} - x_2 \frac{y - v_2}{x - u_2}}{\frac{y - v_1}{x - u_1} - \frac{y - v_2}{x - u_2}},$$

$$v = \frac{x_2 - x_1 + y_1 \frac{x - u_1}{y - v_1} - y_2 \frac{x - u_2}{y - v_2}}{\frac{x - u_1}{y - v_1} - \frac{x - u_2}{y - v_2}}.$$

To utilise this transformation we consider loci with equations of the form

$$2h \frac{y - v_2}{x - u_2} \frac{y - v_1}{x - u_1} + 2g \frac{y - v_2}{x - u_2} + 2f \frac{y - v_1}{x - u_1} + c = 0.\tag{10.11.5}$$

Under the transformation this maps into the locus with equation

$$2h \frac{v - y_2}{u - x_2} \frac{v - y_1}{u - x_1} + 2g \frac{v - y_2}{u - x_2} + 2f \frac{v - y_1}{u - x_1} + c = 0.\tag{10.11.6}$$

On clearing the equation (10.11.5) of fractions we obtain

$$2h(y - v_2)(y - v_1) + 2g(y - v_2)(x - u_1) + 2f(y - v_1)(x - u_2)$$
$$+ c(x - u_1)(x - u_2) = 0,$$

from which we see that W_1 and W_2 are both on this locus. This equation can be re-arranged as

$$cx^2 + 2(g+f)xy + 2hy^2 - (2gv_2 + 2fv_1 + cu_1 + cu_2)x -$$
$$(2gu_1 + 2fu_2 + 2hv_1 + 2hv_2)y + 2hv_1v_2 + 2gu_1v_2 + 2fu_2v_1 + cu_1u_2 = 0. \quad (10.11.7)$$

Similarly we see that Z_1 and Z_2 are on the locus given by (10.11.6), and the equation for it becomes

$$cu^2 + 2(g+f)uv + 2hv^2 - (2gy_2 + 2fy_1 + cx_1 + cx_2)u -$$
$$(2gx_1 + 2fx_2 + 2hy_1 + 2hy_2)v + 2hy_1y_2 + 2gx_1y_2 + 2fx_2y_1 + cx_1x_2 = 0. \quad (10.11.8)$$

We note that W_1 maps to Z_1 and W_2 maps to Z_2 under the transformation in which Z maps to W

To identify all the loci that can occur in (10.11.7) and (10.11.8) would take us beyond the concepts of the present course, so we concentrate on when they represent circles.

Now (10.11.7) is a circle when $c = 2h \neq 0$ and $g = -f$. The equation then becomes

$$x^2 + y^2 + \left[\frac{g}{h}(v_1 - v_2) - u_1 - u_2\right]x + \left[\frac{g}{h}(u_2 - u_1) - v_1 - v_2\right]y$$
$$+ \frac{g}{h}(u_1v_2 - u_2v_1) + u_1u_2 + v_1v_2 = 0. \quad (10.11.9)$$

This is the set of circles which pass through the points W_1 and W_2, a set of coaxal circles. The corresponding equation for the second locus is

$$u^2 + v^2 + \left[\frac{g}{h}(y_1 - y_2) - x_1 - x_2\right]u + \left[\frac{g}{h}(x_2 - x_1) - y_1 - y_2\right]v$$
$$+ \frac{g}{h}(x_1y_2 - x_2y_1) + x_1x_2 + y_1y_2 = 0, \quad (10.11.10)$$

and this gives the set of coaxal circles passing through Z_1 and Z_2.

We can take an arbitrary circle from the first coaxal set and then there is a unique one from the second set corresponding to it. If we take Z_1, Z_2, W_1, W_2 to be concylic we get just one circle and that is the classical case; it occurs when the remaining coefficients in the two equations are pairwise equal.

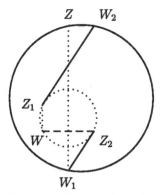

Figure 10.17. Pascal result for two circles.

10.11.3

Instead of using parallelism of lines as the basis of the relation in 10.11.2, we could take instead a fixed line o with equation $lx + my + n = 0$, and let $(Z_1, W_1) \sim (Z_2, W_2)$

if the lines $Z_1 W_2$ and $W_1 Z_2$ meet on o. The results are like those in 10.11.2 and the transformation corresponding to (10.11.4) is

$$\frac{n(y_2 - v) - l(x_2 v - y_2 u)}{n(x_2 - u) + m(x_2 v - y_2 u)} = \frac{n(v_2 - y) - l(u_2 y - v_2 x)}{n(u_2 - x) + m(u_2 y - v_2 x)},$$

$$\frac{n(y_1 - v) - l(x_1 v - y_1 u)}{n(x_1 - u) + m(x_1 v - y_1 u)} = \frac{n(v_1 - y) - l(u_1 y - v_1 x)}{n(u_1 - x) + m(u_1 y - v_1 x)}.$$

Exercises

10.1 Prove the result of Varignon (1731) that if A, B, C, D are the vertices of a convex quadrilateral and

$$P = \mathrm{mp}(A, B), \quad Q = \mathrm{mp}(B, C), \quad R = \mathrm{mp}(C, D), \quad S = \mathrm{mp}(D, A),$$

then P, Q, R, S are the vertices of a parallelogram.

10.2 If $Z_1 \sim z_1, Z_2 \sim z_2$ and $Z_3 \sim z_3$ are non-collinear points show that

$$z_1(\bar{z}_3 - \bar{z}_2) + z_2(\bar{z}_1 - \bar{z}_3) + z_3(\bar{z}_2 - \bar{z}_1) = 4i\delta_{\mathcal{F}}(Z_1, Z_2, Z_3) \neq 0.$$

10.3 Let $A \sim a$, $B \sim b$, $C \sim c$ be non-collinear points and $P \sim p$ a point such that AP, BP, CP meet BC, CA, AB at $D \sim d$, $E \sim e$, $F \sim f$, respectively. Show for sensed ratio that

$$\frac{\overline{BD}}{\overline{DC}} = \frac{p(\bar{b} - \bar{a}) + \bar{p}(a - b) + b\bar{a} - a\bar{b}}{p(\bar{a} - \bar{c}) + \bar{p}(c - a) + a\bar{c} - c\bar{a}},$$

and hence prove Ceva's theorem that

$$\frac{\overline{BD}}{\overline{DC}} \frac{\overline{CE}}{\overline{EA}} \frac{\overline{AF}}{\overline{FB}} = 1.$$

10.4 Let $A \sim a$, $B \sim b$, $C \sim c$ be non-collinear points. Given any point $P \sim p$, show that as $(c - a)/(b - a)$ is non-real there exist unique real numbers y and z such that $p - a = y(b - a) + z(c - a)$, and so $p = xa + yb + zc$ where $x + y + z = 1$. Show that if AP meets BC it is in a point $D \sim d$ such that

$$d = \frac{1}{1+r}b + \frac{r}{1+r}c,$$

where $r = z/y$. Hence prove Ceva's theorem that if $D \in BC$, $E \in CA$, $F \in AB$ are such that AD, BE, CF are concurrent, then

$$\frac{\overline{BD}}{\overline{DC}} \frac{\overline{CE}}{\overline{EA}} \frac{\overline{AF}}{\overline{FB}} = 1.$$

10.5 Let $A \sim a$, $B \sim b$, $C \sim c$ be non-collinear points. If $D \sim d$, $E \sim e$ where

$$d = \frac{1}{1+\lambda}b + \frac{\lambda}{1+\lambda}c, \; e = \frac{1}{1+\mu}c + \frac{\mu}{1+\mu}a,$$

and DE meets AB it is in a point $F \sim f$ where

$$f = \frac{1}{1+\nu}a + \frac{\nu}{1+\nu}b,$$

and $\lambda\mu\nu = -1$.

Prove Menelaus' theorem that if $D \in BC$, $E \in CA$, $F \in AB$ are collinear, then

$$\frac{\overline{BD}}{\overline{DC}} \frac{\overline{CE}}{\overline{EA}} \frac{\overline{AF}}{\overline{FB}} = -1.$$

10.6 Let A, B, C be non-collinear points and take $D \in BC$, $E \in CA$, $F \in AB$ such that

$$\frac{\overline{BD}}{\overline{BC}} = r, \; \frac{\overline{CE}}{\overline{CA}} = s, \; \frac{\overline{AF}}{\overline{AB}} = t.$$

Let l, m, n be, respectively, the lines through D, E, F which are perpendicular to the side-lines BC, CA, AB. Show that l, m, n are concurrent if and only if

$$(1 - 2r)|B, C|^2 + (1 - 2s)|C, A|^2 + (1 - 2t)|A, B|^2 = 0.$$

10.7 If $\mathcal{R}(Z_0)$ is the set of all rotations about the point Z_0, show that $(\mathcal{R}(Z_0), \circ)$ is a commutative group.

10.8 Show that the composition of axial symmetries in two parallel lines is equal to a translation, and conversely that each translation can be expressed in this form.

10.9 Prove that $s_{\phi;Z_0} \circ s_{\theta;Z_0} = r_{2(\phi-\theta);Z_0}$.

10.11 Prove that $s_{\phi;Z_0} \circ r_{\theta;Z_0} = s_{\phi-\frac{1}{2}\theta;Z_0}$. Deduce that any rotation about the point Z_0 can be expressed as the composition of two axial symmetries in lines which pass through Z_0.

10.12 Let $\mathcal{F}_1 \sim \mathcal{F}_2$ if the frame of reference $\mathcal{F}_2$ is positively oriented with respect to $\mathcal{F}_1$. Show that $\sim$ is an equivalence relation.

10.13 Prove the Stewart identity

$$(z_4 - z_1)^2(z_3 - z_2) + (z_4 - z_2)^2(z_1 - z_3) + (z_4 - z_3)^2(z_2 - z_1)$$
$$= -(z_3 - z_2)(z_1 - z_3)(z_2 - z_1).$$

Interpret this trigonometrically.

10.14 Prove DeMoivre's theorem that

$$(\cos \alpha + \imath \sin \alpha)^n = \cos(n\alpha) + \imath \sin(n\alpha),$$

for all positive integers n and all angles $\alpha \in \mathcal{A}^*(\mathcal{F})$, where $\imath$ is the complex number satisfying $\imath^2 = -1$.

10.15 Suppose that l, m, n are distinct parallel lines. Let $Z_1, Z_2, Z_3, Z_4 \in l$ with $Z_1 \neq Z_2$, $Z_3 \neq Z_4$. Suppose that $Z_5, Z_6 \in n$, Z_1Z_5, Z_2Z_5 meet m at Z_7, Z_8, respectively, and Z_3Z_6, Z_4Z_6 meet m at Z_9, Z_{10}, respectively. Prove that then

$$\frac{\overline{Z_9Z_{10}}}{\overline{Z_7Z_8}} = \frac{\overline{Z_3Z_4}}{\overline{Z_1Z_2}}.$$

10.16 If $[Z_1, Z_2, Z_3, Z_4]$ is a parallelogram, W is a point on the diagonal line Z_1Z_3, a line through W parallel to Z_1Z_2 meets Z_1Z_4 and Z_2Z_3 at W_1 and W_2 respectively, and a line through W parallel to Z_1Z_4 meets Z_1Z_2 and Z_3Z_4 at W_3 and W_4, respectively, prove that

$$\delta_{\mathcal{F}}(W, W_4, W_1) = \delta_{\mathcal{F}}(W, W_3, W_2).$$

10.17 If $Z_1 \neq Z_2$ and $\delta_{\mathcal{F}}(Z_1, Z_2, Z_3) = -\delta_{\mathcal{F}}(Z_1, Z_2, Z_4)$, prove that the mid-point of Z_3 and Z_4 is on Z_1Z_2.

10.18 Suppose that Z_1, Z_2, Z_3, Z_4 are points no three of which are collinear. Show that $[Z_1, Z_3] \cap [Z_2, Z_4] \neq \emptyset$ if and only if

$$\frac{\delta_{\mathcal{F}}(Z_1, Z_2, Z_4)}{\delta_{\mathcal{F}}(Z_3, Z_2, Z_4)} < 0 \text{ and } \frac{\delta_{\mathcal{F}}(Z_2, Z_1, Z_3)}{\delta_{\mathcal{F}}(Z_4, Z_1, Z_3)} < 0.$$

11

Position vectors; vector and complex-number methods in geometry

11.1 EQUIPOLLENCE

11.1.1

Definition. An ordered pair (Z_1, Z_2) of points in Π is said to be **equipollent** to the pair (Z_3, Z_4), written symbolically $(Z_1, Z_2) \uparrow (Z_3, Z_4)$, if $\mathrm{mp}(Z_1, Z_4) = \mathrm{mp}(Z_2, Z_3)$. Thus $\uparrow$ is a binary relation in $\Pi \times \Pi$.

Equipollence has the properties:-

(i) *If $Z_1 \equiv (x_1, y_1)$, $Z_2 \equiv (x_2, y_2)$, $Z_3 \equiv (x_3, y_4)$, $Z_4 \equiv (x_4, y_4)$, then $(Z_1, Z_2) \uparrow (Z_3, Z_4)$ if and only if $x_1 + x_4 = x_2 + x_3$, $y_1 + y_4 = y_2 + y_3$, or equivalently $x_2 - x_1 = x_4 - x_3$, $y_2 - y_1 = y_4 - y_3$.*

(ii) *Given any points $Z_1, Z_2, Z_3 \in \Pi$, there is a unique point Z_4 such that $(Z_1, Z_2) \uparrow (Z_3, Z_4)$.*

(iii) *For all $Z_1, Z_2 \in \Pi$, $(Z_1, Z_2) \uparrow (Z_1, Z_2)$.*

(iv) *If $(Z_1, Z_2) \uparrow (Z_3, Z_4)$ then $(Z_3, Z_4) \uparrow (Z_1, Z_2)$.*

(v) *If $(Z_1, Z_2) \uparrow (Z_3, Z_4)$ and $(Z_3, Z_4) \uparrow (Z_5, Z_6)$, then $(Z_1, Z_2) \uparrow (Z_5, Z_6)$.*

(vi) *If $(Z_1, Z_2) \uparrow (Z_3, Z_4)$ then $(Z_1, Z_3) \uparrow (Z_2, Z_4)$.*

(vii) *If $(Z_1, Z_2) \uparrow (Z_3, Z_4)$, then $|Z_1, Z_2| = |Z_3, Z_4|$.*

(viii) *For all $Z_1 \in \Pi$, $(Z_1, Z_1) \uparrow (Z_3, Z_4)$ if and only if $Z_3 = Z_4$.*

(ix) *If $Z_1 \neq Z_2$ and $Z_3 \in l = Z_1 Z_2$, then $(Z_1, Z_2) \uparrow (Z_3, Z_4)$ if and only $Z_4 \in l$, $|Z_1, Z_2| = |Z_3, Z_4|$ and if $\leq_l$ is the natural order for which $Z_1 \leq_l Z_2$, then $Z_3 \leq_l Z_4$.*

(x) *If $Z_1 \neq Z_2$ and $Z_3 \notin Z_1 Z_2$, then $(Z_1, Z_2) \uparrow (Z_3, Z_4)$ if and only if $[Z_1, Z_2, Z_4, Z_3]$ is a parallelogram.*

Proof.
(i) By the mid-point formula,

$$\mathrm{mp}(Z_1, Z_4) \equiv \left(\frac{x_1 + x_4}{2}, \frac{y_1 + y_4}{2} \right), \ \mathrm{mp}(Z_2, Z_3) \equiv \left(\frac{x_2 + x_3}{2}, \frac{y_2 + y_3}{2} \right),$$

and the result follows immediately from this.

(ii) By part (i) it is necessary and sufficient that we choose Z_4 so that $x_4 = x_2 + x_3 - x_1$, $y_4 = y_2 + y_3 - y_1$.

(iii) This is immediate as $x_1 + x_2 = x_1 + x_2$, $y_1 + y_2 = y_1 + y_2$.

(iv) This is immediate as $x_2 + x_3 = x_1 + x_4$, $y_2 + y_3 = y_1 + y_4$.

(v) We are given that

$$x_1 + x_4 \ = \ x_2 + x_3, \ y_1 + y_4 = y_2 + y_3,$$
$$x_3 + x_6 \ = \ x_4 + x_5, \ y_3 + y_6 = y_4 + y_5.$$

By addition $(x_1 + x_6) + (x_3 + x_4) = (x_2 + x_5) + (x_3 + x_4)$, so by cancellation of $x_3 + x_4$ we have $x_1 + x_6 = x_2 + x_5$. Similarly $y_1 + y_6 = y_2 + y_5$ and so the result follows.

(vi) For by (i) above we have $x_1 + x_4 = x_3 + x_2$, $y_1 + y_4 = y_3 + y_2$.

(vii) For by (i) above

$$(x_2 - x_1)^2 + (y_2 - y_1)^2 = (x_4 - x_3)^2 + (y_4 - y_3)^2,$$

and now we apply the distance formula.

(viii) For if $x_1 = x_2, y_1 = y_2$, then (i) above is satisfied if and only if $x_3 = x_4, y_3 = y_4$.

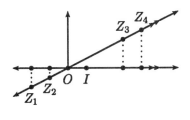
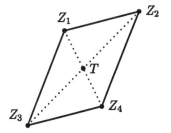

Figure 11.1.

(ix) For suppose first that $(Z_1, Z_2) \uparrow (Z_3, Z_4)$. Then $Z_3 \in l$ and $\mathrm{mp}(Z_3, Z_4) \in l$, so $Z_4 \in l$. By (vii) above we have $|Z_1, Z_2| = |Z_3, Z_4|$. Suppose first that l is not perpendicular to OI and that, as in 6.4.1(ii), the correspondence between $\leq_l$ and the natural order $\leq_{OI}$, under which $O \leq_{OI} I$, is direct. Then $\pi_{OI}(Z_1) \leq_{OI} \pi_{OI}(Z_2)$, so $x_1 \leq x_2$. Then by (i) above $x_3 \leq x_4$, and so by this argument traced in reverse we have $Z_3 \leq_l Z_4$. If the correspondence is indirect, we have $x_2 \leq x_1$, $x_4 \leq x_3$ instead. When l is perpendicular to OI, we project to OJ instead and make use of the y-coordinates.

Conversely suppose that $Z_4 \in l$, $|Z_1, Z_2| = |Z_3, Z_4|$ and $Z_3 \leq_l Z_4$. Now l has parametric equations $x = x_1 + t(x_2 - x_1)$, $y = y_1 + t(y_2 - y_1)$ $(t \in \mathbf{R})$. Suppose that Z_3 and Z_4 have parameters t_3, t_4, respectively, so that

$$x_3 = x_1 + t_3(x_2 - x_1), \quad y_3 = y_1 + t_3(y_2 - y_1),$$
$$x_4 = x_1 + t_4(x_2 - x_1), \quad y_4 = y_1 + t_4(y_2 - y_1).$$

Recall that Z_1, Z_2 have parameters 0 and 1 and $0 < 1$. As in the last paragraph above, if l is not perpendicular to OI and the correspondence between $\leq_l$ and $\leq_{OI}$ is direct, then $x_1 < x_2, x_3 < x_4$; hence $t_3 < t_4$ and we obtain this same conclusion when the correspondence is inverse. When l is perpendicular to OI we project to OJ instead, and use the y-coordinates. Moreover

$$|Z_3, Z_4|^2 = [(t_4 - t_3)(x_2 - x_1)]^2 + [(t_4 - t_3)(y_2 - y_1)]^2 = (t_4 - t_3)^2|Z_1, Z_2|^2.$$

Hence $|t_4 - t_3| = 1$, and so as $t_3 < t_4$ we have $t_4 = 1 + t_3$. Then

$$x_1 + x_4 = 2x_1 + (1 + t_3)(x_2 - x_1), \quad x_2 + x_3 = x_2 + x_1 + t_3(x_2 - x_1)$$

and these are equal. Similarly

$$y_1 + y_4 = 2y_1 + (1 + t_3)(y_2 - y_1), \quad y_2 + y_3 = y_2 + y_1 + t_3(y_2 - y_1)$$

and these are equal. By (i) above we now have $(Z_1, Z_2) \uparrow (Z_3, Z_4)$.

(x) If $[Z_1, Z_2, Z_4, Z_3]$ is a parallelogram, then $\mathrm{mp}(Z_1, Z_4) = \mathrm{mp}(Z_2, Z_3)$. Conversely suppose that $Z_1 \neq Z_2, Z_3 \notin Z_1 Z_2$ and $\mathrm{mp}(Z_1, Z_4) = \mathrm{mp}(Z_2, Z_3)$. Then $Z_1 Z_2, Z_3 Z_4$ have equations

$$-(y_2 - y_1)(x - x_1) + (x_2 - x_1)(y - y_1) = 0,$$
$$-(y_4 - y_3)(x - x_3) + (x_4 - x_3)(y - y_3) = 0.$$

By (i) above, $x_2 - x_1 = x_4 - x_3, y_2 - y_1 = y_4 - y_3$, so these lines are parallel. Similarly $Z_1 Z_3, Z_2 Z_4$ have equations

$$-(y_3 - y_1)(x - x_1) + (x_3 - x_1)(y - y_1) = 0,$$
$$-(y_4 - y_2)(x - x_2) + (x_4 - x_2)(y - y_2) = 0,$$

and by (i) above $x_3 - x_1 = x_4 - x_2, y_3 - y_1 = y_4 - y_2$, so that these lines are parallel. Thus $[Z_1, Z_2, Z_4, Z_3]$ is a parallelogram.

11.2   SUM OF COUPLES, MULTIPLICATION OF A COUPLE BY A SCALAR

11.2.1

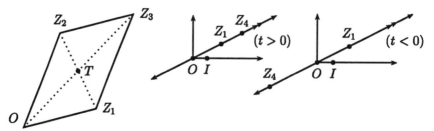

Figure 11.2.

Definition. For $O \in \Pi$, let $\mathcal{V}(\Pi; O)$ be the set of all couples (O, Z) for $Z \in \Pi$. We define the **sum** $(O, Z_1) + (O, Z_2)$ of two couples to be (O, Z_3) where $\text{mp}(O, Z_3) = \text{mp}(Z_1, Z_2)$, so that $(O, Z_1) \uparrow (Z_2, Z_3)$. Thus $+$ is a binary operation in $\mathcal{V}(\Pi; O)$. We define the **product by a number or scalar** $t.(O, Z_1)$, of a number $t \in \mathbf{R}$ and a couple, to be a couple (O, Z_4) as follows. When $Z_1 = O$ we take $Z_4 = O$ for all $t \in \mathbf{R}$. When $Z_1 \neq O$ we take Z_4 to be in the line $l = OZ_1$ and with $|O, Z_4| = |t||O, Z_1|$; furthermore if $\leq_l$ is the natural order for which $O \leq_l Z_1$, we take $O \leq_l Z_4$ when $t \geq 0$, and $Z_4 \leq_l O$ when $t < 0$. Thus product by a number is a function on $\mathbf{R} \times \mathcal{V}(\Pi; O)$ into $\mathcal{V}(\Pi; O)$.

COMMENT. To prove by synthetic means the basic properties of couples listed in 11.2.2 and 11.3.1, would be very laborious in covering all the cases. We establish instead initial algebraic characterizations which allow an effective algebraic approach.

If O is the origin and $Z_1 \equiv (x_1, y_1)$, $Z_2 \equiv (x_2, y_2)$, then

(i) $(O, Z_1) + (O, Z_2) = (O, Z_3)$ where $Z_3 \equiv (x_1 + x_2, y_1 + y_2)$.

(ii) $t.(O, Z_1) = (O, Z_4)$ where $Z_4 \equiv (tx_1, ty_1)$.

Proof.
(i) For this we have $0 + x_3 = x_1 + x_2$, $0 + y_3 = y_1 + y_2$.
(ii) We verify this as follows. Let $(x_4, y_4) = (tx_1, ty_1)$. When $(x_1, y_1) = (0, 0)$ clearly we have $(x_4, y_4) = (x_1, y_1)$. When $(x_1, y_1) \neq (0, 0)$, clearly $Z_4 \in OZ_1$ while

$$|O, Z_4|^2 = (tx_1)^2 + (ty_1)^2 = t^2 |O, Z_1|^2.$$

Now if l is not perpendicular to OI and the correspondence between the natural order $\leq_l$ and the natural order $\leq_{OI}$ on OI, under which $O \leq_{OI} I$, is direct then $x_1 < x_2$. Thus when $t > 0$, we have $tx_1 > 0$ and so $O \leq_l Z_4$; when $t < 0$, we have $tx_1 < 0$ and so $Z_4 \leq_l O$. When the correspondence between the natural orders is inverse, we reach the same conclusion. When l is perpendicular to OI we project to OJ instead.

11.2.2   Vector space over **R**

Definition. A triple $(\mathcal{V}, +, .)$ is said to be a **vector space** over **R** if the following hold:-

(i) First, $+$ is a binary operation in $\mathcal{V}$.

(ii) For all $\underline{a}, \underline{b}, \underline{c} \in \mathcal{V}$, $(\underline{a} + \underline{b}) + \underline{c} = \underline{a} + (\underline{b} + \underline{c})$.

(iii) There is an $\underline{o} \in \mathcal{V}$ such that for all $\underline{a} \in \mathcal{V}$,

$$\underline{a} + \underline{o} = \underline{a}, \quad \underline{o} + \underline{a} = \underline{a}.$$

(iv) Corresponding to each $\underline{a} \in \mathcal{V}$, there is some $-\underline{a} \in \mathcal{V}$ such that

$$(-\underline{a}) + \underline{a} = \underline{o}, \quad \underline{a} + (-\underline{a}) = \underline{o}.$$

(v) For all $\underline{a}, \underline{b} \in \mathcal{V}$, $\underline{a} + \underline{b} = \underline{b} + \underline{a}$.

(vi) Next, $. : \mathbf{R} \times \mathcal{V} \to \mathcal{V}$ is a function.

(vii) For all $\underline{a} \in \mathcal{V}$ and all $t_1, t_2 \in \mathbf{R}$, $t_2.(t_1.\underline{a}) = (t_2 t_1).\underline{a}$.

(viii) For all $\underline{a}, \underline{b} \in \mathcal{V}$ and all $t \in \mathbf{R}$, $t.(\underline{a} + \underline{b}) = t.\underline{a} + t.\underline{b}$.

(ix) For all $\underline{a} \in \mathcal{V}$ and all $t_1, , t_2 \in \mathbf{R}$, $(t_1 + t_2).\underline{a} = t_1.\underline{a} + t_2.\underline{a}$.

(x) For all $\underline{a} \in \mathcal{V}$, $1.\underline{a} = \underline{a}$.

We then have the following result.

$(\mathcal{V}(\Pi; O), +, .)$ *is a vector space over* **R**.
Proof.
(i) This has been covered already in 11.2.1.
(ii) Now $(O, Z_1) + (O, Z_2) = (O, Z_4)$ where $(x_4, y_4) = (x_1 + x_2, y_1 + y_2)$. Then

$$[(O, Z_1) + (O, Z_2)] + (O, Z_3) = (O, Z_4) + (O, Z_3) = (O, Z_5),$$

where

$$(x_5, y_5) = (x_4 + x_3, y_4 + y_3) = ((x_1 + x_2) + x_3, (y_1 + y_2) + y_3).$$

Similarly $(O, Z_2) + (O, Z_3) = (O, Z_6)$ where $(x_6, y_6) = (x_2 + x_3, y_2 + y_3)$, and so

$$(O, Z_1) + [(O, Z_2) + (O, Z_3)] = (O, Z_1) + (O, Z_6) = (O, Z_7)$$

where $(x_7, y_7) = (x_1 + x_6, y_1 + y_6) = (x_1 + (x_2 + x_3), y_1 + (y_2 + y_3))$. Clearly $Z_5 = Z_7$.
(iii) For any $Z_1 \in \Pi$, $(O, Z_1) + (O, O) = (O, Z_2)$ where $(x_2, y_2) = (x_1 + 0, y_1 + 0) = (x_1, y_1)$, so that $Z_2 = Z_1$. Similarly $(O, O) + (O, Z_1) = (O, Z_3)$ where $(x_3, y_3) = (0 + x_1, 0 + y_1) = (x_1, y_1)$, so that $Z_3 = Z_1$.
(iv) Now $(O, Z_1) + (O, Z_2) = (O, Z_3)$, $(O, Z_2) + (O, Z_1) = (O, Z_4)$ where $(x_3, y_3) = (x_1 + x_2, y_1 + y_2)$ and $(x_4, y_4) = (x_2 + x_1, y_2 + y_1)$. Clearly $Z_3 = Z_4$.
(v) If $(x_2, y_2) = (-x_1, -y_1)$, then $(O, Z_1) + (O, Z_2) = (O, Z_3)$ where $(x_3, y_3) = (x_1 - x_1, y_1 - y_1) = (0, 0)$; hence $Z_3 = O$. Similarly $(O, Z_2) + (O, Z_1) = (O, Z_4)$ where $(x_4, y_4) = (-x_1 + x_1, -y_1 + y_1) = (0, 0)$; hence $Z_4 = O$.
(vi) This was covered in 11.2.1.
(vii) For $t_1.(O, Z_1) = (O, Z_2)$ where $(x_2, y_2) = (t_1 x_1, t_1 y_1)$. Then $t_2. (t_1.(O, Z_1)) = t_2.(O, Z_2) = (O, Z_3)$ where $(x_3, y_3) = (t_2(t_1 x_1), t_2(t_1 y_1))$. Also $(t_2 t_1).(O, Z_1) = (O, Z_4)$ where $(x_4, y_4) = ((t_2 t_1) x_1, (t_2 t_1) y_1)$. Thus $Z_3 = Z_4$.
(viii) For $(O, Z_1) + (O, Z_2) = (O, Z_3)$ and $t.[(O, Z_1) + (O, Z_2)] = t.(O, Z_3) = (O, Z_4)$ where $(x_3, y_3) = (x_1 + x_2, y_1 + y_2)$, $(x_4, y_4) = (t(x_1 + x_2), t(y_1 + y_2))$. Also $t.(O, Z_1) = (O, Z_5)$, $t.(O, Z_2) = (O, Z_6)$ where $(x_5, y_5) = (t x_1, t y_1)$, $(x_6, y_6) = (t x_2, t y_2)$. Moreover $(O, Z_5) + (O, Z_6) = (O, Z_7)$ where $(x_7, y_7) = (x_5 + x_6, y_5 + y_6) = (t x_1 + t x_2, t y_1 + t y_2)$. Hence $Z_4 = Z_7$.
(ix) For $t_1.(O, Z_1) = (O, Z_2)$, $t_2.(O, Z_1) = (O, Z_3)$, $(t_1 + t_2).(O, Z_1) = (O, Z_4)$ and $(O, Z_2) + (O, Z_3) = (O, Z_5)$ where $(x_2, y_2) = (t_1 x_1, t_1 y_1)$, $(x_3, y_3) = (t_2 x_1, t_2 y_1)$ and $(x_4, y_4) = ((t_1 + t_2) x_1, (t_1 + t_2) y_1)$. Moreover $(x_5, y_5) = (x_2 + x_3, y_2 + y_3) = (t_1 x_1 + t_2 x_1, t_1 y_1 + t_2 y_1)$. Clearly $Z_4 = Z_5$.
(x) For $1.(O, Z_1) = (O, Z_2)$ where $(x_2, y_2) = (1.x_1, 1.y_1) = (x_1, y_1)$. Thus $Z_2 = Z_1$.

11.3 SCALAR OR DOT PRODUCTS

11.3.1

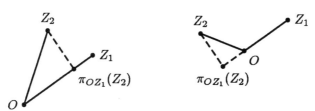

Figure 11.3.

Definitions. We define a **scalar product**, or **dot product**, $(O, Z_1).(O, Z_2)$ as follows. If $Z_1 = O$ then $(O, Z_1).(O, Z_2) = 0$; otherwise $Z_1 \neq O$ and we set

$$(O, Z_1).(O, Z_2) = \begin{cases} |O, Z_1||O, \pi_{OZ_1}(Z_2)|, & \text{if } \pi_{OZ_1}(Z_2) \in [O, Z_1, \\ -|O, Z_1||O, \pi_{OZ_1}(Z_2)|, & \text{if } \pi_{OZ_1}(Z_2) \in OZ_1 \setminus [O, Z_1. \end{cases}$$

Clearly the scalar product is a function on $\mathcal{V}(\Pi; O) \times \mathcal{V}(\Pi; O)$ into **R**.

The **norm** $\|\underline{a}\|$ of a vector $\underline{a} = (O, Z)$ is defined to be the distance $|O, Z|$.

The scalar product has the following properties:-

(i) *If $Z_j \equiv (x_j, y_j)$ for $j = 1, 2$ then $\underline{a}.\underline{b} = (O, Z_1).(O, Z_2) = x_1x_2 + y_1y_2$.*

(ii) *For all $\underline{a}, \underline{b} \in \mathcal{V}(\Pi; O)$, $\underline{a}.\underline{b} = \underline{b}.\underline{a}$.*

(iii) *For all $\underline{a}, \underline{b}, \underline{c} \in \mathcal{V}(\Pi; O)$, $\underline{a}.(\underline{b} + \underline{c}) = \underline{a}.\underline{b} + \underline{a}.\underline{c}$.*

(iv) *For all $\underline{a}, \underline{b} \in \mathcal{V}(\Pi; O)$ and all $t \in \mathbf{R}$, $t.(\underline{a}.\underline{b}) = (t.\underline{a}).\underline{b}$.*

(v) *For all $\underline{a} \neq \underline{o}$, $\underline{a}.\underline{a} > 0$, while $\underline{o}.\underline{o} = 0$.*

(vi) *For all $\underline{a}$, $\|\underline{a}\| = \sqrt{\underline{a}.\underline{a}}$.*

Proof.
(i) If $Z_1 = O$, then $x_1 = y_1 = 0$ so that $x_1x_2 + y_1y_2 = 0$ as required.

Suppose then that $Z_1 \neq O$. Write $l = OZ_1$ and let m be the line through the point O which is perpendicular to l. Define the closed half-plane $\mathcal{H}_5 = \{X : \pi_l(X) \in [O, Z_1\}$ and let $\mathcal{H}_6$ be the other closed half-plane with edge m. Now $l \equiv -y_1x + x_1y = 0$ and $m \equiv x_1x + y_1y = 0$.

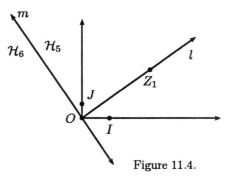

Figure 11.4.

Then as $Z_1 \in \mathcal{H}_5$,

$$\mathcal{H}_5 = \{Z \equiv (x,y) : x_1 x + y_1 y \geq 0\}, \quad \mathcal{H}_6 = \{Z \equiv (x,y) : x_1 x + y_1 y \leq 0\}.$$

But by 6.6.1(ii),

$$\pi_l(Z_2) \equiv \left(x_2 + \frac{y_1}{y_1^2 + x_1^2}(-y_1 x_2 + x_1 y_2),\ y_2 - \frac{x_1}{y_1^2 + x_1^2}(-y_1 x_2 + x_1 y_2) \right)$$

$$= \left(x_1 \frac{x_1 x_2 + y_1 y_2}{x_1^2 + y_1^2},\ y_1 \frac{x_1 x_2 + y_1 y_2}{x_1^2 + y_1^2} \right).$$

Thus

$$|0, Z_1|^2 |0, \pi_l(Z_2)|^2 = (x_1^2 + y_1^2) \left[x_1^2 \frac{(x_1 x_2 + y_1 y_2)^2}{(x_1^2 + y_1^2)^2} + y_1^2 \frac{(x_1 x_2 + y_1 y_2)^2}{(x_1^2 + y_1^2)^2} \right]$$

$$= (x_1 x_2 + y_1 y_2)^2,$$

so that $|0, Z_1 \| 0, \pi_l(Z_2)| = |x_1 x_2 + y_1 y_2|$.
If $Z_2 \in \mathcal{H}_5$ so that

$$(O, Z_1).(O, Z_2) = |0, Z_1 \| 0, \pi_l(Z_2)|,$$

and $x_1 x_2 + y_1 y_2 \geq 0$ so that $|x_1 x_2 + y_1 y_2| = x_1 x_2 + y_1 y_2$, clearly $(O, Z_1).(O, Z_2) = x_1 x_2 + y_1 y_2$.
If $Z_2 \in \mathcal{H}_6 \setminus m$ we have $\pi_l(Z_2) \in l \setminus [O, Z_1$. Then

$$(O, Z_1).(O, Z_2) = -|0, Z_1 \| 0, \pi_l(Z_2)|$$

and $x_1 x_2 + y_1 y_2 \leq 0$, so that $|x_1 x_2 + y_1 y_2| = -(x_1 x_2 + y_1 y_2)$. Clearly again $(O, Z_1).(O, Z_2) = x_1 x_2 + y_1 y_2$.

(ii) Let $\underline{a} = (O, Z_1)$, $\underline{b} = (O, Z_2)$. Then by (i) of the present theorem, $\underline{a}.\underline{b} = x_1 x_2 + y_1 y_2$, $\underline{b}.\underline{a} = x_2 x_1 + y_2 y_1$, and clearly these are equal.

(iii) Let $\underline{a} = (O, Z_1)$, $\underline{b} = (O, Z_2)$, $\underline{c} = (O, Z_3)$. Then by 11.2.1(i) $\underline{b} + \underline{c} = (O, Z_4)$ where $Z_4 \equiv (x_2 + x_3, y_2 + y_3)$. Then by (i) above $\underline{a}.(\underline{b} + \underline{c}) = x_1(x_2 + x_3) + y_1(y_2 + y_3)$, while $\underline{a}.\underline{b} + \underline{a}.\underline{c} = (x_1 x_2 + y_1 y_2) + (x_1 x_3 + y_1 y_3)$, and these are equal.

(iv) Let $\underline{a} = (O, Z_1)$, $\underline{b} = (O, Z_2)$. Then $t.(\underline{a}.\underline{b}) = t(x_1 x_2 + y_1 y_2)$. But by 11.2.1(ii), $t.\underline{a} = (O, Z_4)$ where $Z_4 \equiv (tx_1, ty_1)$ and so $(t.\underline{a}).\underline{b} = (tx_1)x_2 + (ty_1)y_2$, which is equal to the earlier expression.

(v) If $\underline{a} = (O, Z)$ then $\underline{a}.\underline{a} = x^2 + y^2$. This is positive when $(x, y) \neq (0, 0)$, and equal to 0 for $x = y = 0$.

(vi) This follows immediately.

NOTE. Note that 11.2.2(i) to (v) make $(\mathcal{V}, +)$ a **commutative group**. In textbooks on algebra it is proved that there is not a second element which has the property (iii); we shall refer to $\underline{o}$ as the **null vector**. It is also a standard result that for each $\underline{a} \in \mathcal{V}$ there is not a second element with the property (iv); we call $-\underline{a}$ **the inverse** of $\underline{a}$. **Subtraction** $-$ is defined by specifying the difference $\underline{b} - \underline{a} = \underline{b} + (-\underline{a})$; then $-$ is a binary operation on $\mathcal{V}$. If $\underline{a} = (O, Z_1)$, $\underline{b} = (O, Z_2)$, then $-\underline{a} = (O, Z_3)$ where $Z_3 \equiv (-x_1, -y_1)$, and consequently $\underline{b} - \underline{a} = (O, Z_4)$ where $Z_4 \equiv (x_2 - x_1, y_2 - y_1)$. Thus $(O, Z_2) - (O, Z_1) = (O, Z_4)$ if and only if $(Z_1, Z_2) \uparrow (O, Z_4)$.

COMMENT. Now that we have set up our couples we call (O, Z) a **position vector** with respect to the point O, and we adopt the standard notation $\overrightarrow{OZ}$ for (O, Z).

Position vectors can be used for many geometrical purposes instead of Cartesian coordinates, or complex coordinates and complex-valued distances. We would note that by 6.1.1(iv) and 11.2 $Z_0 = \text{mp}(Z_1, Z_2)$ if and only if $\overrightarrow{OZ_0} = \frac{1}{2}(\overrightarrow{OZ_1} + \overrightarrow{OZ_2})$; by 10.1.1(v) that $Z_1 Z_2 \parallel Z_3 Z_4$ if and only if $\overrightarrow{OZ_4} - \overrightarrow{OZ_3} = t(\overrightarrow{OZ_2} - \overrightarrow{OZ_1})$ for some $t \neq 0$ in $\mathbf{R}$, and by 9.7.1(ii) and 6.5.1 Corollary (ii) that $Z_1 Z_2 \perp Z_3 Z_4$ if and only if $(\overrightarrow{OZ_1} - \overrightarrow{OZ_2}).(\overrightarrow{OZ_3} - \overrightarrow{OZ_4}) = 0$. Most importantly, from parametric equations of a line $x = x_1 + t(x_2 - x_1)$, $y = y_1 + t(y_2 - y_1)$ $(t \in \mathbf{R})$, we have that $Z \in Z_1 Z_2$ if and only if

$$\overrightarrow{OZ} = \overrightarrow{OZ_1} + t(\overrightarrow{OZ_2} - \overrightarrow{OZ_1}) = (1 - t)\overrightarrow{OZ_1} + t\overrightarrow{OZ_2} \tag{11.3.1}$$

for some $t \in \mathbf{R}$;

COMMENT. It is usual, in modern treatments, to define vectors to be the equivalence classes for equipollence. This defines free vectors. Position vectors are then defined by taking a specific point O in Π so that we have a pointed plane, and then concentrating on the representatives of the form (O, Z) for the vectors. But if our objective is to introduce position vectors, it is wasteful of effort to set up the free vectors, and in fact the use of free vectors and subsequent specialisation to position vectors can be a confusing route to position vectors.

11.4   COMPONENTS OF A VECTOR

11.4.1   Components

Given non-collinear points Z_1, Z_2, Z_3, we wish to obtain an expression

$$\overrightarrow{Z_1 Z} = p\overrightarrow{Z_1 Z_2} + q\overrightarrow{Z_1 Z_3}.$$

For this we need

$$(x_2 - x_1)p + (x_3 - x_1)q = x - x_1,$$
$$(y_2 - y_1)p + (y_3 - y_1)q = y - y_1.$$

We obtain the solutions

$$p = \frac{\delta_{\mathcal{F}}(Z_1, Z, Z_3)}{\delta_{\mathcal{F}}(Z_1, Z_2, Z_3)}, \quad q = \frac{\delta_{\mathcal{F}}(Z_1, Z_2, Z)}{\delta_{\mathcal{F}}(Z_1, Z_2, Z_3)},$$

and so have

$$\overrightarrow{Z_1 Z} = \frac{\delta_{\mathcal{F}}(Z_1, Z, Z_3)}{\delta_{\mathcal{F}}(Z_1, Z_2, Z_3)}\overrightarrow{Z_1 Z_2} + \frac{\delta_{\mathcal{F}}(Z_1, Z_2, Z)}{\delta_{\mathcal{F}}(Z_1, Z_2, Z_3)}\overrightarrow{Z_1 Z_3}.$$

11.4.2   Areal coordinates

Given non-collinear points Z_1, Z_2, Z_3, the position vector of any point Z of the plane can be expressed in the form $\overrightarrow{OZ} = p\overrightarrow{OZ_1} + q\overrightarrow{OZ_2} + r\overrightarrow{OZ_3}$, with $p + q + r = 1$. This is equivalent to having q, r such that

$$q(x_2 - x_1) + r(x_3 - x_1) = x - x_1,$$
$$q(y_2 - y_1) + r(y_3 - y_1) = y - y_1.$$

These equations have the unique solution

$$q = \frac{\delta_{\mathcal{F}}(Z, Z_3, Z_1)}{\delta_{\mathcal{F}}(Z_1, Z_2, Z_3)}, \quad r = \frac{\delta_{\mathcal{F}}(Z, Z_1, Z_2)}{\delta_{\mathcal{F}}(Z_1, Z_2, Z_3)},$$

and now we take $p = 1 - q - r$ so that by 10.5.4

$$p = \frac{\delta_{\mathcal{F}}(Z, Z_2, Z_3)}{\delta_{\mathcal{F}}(Z_1, Z_2, Z_3)}.$$

For non-collinear points Z_1, Z_2, Z_3, for any Z we write

$$\alpha = \delta_{\mathcal{F}}(Z, Z_2, Z_3), \ \beta = \delta_{\mathcal{F}}(Z, Z_3, Z_1), \ \gamma = \delta_{\mathcal{F}}(Z, Z_1, Z_2),$$

and call (α, β, γ) *areal point coordinates* of Z with respect to (Z_1, Z_2, Z_3). Note that we have

$$p = \frac{\alpha}{\delta_{\mathcal{F}}(Z_1, Z_2, Z_3)}, \ q = \frac{\beta}{\delta_{\mathcal{F}}(Z_1, Z_2, Z_3)}, \ r = \frac{\gamma}{\delta_{\mathcal{F}}(Z_1, Z_2, Z_3)},$$

and $\alpha + \beta + \gamma = \delta_{\mathcal{F}}(Z_1, Z_2, Z_3)$. These were first used by Möbius in 1827.

11.4.3 Cartesian coordinates from areal coordinates

With the notation in 11.4.2, we have

$$(y_2 - y_3)x - (x_2 - x_3)y = 2\alpha - x_2 y_3 + x_3 y_2,$$
$$(y_3 - y_1)x - (x_3 - x_1)y = 2\beta - x_3 y_1 + x_1 y_3,$$

and if we solve these we obtain

$$x = \frac{x_1 \alpha + x_2 \beta + x_3 \gamma}{\delta_{\mathcal{F}}(Z_1, Z_2, Z_3)}, \ y = \frac{y_1 \alpha + y_2 \beta + y_3 \gamma}{\delta_{\mathcal{F}}(Z_1, Z_2, Z_3)}.$$

11.4.4

The representation in 11.4.2 is in fact independent of the origin O. For we have

$$x = px_1 + qx_2 + rx_3, \ y = py_1 + qy_2 + ry_3,$$

and so for any point $Z_0 \equiv_{\mathcal{F}} (x_0, y_0)$,

$$x - x_0 = p(x_1 - x_0) + q(x_2 - x_0) + r(x_3 - x_0),$$
$$y - y_0 = p(y_1 - y_0) + q(y_2 - y_0) + r(y_3 - y_0).$$

But $Z \equiv_{\mathcal{F}'} (x - x_0, y - y_0)$, where $\mathcal{F}' = t_{0, z_0}(\mathcal{F})$. Hence $\overrightarrow{Z_0 Z} = p\overrightarrow{Z_0 Z_1} + q\overrightarrow{Z_0 Z_2} + r\overrightarrow{Z_0 Z_3}$, with $p + q + r = 1$.

NOTATION. Where a vector equation is independent of the origin, as in $\overrightarrow{OZ} = p\overrightarrow{OZ_1} + q\overrightarrow{OZ_2} + r\overrightarrow{OZ_3}$, with $p + q + r = 1$, it is convenient to write this as $Z = pZ_1 + qZ_2 + rZ_3$ with $p + q + r = 1$. In particular, in (11.3.1) we write $Z = (1-t)Z_1 + tZ_2$.

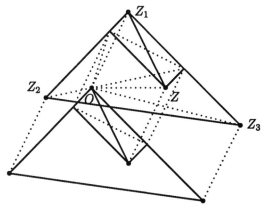

Figure 11.5.

Figure 11.5 caters for when O and Z_1 are taken as origins, a similar diagram would cater for when Z_0 and Z_1 are origins, and then a combination of the two would give the stated result.

11.4.5

We also use the notation $\delta_{\mathcal{F}}(Z_1, Z_2, pZ_4 + qZ_5 + rZ_6)$ for $\delta_{\mathcal{F}}(Z_1, Z_2, Z_3)$ where $\overrightarrow{OZ_3} = p\overrightarrow{OZ_4} + q\overrightarrow{OZ_5} + r\overrightarrow{OZ_6}$ and $p + q + r = 1$. We can then write the conclusion of 10.5.3 as

$$\delta_{\mathcal{F}}(Z_1, Z_2, (1-s)Z_4 + sZ_5) = (1-s)\delta_{\mathcal{F}}(Z_1, Z_2, Z_4) + s\delta_{\mathcal{F}}(Z_1, Z_2, Z_5).$$

The more general result

$$\delta_{\mathcal{F}}(Z_1, Z_2, pZ_4 + qZ_5 + rZ_6) = p\delta_{\mathcal{F}}(Z_1, Z_2, Z_4) + q\delta_{\mathcal{F}}(Z_1, Z_2, Z_5) + r\delta_{\mathcal{F}}(Z_1, Z_2, Z_6),$$

where $p + q + r = 1$, can be deduced from this. For

$$
\begin{aligned}
& \delta_{\mathcal{F}}(Z_1, Z_2, pZ_4 + qZ_5 + rZ_6) \\
= {}& \delta_{\mathcal{F}}\left(Z_1, Z_2, pZ_4 + (1-p)\left(\frac{q}{1-p}Z_5 + \frac{r}{1-p}Z_6\right)\right) \\
= {}& p\delta_{\mathcal{F}}(Z_1, Z_2, Z_4) + (1-p)\delta_{\mathcal{F}}\left(Z_1, Z_2, \frac{q}{1-p}Z_5 + \frac{r}{1-p}Z_6\right) \\
= {}& p\delta_{\mathcal{F}}(Z_1, Z_2, Z_4) + (1-p)\left[\frac{q}{1-p}\delta_{\mathcal{F}}(Z_1, Z_2, Z_5) + \frac{r}{1-p}\delta_{\mathcal{F}}(Z_1, Z_2, Z_6)\right].
\end{aligned}
$$

In this we have used the fact that

$$\frac{q}{1-p} + \frac{r}{1-p} = \frac{q+r}{1-p} = \frac{1-p}{1-p} = 1.$$

11.5   VECTOR METHODS IN GEOMETRY

There is an informative account of many of the results of this chapter contained in Coxeter and Greitzer [5], dealt with by the methods of pure geometry.

Some results are very basic, involving just collinearities or concurrencies, or ratio results. We start by showing how vector notation can be used to prove such results in a very straightforward fashion.

11.5.1 Menelaus' theorem, c.100A.D.

For non-collinear points Z_1, Z_2 and Z_3, let $Z_4 \in Z_2 Z_3$, $Z_5 \in Z_3 Z_1$ and $Z_6 \in Z_1 Z_2$. Then Z_4, Z_5 and Z_6 are collinear if and only if

$$\frac{\overline{Z_2 Z_4}}{\overline{Z_4 Z_3}} \frac{\overline{Z_3 Z_5}}{\overline{Z_5 Z_1}} \frac{\overline{Z_1 Z_6}}{\overline{Z_6 Z_2}} = -1.$$

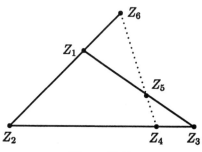

Figure 11.6.

Proof. Let $Z_4 = (1-r)Z_2 + rZ_3$, $Z_5 = (1-s)Z_3 + sZ_1$, $Z_6 = (1-t)Z_1 + tZ_2$.

Since Z_4, Z_5 and Z_6 are collinear, we have that $Z_6 = (1-u)Z_4 + uZ_5$, for some real number u. Then

$$(1-t)Z_1 + tZ_2 = (1-u)[(1-r)Z_2 + rZ_3] + u[(1-s)Z_3 + sZ_1].$$

As the coefficients on each side add to 1, by the uniqueness in 11.4.2 we can equate coefficients and thus obtain

$$1 - t = su, \ t = (1-u)(1-r), \ r(1-u) = -u(1-s).$$

On eliminating u we obtain

$$\frac{r}{1-s}\frac{s}{1-t}\frac{t}{1-r} = -\frac{u}{1-u}\frac{1}{u}(1-u) = -1,$$

and so

$$\frac{r}{1-r}\frac{s}{1-s}\frac{t}{1-t} = -1.$$

This yields the stated result.

This is known as MENELAUS' THEOREM.

11.5.2 Ceva's theorem and converse, 1678

For non-collinear points Z_1, Z_2 and Z_3, let $Z_4 \in Z_2 Z_3$, $Z_5 \in Z_3 Z_1$ and $Z_6 \in Z_1 Z_2$. If $Z_1 Z_4, Z_2 Z_5$ and $Z_3 Z_6$ are concurrent, then

$$\frac{\overline{Z_2 Z_4}}{\overline{Z_4 Z_3}} \frac{\overline{Z_3 Z_5}}{\overline{Z_5 Z_1}} \frac{\overline{Z_1 Z_6}}{\overline{Z_6 Z_2}} = 1. \tag{11.5.1}$$

Proof. Denoting the point of concurrency by Z_0, we have

$$\begin{aligned} Z_4 &= (1-u)Z_0 + uZ_1 = (1-r)Z_2 + rZ_3, \\ Z_5 &= (1-v)Z_0 + vZ_2 = (1-s)Z_3 + sZ_1, \\ Z_6 &= (1-w)Z_0 + wZ_3 = (1-t)Z_1 + tZ_2, \end{aligned}$$

for some $u, v, w, r, s, t \in \mathbf{R}$. Then

$$Z_0 = -\frac{u}{1-u}Z_1 + \frac{1-r}{1-u}Z_2 + \frac{r}{1-u}Z_3,$$

$$Z_0 = \frac{s}{1-v}Z_1 - \frac{v}{1-v}Z_2 + \frac{1-s}{1-v}Z_3,$$

$$Z_0 = \frac{1-t}{1-w}Z_1 + \frac{t}{1-w}Z_2 - \frac{w}{1-w}Z_3.$$

On equating the coefficients of Z_1, Z_2 and Z_3, in turn, we obtain

$$-\frac{u}{1-u} = \frac{s}{1-v} = \frac{1-t}{1-w},$$

$$\frac{1-r}{1-u} = -\frac{v}{1-v} = \frac{t}{1-w},$$

$$\frac{r}{1-u} = \frac{1-s}{1-v} = -\frac{w}{1-w}.$$

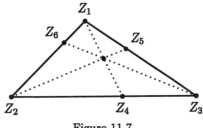

Figure 11.7.

From this

$$\frac{s}{1-t} = \frac{1-v}{1-w}, \ \frac{t}{1-r} = \frac{1-w}{1-u}, \ \frac{r}{1-s} = \frac{1-u}{1-v},$$

and so by multiplication

$$\frac{s}{1-t}\frac{t}{1-r}\frac{r}{1-s} = 1.$$

Thus we obtain our conclusion. This is known as CEVA'S THEOREM.
   In fact we also have that

$$\frac{u}{s} = -\frac{1-u}{1-v}, \ \frac{v}{t} = -\frac{1-v}{1-w}, \ \frac{w}{r} = -\frac{1-w}{1-u},$$

which gives $uvw = -rst$. This is

$$\frac{\overline{Z_0 Z_4}}{\overline{Z_0 Z_1}}\frac{\overline{Z_0 Z_5}}{\overline{Z_0 Z_2}}\frac{\overline{Z_0 Z_6}}{\overline{Z_0 Z_3}} = -\frac{\overline{Z_2 Z_4}}{\overline{Z_2 Z_3}}\frac{\overline{Z_3 Z_5}}{\overline{Z_3 Z_1}}\frac{\overline{Z_1 Z_6}}{\overline{Z_1 Z_2}}.$$

CONVERSE of CEVA'S THEOREM. *Conversely, for non-collinear points* Z_1, Z_2 *and* Z_3*, let* $Z_4 \in Z_2 Z_3$*,* $Z_5 \in Z_3 Z_1$ *and* $Z_6 \in Z_1 Z_2$*. If (11.5.1) holds and* $Z_2 Z_5$ *and* $Z_3 Z_6$ *meet at a point* Z_0*, then* $Z_1 Z_4$ *also passes through* Z_0*.*
   To start our proof we note that we have

$$Z_5 = (1-v)Z_0 + vZ_2 = (1-s)Z_3 + sZ_1,$$

$$Z_6 = (1-w)Z_0 + wZ_3 = (1-t)Z_3 + tZ_1.$$

Hence

$$Z_0 = \frac{s}{1-v}Z_1 - \frac{v}{1-v}Z_2 + \frac{1-s}{1-v}Z_3, \ Z_0 = \frac{1-t}{1-w}Z_1 + \frac{t}{1-w}Z_2 - \frac{w}{1-w}Z_3.$$

It follows that

$$\frac{s}{1-v} = \frac{1-t}{1-w}, \quad -\frac{v}{1-v} = \frac{t}{1-w}, \quad \frac{1-s}{1-v} = -\frac{w}{1-w},$$

from which

$$\frac{s}{1-t} = \frac{1-v}{1-w}, \quad (1-s)t = vw.$$

On eliminating s between these, we obtain $(1-v)t^2 + (v-w)t - vw(1-w) = 0$. We then obtain two pairs of solutions, $t = w$, $s = 1 - v$, and

$$t = -v\frac{1-w}{1-v}, \quad s = \frac{1-vw}{1-w}.$$

The first pair of solutions leads to $v = w = 0$ and so $Z_5 = Z_6 = Z_0 = Z_1$, which we regard as a degenerate case.

With $Z_4 = (1-r)Z_2 + rZ_3$, we are given that

$$\frac{1-r}{r} = \frac{st}{(1-s)(1-t)},$$

and so have

$$Z_4 = \frac{st}{st + (1-s)(1-t)}Z_2 + \frac{(1-s)(1-t)}{st + (1-s)(1-t)}Z_3.$$

With the second pair of solutions above, we obtain that

$$Z_0 = \frac{1-vw}{(1-v)(1-w)}Z_1 - \frac{v(1-w)}{(1-v)(1-w)}Z_2 \frac{w(1-v)}{(1-v)(1-w)}Z_3,$$

and also that

$$Z_4 = \frac{v(1-w)}{v+w-2vw}Z_2 + \frac{w(1-v)}{v+w-2vw}Z_3,$$

so that

$$Z_0 = \frac{1-vw}{(1-v)(1-w)}Z_1 - \frac{v+w-2vw}{(1-v)(1-w)}Z_4.$$

As the sum of the coefficients of Z_1 and Z_4 is equal to 1, Z_1Z_4 passes through Z_0. This proves the result.

To obtain a formula for Z_0 we note that on solving the second pair of solutions above for v and w, we obtain the pair of solutions

$$v = 1, \ w = 1; \quad v = -\frac{st}{1-t}, \quad w = -\frac{(1-s)(1-t)}{s}.$$

To see this, note that

$$1 - w = -\frac{1-v}{v}t, \quad w = 1 + \frac{1-v}{v}t,$$

so that

$$\frac{1 - v\left(1 + \frac{1-v}{v}t\right)}{\frac{1-v}{v}t} = s, \quad \text{i.e.} \quad \frac{1 - v - (1-v)t}{-\frac{1-v}{v}t} = s.$$

Thus either $v = 1$ and consequently $w = 1$, or

$$\frac{1-t}{-\frac{t}{v}} = s, \text{ i.e. } v = \frac{st}{1-t},$$

and hence

$$w = -\frac{(1-s)(1-t)}{s}.$$

The first pair lead to $Z_5 = Z_2$, $Z_6 = Z_3$, another degenerate case, while the second pair lead to

$$Z_0 = \frac{s(1-t)}{1-t+st}Z_1 + \frac{st}{1-t+st}Z_2 + \frac{(1-s)(1-t)}{1-t+st}Z_3. \qquad (11.5.2)$$

Because of the condition (11.5.1) the coefficients in (11.5.2) could be given in several different forms.

11.5.3   Desargues' perspective theorem, 1648

Let (Z_1, Z_2, Z_3) and (Z_4, Z_5, Z_6) be two pairs of non-collinear points. Let Z_2Z_3 and Z_5Z_6 meet at W_1, Z_3Z_1 and Z_6Z_4 meet at W_2, and Z_1Z_2 and Z_4Z_5 meet at W_3. Then W_1, W_2, W_3 are collinear if and only if Z_1Z_4, Z_2Z_5, Z_3Z_6 are concurrent.

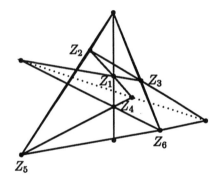

Figure 11.8.

Proof.   Suppose that Z_1Z_4, Z_2Z_5, Z_3Z_6 meet at a point Z_0. Then

$$\begin{aligned}
Z_4 &= (1-u)Z_0 + uZ_1, \\
Z_5 &= (1-v)Z_0 + vZ_2, \\
Z_6 &= (1-w)Z_0 + wZ_3,
\end{aligned}$$

for some $u, v, w, \in \mathbf{R}$.

On eliminating Z_0 between the second and third of these, we obtain that

$$(1-w)Z_5 - (1-v)Z_6 = v(1-w)Z_2 - w(1-v)Z_3,$$

from which we obtain that

$$\frac{1-w}{v-w}Z_5 - \frac{1-v}{v-w}Z_6 = \frac{v(1-w)}{v-w}Z_2 - \frac{w(1-v)}{v-w}Z_3.$$

Now the sum of the coefficients of Z_5 and Z_6 is equal to 1, so the left-hand side represents a point on the line Z_5Z_6. Similarly, the sum of the coefficients of Z_2 and Z_3 is equal to 1, so the right-hand side represents a point on the line Z_2Z_3. Thus this must be the point W_1.

By a similar argument based on the third and first lines, we find that

$$\frac{1-u}{w-u}Z_6 - \frac{1-w}{w-u}Z_4 = \frac{w(1-u)}{w-u}Z_3 - \frac{u(1-w)}{w-u}Z_1$$

must be the point W_2, and by a similar argument based on the first and second lines, we find that

$$\frac{1-v}{u-v}Z_4 - \frac{1-u}{u-v}Z_5 = \frac{u(1-v)}{u-v}Z_1 - \frac{v(1-u)}{u-v}Z_2$$

must be the point W_3.

Then by repeated use of 10.5.3 and 11.4.5

$$\delta_{\mathcal{F}}(W_1, W_2, W_3)$$

$$= \delta_{\mathcal{F}}\left(\frac{v(1-w)}{v-w}Z_2 - \frac{w(1-v)}{v-w}Z_3, -\frac{u(1-w)}{w-u}Z_1 + \frac{w(1-u)}{w-u}Z_3,\right.$$

$$\left.\frac{u(1-v)}{u-v}Z_1 - \frac{v(1-u)}{u-v}Z_2\right)$$

$$= \left[\frac{v(1-w)}{v-w}\frac{w(1-u)}{w-u}\frac{u(1-v)}{u-v} - \frac{w(1-v)}{v-w}\frac{-u(1-w)}{w-u}\frac{-v(1-u)}{u-v}\right]\delta_{\mathcal{F}}(Z_1, Z_2, Z_3)$$

$$= 0.$$

This shows that W_1, W_2 and W_3 are collinear.

This is known as DESARGUES' PERSPECTIVE OR TWO-TRIANGLE THEOREM. Conversely, let

$$W_1 = (1-l)Z_2 + lZ_3 = (1-m)Z_5 + mZ_6,$$
$$W_2 = (1-p)Z_3 + pZ_1 = (1-q)Z_6 + qZ_4,$$
$$W_3 = (1-r)Z_1 + rZ_2 = (1-s)Z_4 + sZ_5.$$

From the third of these we deduce that $(1-r)Z_1 - (1-s)Z_4 = sZ_5 - rZ_2$, and from this

$$\frac{1-r}{s-r}Z_1 - \frac{1-s}{s-r}Z_4 = \frac{s}{s-r}Z_5 - \frac{r}{s-r}Z_2,$$

so that this must be the point of intersection of Z_1Z_4 and Z_2Z_5.

By a similar argument, we deduce from the second equation that

$$\frac{1-l}{m-l}Z_2 - \frac{1-m}{m-l}Z_5 = \frac{m}{m-l}Z_6 - \frac{l}{m-l}Z_3,$$

and so this must be the point of intersection of Z_2Z_5 and Z_3Z_6. By a similar argument, we deduce from the first equation that

$$\frac{1-p}{q-p}Z_3 - \frac{1-p}{q-p}Z_6 = \frac{q}{q-p}Z_4 - \frac{p}{q-p}Z_1,$$

and so this must be the point of intersection of Z_3Z_6 and Z_1Z_4.

We are given now that W_1, W_2 and W_3 are collinear, so that $W_3 = (1-t)W_1 + tW_2$, for some $t \in \mathbf{R}$. Then

$$(1-t)[(1-l)Z_2 + lZ_3] + t[(1-p)Z_3 + pZ_1] = (1-r)Z_1 + rZ_2,$$
$$(1-t)[(1-m)Z_5 + mZ_6] + t[(1-q)Z_6 + qZ_4] = (1-s)Z_4 + sZ_5.$$

Since the points Z_1, Z_2, Z_3 are not collinear we can equate the coefficients in the first line here, and obtain that

$$pt = 1 - r, \ (1 - t)(1 - l) = r, \ (1 - t)l + t(1 - p) = 0,$$

and since the points Z_4, Z_5, Z_6 are not collinear we can equate the coefficients in the second line, and obtain that

$$qt = 1 - s, \ (1 - t)(1 - m) = s, \ (1 - t)m + t(1 - q) = 0.$$

Now for $Z_2 Z_5$ and $Z_3 Z_6$ to meet $Z_1 Z_4$ in the same point, we need to have

$$\frac{1 - r}{s - r} = -\frac{p}{q - p},$$

and from this

$$\frac{1 - r}{p} = -\frac{s - r}{q - p}.$$

But we have from above

$$\frac{1 - r}{p} = \frac{1 - s}{q},$$

as a common value of t, and so need

$$\frac{1 - s}{q} = -\frac{s - r}{q - p}$$

or equivalently $q(1 - r) = p(1 - s)$, and we have already noted that this is so.

It follows that $Z_1 Z_4, Z_2 Z_5$ and $Z_3 Z_6$ are concurrent.

11.5.4   Pappus' theorem, c.300A.D.

Let the points Z_1, Z_2, Z_3 lie on one line, and the points Z_4, Z_5, Z_6 lie on a second line, these two lines intersecting at some point Z_0. Suppose that $Z_2 Z_6$ and $Z_5 Z_3$ meet at W_1, $Z_3 Z_4$ and $Z_6 Z_1$ meet at W_2, and $Z_1 Z_5$ and $Z_4 Z_2$ meet at W_3. Then the points W_1, W_2, W_3 are collinear.

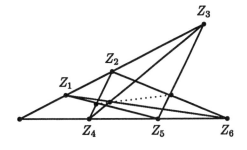

Figure 11.9.

Proof. We have that

$$Z_2 = (1 - p)Z_0 + pZ_1, \ Z_3 = (1 - q)Z_0 + qZ_1,$$
$$Z_5 = (1 - u)Z_0 + uZ_4, \ Z_6 = (1 - v)Z_0 + vZ_4,$$

for some $p, q, u, v \in \mathbf{R}$. On eliminating Z_0 from the equations for Z_2 and Z_5, we find that

$$(1 - u)Z_2 - (1 - p)Z_5 = p(1 - u)Z_1 - u(1 - p)Z_4,$$

and so

$$\frac{1-u}{1-pu}Z_2 + \frac{u(1-p)}{1-pu}Z_4 = \frac{p(1-u)}{1-pu}Z_1 + \frac{1-p}{1-pu}Z_5.$$

This must be the point W_3 then. Similarly, on eliminating Z_0 from the equations for Z_3 and Z_6 we have that

$$(1-v)Z_3 - (1-q)Z_6 = q(1-v)Z_1 - (1-q)vZ_4,$$

and so

$$\frac{1-v}{1-qv}Z_3 + \frac{(1-q)v}{1-qv}Z_4 = \frac{q(1-v)}{1-qv}Z_1 + \frac{1-q}{1-qv}Z_6.$$

This must be the point W_2 then.

Now from the equations for Z_2 and Z_3 we have that $pZ_3 - qZ_2 = (p-q)Z_0$, and from the equations for Z_5 and Z_6 we have that $uZ_6 - vZ_5 = (u-v)Z_0$. On combining these, we have that

$$(v-u)(pZ_3 - qZ_2) = (q-p)(uZ_6 - vZ_5).$$

From this we have that

$$\frac{u(q-p)}{qv-pu}Z_6 + \frac{q(v-u)}{qv-pu}Z_2 = \frac{p(v-u)}{qv-pu}Z_3 + \frac{(q-p)v}{qv-pu}Z_5.$$

This must then be the point W_1.

However, the left-hand sides of the representations for W_1, W_2 and W_3 contain four points Z_1, Z_2, Z_5, Z_6 and we wish to reduce this to three non-collinear points. For this purpose we eliminate Z_5. From the equations for Z_5 and Z_6 we have that $uZ_6 - vZ_5 = (u-v)Z_0$, while from the equation for Z_2 we have $Z_2 - pZ_1 = (1-p)Z_0$. Combining these gives

$$Z_5 = \frac{u}{v}Z_6 - \frac{(u-v)}{v(1-p)}(Z_2 - pZ_1).$$

On substitution, this gives that

$$W_3 = \frac{pu(1-v)}{v(1-pu)}Z_1 + \frac{v-u}{v(1-pu)}Z_2 + \frac{u(1-p)}{v(1-pu)}Z_6.$$

We note that the sum of the coefficients for each of W_1, W_2, W_3 in terms of Z_1, Z_2 and Z_6 is equal to 1, and so by repeated use of 10.5.3 and 11.4.5 we have that $\delta_{\mathcal{F}}(W_1, W_2, W_3)$ is equal to

$$\delta_{\mathcal{F}}\left(\frac{q(v-u)}{qv-pu}Z_2 + \frac{u(q-p)}{qv-pu}Z_6, \frac{q(1-v)}{1-qv}Z_1 + \frac{1-q}{1-qv}Z_6,\right.$$

$$\left.\frac{pu(1-v)}{v(1-pu)}Z_1 + \frac{v-u}{v(1-pu)}Z_2 + \frac{u(1-p)}{v(1-pu)}Z_6\right)$$

$$= \frac{q(v-u)}{qv-pu}\frac{q(1-v)}{1-qv}\frac{u(1-p)}{v(1-pu)}\delta_{\mathcal{F}}(Z_2, Z_1, Z_6) + \frac{q(v-u)}{qv-pu}\frac{1-q}{1-qv}\frac{pu(1-v)}{v(1-pu)}\delta_{\mathcal{F}}(Z_2, Z_6, Z_1)$$

$$+ \frac{u(q-p)}{qv-pu}\frac{q(1-v)}{1-qv}\frac{v-u}{v(1-pu)}\delta_{\mathcal{F}}(Z_6, Z_1, Z_2)$$

$$= \frac{qu(1-v)(v-u)}{v(qv-pu)(1-qv)((1-pu)}[-q(1-p)+p(1-q)+q-p]\delta_{\mathcal{F}}(Z_1, Z_2, Z_6) = 0.$$

This shows that W_1, W_2 and W_3 are collinear.

This is known as PAPPUS' THEOREM.

11.5.5   Centroid of a triangle

If Z_4, Z_5, Z_6 are the mid-points of $\{Z_2, Z_3\}$, $\{Z_3, Z_1\}$, $\{Z_1, Z_2\}$, respectively, then with the notation of 11.5.2 we have that $r = s = t = \frac{1}{2}$, and the condition (11.5.1) in the converse of Ceva's theorem holds. Note that $[Z_3, Z_6]$ is a cross-bar for the interior region $\mathcal{IR}(|Z_3 Z_2 Z_1)$ and so $[Z_2, Z_5$ meets $[Z_3, Z_6]$ in a point Z_0, which is thus on both $Z_2 Z_5$ and $Z_3 Z_6$. It follows that it is also on $Z_1 Z_4$. *Thus the lines joining the vertices of a triangle to the mid-points of the opposite sides are concurrent.* The point of concurrence Z_0 is called the **centroid** of the triangle, and for it by 11.5.2 we have

$$Z_0 = \tfrac{1}{3}Z_1 + \tfrac{1}{3}Z_2 + \tfrac{1}{3}Z_3.$$

11.5.6   Orthocentre of a triangle

Let Z_4, Z_5, Z_6 be the feet of the perpendiculars from Z_1 to $Z_2 Z_3$, Z_2 to $Z_3 Z_1$, Z_3 to $Z_1 Z_2$, respectively. Then with the notation of 11.5.2 we have that

$$r = \frac{c}{a}\cos\beta, \; s = \frac{a}{b}\cos\gamma, \; t = \frac{b}{c}\cos\alpha.$$

Hence

$$1 - r = \frac{a - c\cos\beta}{a} = \frac{c\cos\beta + b\cos\gamma - c\cos\beta}{a} = \frac{b}{a}\cos\gamma.$$

Similarly

$$1 - s = \frac{c}{b}\cos\alpha, \; 1 - t = \frac{a}{c}\cos\beta.$$

It follows that the condition (11.5.1) in the converse of Ceva's theorem is true.

Repeating an argument that we used in 7.2.3, suppose now that m and n are any lines which are perpendicular to $Z_3 Z_1$ and $Z_1 Z_2$, respectively. If we had $m \parallel n$, then we would have $m \perp Z_3 Z_1$, $n \parallel m$ so that $n \perp Z_3 Z_1$; but already $n \perp Z_1 Z_2$ so $Z_3 Z_1 \parallel Z_1 Z_2$; as Z_1, Z_2, Z_3 are not collinear, this gives a contradiction.

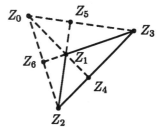

Figure 11.10. Orthocentre of triangle.

Thus m is not parallel to n and so these lines meet in a unique point Z_0. In particular the lines $Z_2 Z_5, Z_3 Z_6$ must meet in a unique point Z_0 and then by the converse of Ceva's theorem, $Z_1 Z_4$ will pass through Z_0. *Thus the lines through the vertices of a triangle which are perpendicular to the opposite side-lines are concurrent.* The point of concurrence Z_0 is called the **orthocentre** of the triangle.

By (11.5.2) we thus have

$$Z_0 =$$

$$\frac{\frac{a}{b}\cos\gamma\frac{a}{c}\cos\beta}{\frac{a}{c}\cos\beta + \frac{a}{c}\cos\gamma\cos\alpha}Z_1 + \frac{\frac{a}{c}\cos\gamma\cos\alpha}{\frac{a}{c}\cos\beta + \frac{a}{c}\cos\gamma\cos\alpha}Z_2 + \frac{\frac{a}{b}\cos\alpha\cos\beta}{\frac{a}{c}\cos\beta + \frac{a}{c}\cos\gamma\cos\alpha}Z_3$$

$$= \frac{a}{b}\frac{\cos\beta\cos\gamma}{\cos\beta + \cos\gamma\cos\alpha}Z_1 + \frac{\cos\gamma\cos\alpha}{\cos\beta + \cos\gamma\cos\alpha}Z_2 + \frac{c}{b}\frac{\cos\alpha\cos\beta}{\cos\beta + \cos\gamma\cos\alpha}Z_2.$$

We could also proceed in this special case as follows. The argument is laid out for the case in the diagram, with β and γ acute angles and (Z_1, Z_2, Z_3) positive in orientation. The other cases can be treated similarly.

Now by 5.2.2 applied to $[Z_6, Z_2, Z_3]$, $|\angle Z_0 Z_2 Z_4|^\circ = |\angle Z_6 Z_2 Z_3|^\circ = 90 - |\gamma|^\circ$ so that

$$\frac{|Z_0, Z_4|}{|Z_2, Z_4|} = \tan\angle Z_0 Z_2 Z_4 = \cot\gamma.$$

But $|Z_2, Z_4| = c\cos\beta$ and so $|Z_0, Z_4| = c\cos\beta\cot\gamma$. Thus $\delta_{\mathcal{F}}(Z_0, Z_2, Z_3) = \frac{1}{2}ac\cos\beta\cot\gamma$ and since $\delta_{\mathcal{F}}(Z_1, Z_2, Z_3) = \frac{1}{2}ac\sin\beta$, we have that

$$\frac{\delta_{\mathcal{F}}(Z_0, Z_2, Z_3)}{\delta_{\mathcal{F}}(Z_1, Z_2, Z_3)} = \cot\beta\cot\gamma.$$

As similar results hold in the other two cases, we have by 11.4.2 that

$$Z_0 = \cot\beta\cot\gamma Z_1 + \cot\gamma\cot\alpha Z_2 + \cot\alpha\cot\beta Z_3.$$

That the sum of the coefficients is equal to 1 follows from the identity

$$\tan\alpha + \tan\beta + \tan\gamma = \tan\alpha\tan\beta\tan\gamma$$

for the angles of a triangle. For, using the notation of 10.8.1, $|\alpha|^\circ + |\beta|^\circ + |\gamma|^\circ = 180$, and so $\alpha_{\mathcal{F}} + \beta_{\mathcal{F}} + \gamma_{\mathcal{F}} = 180_{\mathcal{F}}$ so that

$$-\tan\gamma_{\mathcal{F}} = \tan(180_{\mathcal{F}} - \gamma_{\mathcal{F}}) = \tan(\alpha_{\mathcal{F}} + \beta_{\mathcal{F}})$$

$$= \frac{\tan\alpha_{\mathcal{F}} + \tan\beta_{\mathcal{F}}}{1 - \tan\alpha_{\mathcal{F}}\tan\beta_{\mathcal{F}}},$$

whence the result follows by multiplying across and rearranging. This formula fails in the case of a right-angled triangle.

From our two methods we have two formulae for Z_0, but we further note that

$$\cos(\alpha_{\mathcal{F}} + \gamma_{\mathcal{F}}) = \cos(180_{\mathcal{F}} - \beta_{\mathcal{F}}),$$

$$\cos\alpha_{\mathcal{F}}\cos\gamma_{\mathcal{F}} - \sin\alpha_{\mathcal{F}}\sin\gamma_{\mathcal{F}} = -\cos\beta_{\mathcal{F}},$$

$$\cos\alpha\cos\gamma + \cos\beta = \sin\alpha\sin\gamma.$$

On using this with the sine rule, the two formulae for the orthocentre are reconciled.

11.5.7   Incentre of a triangle

Let Z_1, Z_2, Z_3 be non-collinear points. By 5.5.1 the mid-line of $|\underline{Z_2 Z_1 Z_3}$ will meet $[Z_2, Z_3]$ in a point Z_4 where $Z_4 = (1-r)Z_2 + rZ_3$, and

$$\frac{r}{1-r} = \frac{c}{b}.$$

By similar arguments the mid-line of $|\underline{Z_3 Z_2 Z_1}$ will meet $[Z_3, Z_1]$ in a point Z_5 where $Z_5 = (1-s)Z_3 + sZ_1$, and

$$\frac{s}{1-s} = \frac{a}{c},$$

and the mid-line of $|\underline{Z_1 Z_3 Z_2}$ will meet $[Z_1, Z_2]$ in a point Z_6 where $Z_6 = (1-t)Z_1 + tZ_2$, and

$$\frac{t}{1-t} = \frac{b}{a}.$$

The product of these three ratios is clearly equal to 1 so (11.5.1) is satisfied. By the cross-bar theorem, $[Z_2, Z_5\rangle$ will meet $[Z_3, Z_6\rangle$ in a point Z_0 and so $Z_2 Z_5, Z_3 Z_6$ meet in Z_0. It follows that $Z_1 Z_4$ also passes through the point Z_0.

Thus the mid-lines of the angle-supports $|\underline{Z_2 Z_1 Z_3}$, $|\underline{Z_3 Z_2 Z_1}, |\underline{Z_1 Z_3 Z_2}$ *for a triangle* $[Z_1, Z_2, Z_3]$ *are concurrent.* The perpendicular distances from this point Z_0 to the side-lines of the triangle are equal by Ex.4.4, so the circle with Z_0 as centre and length of radius these common perpendicular distances will pass through the feet of these perpendiculars.

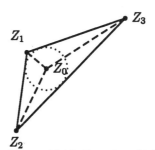

Figure 11.11. Incentre of triangle.

   This circle is called the **incircle** for the triangle; its centre Z_0 is called the **incentre** of the triangle. The three side-lines are tangents to the circle with the points of contact being the feet of the perpendiculars. For the incentre, by (11.5.2) we have the formula

$$Z_0 = \frac{a}{a+b+c}Z_1 + \frac{b}{a+b+c}Z_2 + \frac{c}{a+b+c}Z_3.$$

11.6   MOBILE COORDINATES

In standard vector notation, the vector product or cross product takes us out of the plane Π and into solid geometry. Sensed-area gives us half of the magnitude of the vector product and we use that instead. Without the vector product, however, we have not got orientation of the plane Π by vector means. We go on to supply this lack.

   However the standard vector operations can be awkward in dealing with perpendicularity and distance, and can involve quite a bit of trigonometry, so we also set out a method of reducing unwieldy calculations.

11.6.1   Grassmann's supplement of a vector

Given any $Z \neq O$, we show that there is a unique W such that

$$|O, W| = |O, Z|, \ OW \perp OZ, \ \delta_{\mathcal{F}}(O, Z, W) > 0.$$

Proof. With $Z \equiv (x, y)$, $W \equiv (u, v)$ these require

$$x^2 + y^2 = u^2 + v^2, \ ux + vy = 0, \ xv - yu > 0.$$

By the middle one of these

$$\begin{vmatrix} x & y \\ -v & u \end{vmatrix} = 0,$$

so the rows of this are linearly dependent. Thus we have $r(x, y) + s(-v, u) = (0, 0)$, for some $(r, s) \neq (0, 0)$. We cannot have $r = 0$ as that would imply $W = O$ and so $Z = O$. Then

$$x = \frac{s}{r}v, \ y = -\frac{s}{r}u,$$

so by the first property above

$$x^2 + y^2 = \frac{s^2}{r^2}(x^2 + y^2).$$

Thus we have either $s/r = 1$, so that $u = -y$, $v = x$, for which $2\delta_{\mathcal{F}}(O, Z, W) = x^2 + y^2 > 0$, or we have $s/r = -1$, so that $u = y$, $v = -x$, for which $2\delta_{\mathcal{F}}(O, Z, W) = -(x^2 + y^2) < 0$. Thus the unique solution is $u = -y$, $v = x$.

For any $Z \in \Pi$, we define $\overrightarrow{OZ}^{\perp} = \overrightarrow{OW}$ where $Z \equiv (x, y)$, $W \equiv (-y, x)$, and call this the **Grassmann supplement** of $\overrightarrow{OZ}$. This clearly has the properties

$$(\overrightarrow{OZ_1} + \overrightarrow{OZ_2})^{\perp} = \overrightarrow{OZ_1}^{\perp} + \overrightarrow{OZ_2}^{\perp},$$
$$(k\overrightarrow{OZ})^{\perp} = k(\overrightarrow{OZ})^{\perp},$$
$$(\overrightarrow{OZ}^{\perp})^{\perp} = -\overrightarrow{OZ}.$$

11.6.2

In $\mathcal{F}$ we take $|O, I| = |O, J| = 1$. If $|O, Z| = 1$ and θ is the angle in $\mathcal{A}_{\mathcal{F}}$ with support $|IOZ$, then we recall from 9.2.2 that $Z \equiv (x, y)$ where $x = \cos\theta$, $y = \sin\theta$. As $I \equiv (1, 0)$, $J \equiv (0, 1)$, we note that

$$\overrightarrow{OI}^{\perp} = \overrightarrow{OJ}, \ \overrightarrow{OZ} = \cos\theta\overrightarrow{OI} + \sin\theta\overrightarrow{OJ}, \ \overrightarrow{OZ}^{\perp} = -\sin\theta\overrightarrow{OI} + \cos\theta\overrightarrow{OJ}.$$

Suppose that we also have $\overrightarrow{OW} = \cos\phi\overrightarrow{OI} + \sin\phi\overrightarrow{OJ}$. Then by 11.4.1 we have $\overrightarrow{OW} = r\overrightarrow{OZ} + s\overrightarrow{OZ}^{\perp}$ where

$$r = \cos\phi\cos\theta + \sin\phi\sin\theta, \ s = \sin\phi\cos\theta - \cos\phi\sin\theta.$$

By the addition formulae we recognise that $r = \cos\angle_{\mathcal{F}}ZOW$, $s = \sin\angle_{\mathcal{F}}ZOW$.

11.6.3 Handling a triangle

Although for a triangle $[Z_1, Z_2, Z_3]$, we have the vector form for the centroid as $\frac{1}{3}(Z_1 + Z_2 + Z_3)$, and for the incentre as

$$\frac{a}{a+b+c}Z_1 + \frac{b}{a+b+c}Z_2 + \frac{c}{a+b+c}Z_3,$$

where as usual $a = |Z_2, Z_3|$, $b = |Z_3, Z_1|$, $c = |Z_1, Z_2|$, neither this formula for the incentre, nor the more awkward formula for the orthocentre, are convenient for applications and generalisation. In 7.2.3 we noted that a unique circle passes through the vertices of a triangle $[Z_1, Z_2, Z_3]$. It is called the **circumcircle** of this triangle and its centre is called the **circumcentre**. It is possible to find an expression for the circumcentre in terms of the vertices as in 11.4.4 but it is tedious to cover all the cases. For these reasons we consider the following way of representing any triangle.

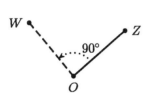

Figure 11.12. Grassmann supplement.

$$w_1 - z_2 = p(z_3 - z_2);$$
$$z_1 - w_1 = qi(z_3 - z_2)$$

Given non-collinear points Z_1, Z_2, Z_3, by 11.4.1 we can express

$$\overrightarrow{OZ_1} = \overrightarrow{OZ_2} + p_1(\overrightarrow{OZ_3} - \overrightarrow{OZ_2}) + q_1(\overrightarrow{OZ_3} - \overrightarrow{OZ_2})^{\perp},$$

for unique non-zero p_1 and q_1. We could work exclusively with material in this form but the manipulations are simpler if we use complex coordinates as well. Then

$$x_1 - x_2 = p_1(x_3 - x_2) - q_1(y_3 - y_2), \quad y_1 - y_2 = p_1(y_3 - y_2) + q_1(x_3 - x_2),$$

so with $Z_1 \sim z_1, Z_2 \sim z_2, Z_3 \sim z_3$ we have $z_1 - z_2 = (p_1 + q_1 i)(z_3 - z_2)$. We coin the name **mobile coordinates** of the point Z_1 with respect to (Z_2, Z_3) and $\mathcal{F}$, for the pair (p_1, q_1).

It follows immediately that $|Z_1, Z_2| = \sqrt{p_1^2 + q_1^2}|Z_2, Z_3|$, and as $z_1 - z_3 = (p_1 - 1 + q_1 i)(z_3 - z_2)$ we also have $|Z_3, Z_1| = \sqrt{(p_1 - 1)^2 + q_1^2}|Z_2, Z_3|$. From

$$\frac{z_3 - z_1}{z_2 - z_1} = \frac{p_1 - 1 + q_1 i}{p_1 + q_1 i}, \quad \frac{z_1 - z_2}{z_3 - z_2} = p_1 + q_1 i, \quad \frac{z_2 - z_3}{z_1 - z_3} = \frac{1}{1 - p_1 + q_1 i},$$

with $\alpha = \angle_{\mathcal{F}} Z_2 Z_1 Z_3$, $\beta = \angle_{\mathcal{F}} Z_3 Z_2 Z_1$, $\gamma = \angle_{\mathcal{F}} Z_1 Z_3 Z_2$, we have that

$$\operatorname{cis}\alpha = \frac{(p_1 - 1 + q_1 i)(p_1 - q_1 i)}{\sqrt{p_1^2 + q_1^2}\sqrt{(p_1 - 1)^2 + q_1^2}}, \quad \operatorname{cis}\beta = \frac{p_1 + q_1 i}{\sqrt{p_1^2 + q_1^2}}, \quad \operatorname{cis}\gamma = \frac{1 - p_1 + q_1 i}{\sqrt{(1 - p_1)^2 + q_1^2}}.$$

We also have that

$$p_1 + i q_1 = \sqrt{p_1^2 + q_1^2}\operatorname{cis}\beta = \frac{c}{a}\operatorname{cis}\beta,$$

$$1 - p_1 - i q_1 = \sqrt{(p_1 - 1)^2 + q_1^2}\operatorname{cis}\gamma = \frac{b}{a}\operatorname{cis}\gamma,$$

$$p_1(p_1 - 1) + q_1^2 + i q_1 = \sqrt{p_1^2 + q_1^2}\sqrt{(p_1 - 1)^2 + q_1^2}\operatorname{cis}\alpha = \frac{bc}{a^2}\operatorname{cis}\alpha. \quad (11.6.1)$$

Thus we have in terms of p_1 and q_1, the ratios of the lengths of the sides and cosines and sines of the angles. Moreover, it is easily calculated that

$$\delta_{\mathcal{F}}(Z_1, Z_2, Z_3) = \frac{q_1}{2}|Z_2, Z_3|^2,$$

so the orientation of this triple is determined by the sign of q_1.

Note too that if

$$z - z_2 = (p + q\imath)(z_3 - z_2), \ z' - z_2 = (p' + q'\imath)(z_3 - z_2),$$

then

$$|z' - z| = |p' - p + (q' - q)\imath||z_3 - z_2|,$$

and so

$$|Z, Z'| = \sqrt{(p' - p)^2 + (q' - q)^2}|Z_2, Z_3| = a\sqrt{(p' - p)^2 + (q' - q)^2}. \qquad (11.6.2)$$

11.6.4   Circumcentre of a triangle

Looking first for the circumcentre, we note that points Z on the perpendicular bisector of $[Z_2, Z_3]$ have complex coordinates of the form $z = z_2 + (\frac{1}{2} + q\imath)(z_3 - z_2)$, where $q \in \mathbf{R}$. But $z_1 - z_2 = (p_1 + q_1\imath)(z_3 - z_2)$ and so

$$z = z_2 + \frac{\frac{1}{2} + q\imath}{p_1 + q_1\imath}(z_1 - z_2).$$

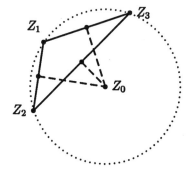

Figure 11.13. Circumcentre of triangle.

From this we have that

$$z - \tfrac{1}{2}(z_1 + z_2) = \left(\frac{\frac{1}{2} + q\imath}{p_1 + q_1\imath} - \frac{1}{2}\right)(z_1 - z_2).$$

To have $Z_3Z \perp Z_1Z_2$ also we need the coefficient of $z_1 - z_2$ in this to be purely imaginary. The coefficient is equal to

$$\frac{\frac{1}{2}p_1 + qq_1 + (qp_1 - \frac{1}{2}q_1)\imath}{p_1^2 + q_1^2} - \frac{1}{2},$$

and so we need

$$\tfrac{1}{2}p_1 + qq_1 = \tfrac{1}{2}(p_1^2 + q_1^2), \text{ i.e. } q = \frac{p_1^2 - p_1 + q_1^2}{2q_1}.$$

Thus the circumcentre has complex coordinate

$$z_0 = z_2 + \frac{1}{2}\left[1 + \frac{p_1^2 - p_1 + q_1^2}{q_1}\imath\right](z_3 - z_2).$$

From this we have that

$$z_0 - \tfrac{1}{2}(z_2 + z_3) = \frac{1}{2}\left[-1 + \frac{p_1^2 - p_1 + q_1^2}{q_1}i\right](z_3 - z_2)$$

$$= \frac{i}{2q_1}(p_1 - q_1 i)(p_1 - 1 + q_1 i)(z_3 - z_2),$$

and from this can conclude that the length of radius of the circumcircle is

$$\frac{a}{2|q_1|}\sqrt{p_1^2 + q_1^2}\sqrt{(p_1 - 1)^2 + q_1^2}.$$

In fact we can deduce from this material a formula for the circumcentre in terms of areal coordinates. For by (11.6.1)

$$p_1^2 - p_1 + q_1^2 = \frac{bc}{a^2}\cos\alpha, \quad q_1 = \frac{bc}{a^2}\sin\alpha,$$

so that $z = z_2 + \tfrac{1}{2}(1 + \cot\alpha\, i)(z_3 - z_2)$. Then by 11.6.3 we have

$$x - x_2 = \tfrac{1}{2}(x_3 - x_2) - \tfrac{1}{2}\cot\alpha(y_3 - y_2), \quad y - y_2 = \tfrac{1}{2}\cot\alpha(x_3 - x_2) + \tfrac{1}{2}(y_3 - y_2).$$

From this we have that

$$\delta_{\mathcal{F}}(Z, Z_2, Z_3) = \delta_{\mathcal{F}}(Z - Z_2, O, Z_3 - Z_2)$$
$$= \tfrac{1}{4}\cot\alpha\left[(x_3 - x_2)^2 + (y_3 - y_2)^2\right] = \tfrac{1}{4}\cot\alpha\, a^2.$$

As $\delta_{\mathcal{F}}(Z_1, Z_2, Z_3) = \tfrac{1}{2}q_1 a^2 = \tfrac{1}{2}bc\sin\alpha$ we have

$$\frac{\delta_{\mathcal{F}}(Z, Z_2, Z_3)}{\delta_{\mathcal{F}}(Z_1, Z_2, Z_3)} = \frac{1}{2}\frac{a^2\cot\alpha}{bc\sin\alpha},$$

and by use of the sine rule this is seen to be equal to

$$\frac{1}{2}\frac{\cos\alpha}{\sin\beta\sin\gamma}.$$

By cyclic rotation we can write down the other two coefficients and so have

$$Z = \frac{1}{2}\frac{\cos\alpha}{\sin\beta\sin\gamma}Z_1 + \frac{1}{2}\frac{\cos\beta}{\sin\gamma\sin\alpha}Z_2 + \frac{1}{2}\frac{\cos\gamma}{\sin\alpha\sin\beta}Z_3.$$

That the sum of the coefficients here is equal to 1 follows from the identity

$$\sin 2\alpha + \sin 2\beta + \sin 2\gamma = 4\sin\alpha\sin\beta\sin\gamma,$$

for the angles of a triangle. For

$$\sin 2\alpha_{\mathcal{F}} + \sin 2\beta_{\mathcal{F}} + \sin 2\gamma_{\mathcal{F}}$$
$$= 2\sin(\alpha_{\mathcal{F}} + \beta_{\mathcal{F}})\cos(\alpha_{\mathcal{F}} - \beta_{\mathcal{F}}) + 2\sin\gamma_{\mathcal{F}}\cos\gamma_{\mathcal{F}}$$
$$= 2\sin\gamma_{\mathcal{F}}[\cos(\alpha_{\mathcal{F}} - \beta_{\mathcal{F}}) + \cos\gamma_{\mathcal{F}}]$$
$$= 2\sin\gamma_{\mathcal{F}}[\cos(\alpha_{\mathcal{F}} - \beta_{\mathcal{F}}) - \cos(\alpha_{\mathcal{F}} + \beta_{\mathcal{F}})]$$
$$= 2\sin\gamma_{\mathcal{F}}.2\sin\alpha_{\mathcal{F}}\sin\beta_{\mathcal{F}}.$$

11.6.5   Other distinguished points for a triangle

For the centroid Z_0 of $[Z_1, Z_2, Z_3]$ we have by 11.5.5 and 11.4.5 that $z_0 = \frac{1}{3}(z_1 + z_2 + z_3)$ and so

$$z_0 - z_2 = \tfrac{1}{3}(z_1 + z_2 + z_3) - z_2 = \tfrac{1}{3}(z_1 - z_2) + \tfrac{1}{3}(z_3 - z_2)$$
$$= \tfrac{1}{3}(p_1 + q_1 i)(z_3 - z_2) + \tfrac{1}{3}(z_3 - z_2) = \tfrac{1}{3}(p_1 + 1 + q_1 i)(z_3 - z_2).$$

This gives *the complex coordinate of the centroid.*

We next turn to the orthocentre of this triangle. Points Z on the line through Z_1 perpendicular to $Z_2 Z_3$ have complex coordinates of the form

$$z = z_2 + (p_1 + qi)(z_3 - z_2) = z_3 + (p_1 - 1 + qi)(z_3 - z_2)$$
$$= z_3 + \frac{p_1 - 1 + qi}{p_1 + q_1 i}(z_1 - z_2).$$

For Z_Z to be also perpendicular to $Z_1 Z_2$ we also need the coefficient of $z_1 - z_2$ here to be purely imaginary. Thus we need

$$(p_1 - 1)p_1 + qq_1 = 0 \quad \text{i.e.} \quad q = -\frac{(p_1 - 1)p_1}{q_1},$$

and so obtain

$$z_2 + p_1\left(1 - \frac{p_1 - 1}{q_1}i\right)(z_3 - z_2),$$

as *the complex coordinate of the orthocentre.*

It takes more of an effort to deal with the incentre. The mid-point of the points with complex coordinates

$$z_2 + \frac{1}{|Z_2, Z_3|}(z_3 - z_2), \; z_2 + \frac{1}{|Z_2, Z_1|}(z_1 - z_2),$$

has complex coordinate

$$z_2 + \frac{1}{2|Z_2, Z_3|}\left(z_3 - z_2 + \frac{1}{\sqrt{p_1^2 + q_1^2}}(z_1 - z_2)\right)$$
$$= z_2 + \frac{1}{2|Z_2, Z_3|}\left(z_3 - z_2 + \frac{p_1 + q_1 i}{\sqrt{p_1^2 + q_1^2}}(z_3 - z_2)\right)$$
$$= z_2 + \frac{1}{2|Z_2, Z_3|}\left(1 + \frac{p_1 + q_1 i}{\sqrt{p_1^2 + q_1^2}}\right)(z_3 - z_2).$$

Points on the midline of $|Z_1 Z_2 Z_3$ then have complex coordinates of the form

$$z_2 + \frac{r}{2|Z_2, Z_3|}\left(1 + \frac{p_1 + q_1 i}{\sqrt{p_1^2 + q_1^2}}\right)(z_3 - z_2).$$

By a similar argument, points on the midline of $\lfloor Z_2\,Z_3\,Z_1$ have complex coordinates of the form

$$z_3 + \frac{s}{2|\overrightarrow{Z_2,Z_3}|}\left(\frac{p_1-1+q_1\imath}{\sqrt{(p_1-1)^2+q_1^2}}\right)(z_3-z_2)$$

$$= z_2 + \left[1 + \frac{s}{2|\overrightarrow{Z_2,Z_3}|}\left(\frac{p_1-1+q_1\imath}{\sqrt{(p_1-1)^2+q_1^2}}\right)\right](z_3-z_2).$$

For a point of intersection we need

$$\frac{r}{2|\overrightarrow{Z_2,Z_3}|}\left(1+\frac{p_1+q_1\imath}{\sqrt{p_1^2+q_1^2}}\right) = 1 + \frac{s}{2|\overrightarrow{Z_2,Z_3}|}\left(\frac{p_1-1+q_1\imath}{\sqrt{(p_1-1)^2+q_1^2}}\right).$$

On solving for r and s, we obtain *for the incentre the complex coordinate*

$$z_2 + \frac{p_1+q_1\imath+\sqrt{p_1^2+q_1^2}}{1+\sqrt{p_1^2+q_1^2}+\sqrt{(p_1-1)^2+q_1^2}}(z_3-z_2).$$

11.6.6   Euler line of a triangle

With the notation Z_7, Z_8, Z_9 for the centroid, circumcentre and orthocentre, respectively, of a triangle $[Z_1, Z_2, Z_3]$ we have the formulae

$$z_7 - z_2 = \tfrac{1}{3}(p_1+1+q_1\imath)(z_3-z_2),$$

$$z_8 - z_2 = \frac{1}{2}\left(1+\frac{p_1^2-p_1+q_1^2}{q_1}\right)(z_3-z_2),$$

$$z_9 - z_2 = p_1\left(1-\frac{p_1-1}{q_1}\imath\right)(z_3-z_2).$$

It is straightforward to check that

$$\tfrac{2}{3}z_8 + \tfrac{1}{3}z_9 = z_2 + \tfrac{2}{3}(z_8-z_2)+\tfrac{1}{3}(z_9-z_2) = z_7,$$

and so $Z_7 \in Z_8 Z_9$.

Thus we have shown that *the centroid, circumcentre and orthocentre of any triangle are collinear.* This is a result due to Euler, after whom this line of collinearity is named the **Euler line** of the triangle.

11.6.7   Similar triangles

For any two triangles $[Z_1, Z_2, Z_3]$ and $[Z_4, Z_5, Z_6]$ we have

$$\overrightarrow{OZ_1} = \overrightarrow{OZ_2} + p_1(\overrightarrow{OZ_3}-\overrightarrow{OZ_2})+q_1(\overrightarrow{OZ_3}-\overrightarrow{OZ_2})^\perp,$$
$$\overrightarrow{OZ_4} = \overrightarrow{OZ_5} + p_4(\overrightarrow{OZ_6}-\overrightarrow{OZ_5})+q_4(\overrightarrow{OZ_6}-\overrightarrow{OZ_5})^\perp,$$

or equivalently $z_1 = z_2 + (p_1+q_1\imath)(z_3-z_2)$, $z_4 = z_5 + (p_4+q_4\imath)(z_6-z_5)$, where p_1, q_1, p_4, q_4 are non-zero real numbers. Then these triangles are similar in the correspondence $(Z_1, Z_2, Z_3) \rightarrow (Z_4, Z_5, Z_6)$ if and only if $p_4 = p_1$, $q_4 = \pm q_1$.

Proof. First suppose that $p_4 = p_1$, $q_4 = q_1$ so that $z_1 = z_2 + (p_1 + q_1 i)(z_3 - z_2)$, $z_4 = z_5 + (p_1 + q_1 i)(z_6 - z_5)$. Then we have that

$$|\measuredangle_\mathcal{F} Z_3 Z_2 Z_1|^\circ = |\measuredangle_\mathcal{F} Z_6 Z_5 Z_4|^\circ, \quad |\measuredangle_\mathcal{F} Z_1 Z_3 Z_2|^\circ = |\measuredangle_\mathcal{F} Z_4 Z_6 Z_5|^\circ.$$

It follows by 5.3.2 that the measures of the corresponding angles of these triangles are equal, and so the triangles are similar, with the lengths of corresponding sides proportional. Moreover, the triples (Z_1, Z_2, Z_3) and (Z_4, Z_5, Z_6) are similarly oriented.

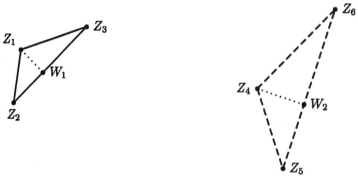

Figure 11.14. Similar triangles.

Next suppose that $z_1 = z_2 + (p_1 + q_1 i)(z_3 - z_2)$, $z_4 = z_5 + (p_1 - q_1 i)(z_6 - z_5)$. By an analogous argument the triangles are still similar, and now the triples (Z_1, Z_2, Z_3) and (Z_4, Z_5, Z_6) are oppositely oriented.

Conversely, suppose that $[Z_1, Z_2, Z_3]$ and $[Z_4, Z_5, Z_6]$ are similar triangles in the correspondence $(Z_1, Z_2, Z_3) \to (Z_4, Z_5, Z_6)$. Let W_1 be the foot of the perpendicular from Z_1 to $Z_2 Z_3$, and from parametric equations of $Z_2 Z_3$ choose $p_1 \in \mathbf{R}$ so that $w_1 = z_2 + p_1(z_3 - z_2)$. Then

$$\frac{|Z_2, W_1|}{|Z_2, Z_3|} = |p_1|,$$

and p_1 is positive or negative according or not as W_1 is on the same or opposite side of Z_2 as Z_3 is on the line $Z_2 Z_3$, that is according as the wedge-angle $\angle Z_3 Z_2 Z_1$ is acute or obtuse. As $W_1 Z_1 \perp Z_2 Z_3$ we can find $q_1 \in \mathbf{R}$ so that $z_1 - w_1 = q_1 i(z_3 - z_2)$. Then

$$\frac{|Z_1, W_1|}{|Z_2, Z_3|} = |q_1|,$$

and q_1 is positive or negative according as (Z_1, Z_2, Z_3) is positively or negatively oriented.

As the lengths of the sides of the two triangles are proportional, we have

$$|Z_5, Z_6| = k|Z_2, Z_3|, \quad |Z_6, Z_4| = k|Z_3, Z_1|, \quad |Z_4, Z_5| = k|Z_1, Z_2|,$$

for some $k > 0$. Let W_2 be the foot of the perpendicular from Z_4 to the line $Z_5 Z_6$. Then the triangles $[Z_1, Z_2, W_1]$ and $[Z_4, Z_5, W_2]$ are similar, so we have that

$$\frac{|Z_5, W_2|}{|Z_2, W_1|} = \frac{|Z_4, Z_5|}{|Z_1, Z_2|} = k.$$

It follows that

$$|Z_5, W_2| = k|Z_2, W_1| = k|p_1||Z_2, Z_3| = |p_1||Z_5, Z_6|.$$

But W_2 is on the same side of the point Z_5 on the line $Z_5 Z_6$ as Z_6 is if the wedge-angle $\angle Z_6 Z_5 Z_4$ is acute, and on the opposite side if this angle is obtuse. Hence we have that $w_2 - z_5 = p_1(z_6 - z_5)$.

As $Z_4 W_2 \perp Z_5 Z_6$, we have $z_4 - w_2 = ji(z_6 - z_5)$ for some $j \in \mathbb{R}$, and then $|Z_4, W_2| = |j||Z_5, Z_6|$. But

$$\frac{|Z_4, W_2|}{|Z_1, W_1|} = \frac{|Z_4, Z_5|}{|Z_1, Z_2|} = \frac{|Z_5, Z_6|}{|Z_2, Z_3|},$$

so

$$\frac{|Z_4, W_2|}{|Z_5, Z_6|} = \frac{|Z_1, W_1|}{|Z_2, Z_3|} = |q_1|.$$

Hence $j = \pm q_1$ and we are to take the plus if (Z_1, Z_2, Z_3) and (Z_4, Z_5, Z_6) have the same orientation, the minus if the opposite orientation.

Thus our mobile coordinates (p_1, q_1) are intimately connected with similarity of triangles.

11.6.8   Similar triangles erected on the sides of a triangle

Given an arbitrary triangle $[Z_1, Z_2, Z_3]$, if we consider points Z_4, Z_5 and Z_6 defined by

$$z_4 = z_2 + (p_1 + q_1 i)(z_3 - z_2), \quad z_5 = z_3 + (p_1 + q_1 i)(z_1 - z_3),$$
$$z_6 = z_1 + (p_1 + q_1 i)(z_2 - z_1),$$

for some non-zero real numbers p_1 and q_1, then *we have triangles erected on the sides* $[Z_2, Z_3]$, $[Z_3, Z_1]$ *and* $[Z_1, Z_2]$, *respectively, which are similar to each other and have the same orientation as each other.* By addition we note that $\frac{1}{3}(z_4 + z_5 + z_6) = \frac{1}{3}(z_1 + z_2 + z_3)$, and so *the triangle* $[Z_4, Z_5, Z_6]$ *has the same centroid as the original triangle* $[Z_1, Z_2, Z_3]$.

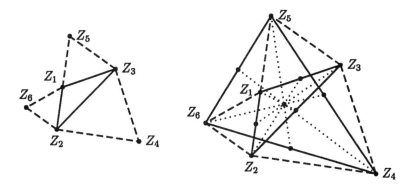

Figure 11.15. Similar triangles on sides of triangle.

Further, if we let Z_7, Z_8, Z_9 be the centroids of the triangles $[Z_2, Z_3, Z_4]$, $[Z_3, Z_1, Z_5]$ and $[Z_1, Z_2, Z_6]$, respectively, we have that $z_7 = \frac{1}{3}(z_2 + z_3 + z_4)$, $z_8 = \frac{1}{3}(z_3 + z_1 + z_5)$, $z_9 = \frac{1}{3}(z_1 + z_2 + z_6)$; it follows that *the centroid of* $[Z_7, Z_8, Z_9]$ *is also the centroid of the original triangle.*

11.6.9 Circumcentres of similar triangles on sides of triangle

In a more complicated fashion than in the last subsection, for an arbitrary triangle $[Z_1, Z_2, Z_3]$ suppose that we take points Z_4, Z_5 and Z_6 so that

$$z_4 = z_2 + (p_1 + q_1 i)(z_3 - z_2), \quad z_3 = z_1 + (p_1 + q_1 i)(z_5 - z_1),$$
$$z_2 = z_6 + (p_1 + q_1 i)(z_1 - z_6),$$

so that *we have similar triangles once again on the sides of the original triangle but now in the correspondences* $(Z_2, Z_3, Z_4) \to (Z_1, Z_5, Z_3) \to (Z_6, Z_1, Z_2)$. We let Z_7, Z_8, Z_9 be the circumcentres of these three similar triangles, so that we have

$$z_7 - z_2 = \frac{1}{2}\left(1 + \frac{p_1^2 - p_1 + q_1^2}{q_1}i\right)(z_3 - z_2),$$

$$z_8 - z_1 = \frac{1}{2}\left(1 + \frac{p_1^2 - p_1 + q_1^2}{q_1}i\right)(z_5 - z_1),$$

$$z_9 - z_6 = \frac{1}{2}\left(1 + \frac{p_1^2 - p_1 + q_1^2}{q_1}i\right)(z_1 - z_6).$$

But

$$z_5 - z_1 = \frac{1}{p_1 + q_1 i}(z_3 - z_1),$$

while $z_2 - z_1 = (1 - p_1 - q_1 i)(z_6 - z_1)$, so that

$$z_6 - z_1 = \frac{1}{1 - p_1 - q_1 i}(z_2 - z_1).$$

Also

$$z_9 - z_1 = z_6 - z_1 - \frac{1}{2}\left(1 + \frac{p_1^2 - p_1 + q_1^2}{q_1}i\right)(z_6 - z_1) = \left(\frac{1}{2} - \frac{p_1^2 - p_1 + q_1^2}{2q_1}i\right)(z_6 - z_1).$$

On combining these we have

$$z_8 - z_1 = \frac{1}{2}\left(1 + \frac{p_1^2 - p_1 + q_1^2}{q_1}i\right)\frac{1}{p_1 + q_1 i}(z_3 - z_1),$$

$$z_9 - z_1 = \left(\frac{1}{2} - \frac{p_1^2 - p_1 + q_1^2}{2q_1}i\right)\frac{1}{1 - p_1 - q_1 i}(z_2 - z_1).$$

Then

$$z_9 + (p_1 + q_1 i)(z_8 - z_9)$$
$$= (1 - p_1 - q_1 i)(z_9 - z_1) + (p_1 + q_1 i)(z_8 - z_1) + z_1$$
$$= \left(\frac{1}{2} - \frac{p_1^2 - p_1 + q_1^2}{2q_1}i\right)(z_2 - z_1) + \left(\frac{1}{2} + \frac{p_1^2 - p_1 + q_1^2}{2q_1}i\right)(z_3 - z_1) + z_1$$
$$= \tfrac{1}{2}(z_2 + z_3) + \frac{p_1^2 - p_1 + q_1^2}{2q_1}i(z_3 - z_2) = z_7.$$

It follows that *the triangle* $[Z_7, Z_8, Z_9]$ *is also similar to the similar triangles above, in the correspondence* $(Z_2, Z_3, Z_4) \to (Z_9, Z_8, Z_7)$.

In the particular case when $p_1 = 1/2$, $q_1 = \sqrt{3}/2$, the similar triangles are all **equilateral** triangles, that is all three sides have equal lengths. In this case the last result is known as NAPOLEON'S THEOREM. It is easier to prove than the more general case, as for it we can work just with centroids.

11.6.10 The nine-point circle

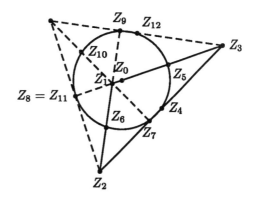

Given the notation of 11.6.3, 11.6.4 and 11.6.5, let $Z_4 = \mathrm{mp}(Z_2, Z_3)$, $Z_5 = \mathrm{mp}(Z_3, Z_1)$ and $Z_6 = \mathrm{mp}(Z_1, Z_2)$. We first seek the circumcircle of the triangle $[Z_4, Z_5, Z_6]$ with vertices these mid-points of the sides of the original triangle.

Figure 11.16. Nine-point circle.

Now

$$z_4 = \tfrac{1}{2}(z_2 + z_3) = z_2 + \tfrac{1}{2}(z_3 - z_2),$$
$$z_5 = \tfrac{1}{2}(z_3 + z_1) = z_2 + \tfrac{1}{2}(p_1 + 1 + q_1\imath)(z_3 - z_2),$$
$$z_6 = \tfrac{1}{2}(z_1 + z_2) = z_2 + \tfrac{1}{2}(p_1 + q_1\imath)(z_3 - z_2).$$

Then

$$z_6 - z_5 = -\tfrac{1}{2}(z_3 - z_2),$$

so that

$$z_4 - z_5 = -\tfrac{1}{2}(p_1 + q_1\imath)(z_3 - z_2) = (p_1 + q_1\imath)(z_6 - z_5).$$

It follows from 11.6.4 that the circumcentre of $[Z_4, Z_5, Z_6]$ has complex coordinate

$$z_5 + \frac{\imath}{2q_1}(p_1 + q_1\imath)(p_1 - 1 - q_1\imath)(z_6 - z_5),$$

and this simplifies to

$$z_2 + \left(\frac{1}{2}(p_1 + 1 + q_1\imath) - \frac{\imath}{4q_1}(p_1 + q_1\imath)(p_1 - 1 - q_1\imath)\right)(z_3 - z_2)$$
$$= z_2 + \frac{1}{4}\left(1 + 2p_1 + q_1\imath + \frac{p_1(1 - p_1)}{q_1}\imath\right)(z_3 - z_2).$$

We denote this point by Z_0.

Next let Z_7, Z_8, Z_9 be the feet of the perpendiculars from the vertices Z_1, Z_2, Z_3 onto the opposite side-lines Z_2Z_3, Z_3Z_1, Z_1Z_2, respectively. We now show that the circumcentre of $[Z_6, Z_4, Z_7]$ is Z_0 also. For this we note that

$$z_4 = z_2 + \tfrac{1}{2}(z_3 - z_2), \quad z_6 = z_2 + \tfrac{1}{2}(p_1 + q_1\imath)(z_3 - z_2), \quad z_7 = z_2 + p_1(z_3 - z_2).$$

From this

$$z_7 - z_4 = (p_1 - \tfrac{1}{2})(z_3 - z_2), \quad z_6 - z_4 = \frac{p_1 - 1 + q_1 \imath}{2p_1 - 1}(z_7 - z_4),$$

and hence

$$z_6 = z_4 + \frac{1}{2p_1 - 1}(p_1 - 1 + q_1 \imath)(z_7 - z_4).$$

By the formula of 11.6.4 for the circumcentre, we know that the incentre of $[Z_6, Z_4, Z_7]$ has complex coordinate

$$z_4 + \frac{\imath}{2q_1/(2p_1 - 1)}\left(\frac{p_1 - 1}{2p_1 - 1} + \frac{q_1}{2p_1 - 1}\imath\right)\left(\frac{p_1 - 1}{2p_1 - 1} - 1 - \frac{q_1}{2p_1 - 1}\imath\right)(z_7 - z_4),$$

and this simplifies to z_0.

Thus Z_7 lies on the circumcircle of the triangle $[Z_4, Z_5, Z_6]$, and as this argument also applies to the other two sides of $[Z_1, Z_2, Z_3]$, so do Z_8 and Z_9. This shows that *the feet of the perpendiculars from the vertices of the triangle onto the opposite sidelines also lie on the above circle.*

Finally, let Z_{10} be the mid-point of the orthocentre and the vertex Z_1 in the original triangle, Z_{11} be the mid-point of the orthocentre and the vertex Z_2, and Z_{12} be the mid-point of the orthocentre and the vertex Z_3. We seek the circumcentre of the triangle $[Z_{11}, Z_4, Z_7]$. Now $z_4 = z_2 + \tfrac{1}{2}(z_3 - z_2)$, $z_7 = z_2 + p_1(z_3 - z_2)$, and by the formula in 11.6.5 for the orthocentre,

$$z_{11} = z_2 + \frac{p_1}{2}\left(1 - \frac{p_1 - 1}{q_1}\imath\right)(z_3 - z_2).$$

From these we have that

$$z_{11} = z_4 + \frac{p_1 - 1}{2p_1 - 1}\left(1 - \frac{p_1}{q_1}\imath\right)(z_7 - z_4).$$

Then the circumcentre of $[Z_{11}, Z_4, Z_7]$ has complex coordinate

$$z_4 + \frac{\imath}{-2p_1(p_1 - 1)/(2p_1 - 1)q_1}\frac{p_1 - 1}{2p_1 - 1}\left(1 - \frac{p_1}{q_1}\imath\right)\left[\frac{p_1 - 1}{2p_1 - 1} - 1 + \frac{p_1(p_1 - 1)}{q_1(2p_1 - 1)}\imath\right](z_7 - z_4),$$

and it can be checked that this reduces to z_0. Hence Z_{11} lies on the circle through the mid-points of the sides of $[Z_1, Z_2, Z_3]$, and thus, as our argument applies equally well to the other two sides of the original triangle, so do Z_{12} and Z_{10}. This shows that *the mid-point of the orthocentre and each vertex also lies on the above circle.*

Thus we have identified nine points on this circle, which is named from this property.

11.7

NOTE. The advantage of mobile coordinates is that they located a point with respect to a triangle using just two instead of three numbers and they also behave like re-scaled Cartesian coordinates. None the less they can lead to unwieldy expressions as in this section and it is a good idea when possible to check the algebraic manipulations using a computer software programme.

11.7.1 Feuerbach's theorem, 1822

Our formula in 11.6.5 for the incentre of a triangle is very awkward to apply because of the complicated term in the denominator. However by eliminating the surds in the denominator in two steps, by multiplying above and below by a conjugate surd of the denominator, we obtain the more convenient formulation that

$$\frac{p_1 + \sqrt{p_1^2 + q_1^2}}{1 + \sqrt{p_1^2 + q_1^2} + \sqrt{(p_1 - 1)^2 + q_1^2}} = \frac{1}{2} + \frac{1}{2}\sqrt{p_1^2 + q_1^2} - \frac{1}{2}\sqrt{(p_1 - 1)^2 + q_1^2}. \quad (11.7.1)$$

In fact once we know the form of this we can establish it more directly and easily by noting that

$$\frac{1}{2}\left[1 + \sqrt{p_1^2 + q_1^2} + \sqrt{(p_1 - 1)^2 + q_1^2}\right]\left[1 + \sqrt{p_1^2 + q_1^2} - \sqrt{(p_1 - 1)^2 + q_1^2}\right]$$

$$= \frac{1}{2}\left[\left(1 + \sqrt{p_1^2 + q_1^2}\right)^2 - \left((p_1 - 1)^2 + q_1^2\right)\right]$$

$$= p_1 + \sqrt{p_1^2 + q_1^2}.$$

We note that the right-hand side in (11.7.1) must be positive.

Recalling from 11.6.5 that the incentre Z_{13} has complex coordinate

$$z_{13} = z_2 + \frac{p_1 + \sqrt{p_1^2 + q_1^2} + q_1 \imath}{1 + \sqrt{p_1^2 + q_1^2} + \sqrt{(p_1 - 1)^2 + q_1^2}}(z_3 - z_2),$$

we re-write this as

$$z_{13} - z_2$$

$$= \frac{1}{2}\left[1 + \sqrt{p_1^2 + q_1^2} - \sqrt{(p_1 - 1)^2 + q_1^2}\right]\left[1 + \frac{q_1 \imath}{p_1 + \sqrt{p_1^2 + q_1^2}}\right](z_3 - z_2)$$

$$= \frac{1}{2q_1}\left[1 + \sqrt{p_1^2 + q_1^2} - \sqrt{(p_1 - 1)^2 + q_1^2}\right]\left[q_1 + \left(\sqrt{p_1^2 + q_1^2} - p_1\right)\imath\right](z_3 - z_2).$$

From this, the foot of the perpendicular from the incentre Z_{13} to the line $Z_2 Z_3$ has complex coordinate

$$z_2 + \frac{1}{2q_1}\left[1 + \sqrt{p_1^2 + q_1^2} - \sqrt{(p_1 - 1)^2 + q_1^2}\right]q_1(z_3 - z_2)$$

and so *the length of radius of the incircle is equal to*

$$\frac{a}{2|q_1|}\left[1+\sqrt{p_1^2+q_1^2}-\sqrt{(p_1-1)^2+q_1^2}\right]\left[\sqrt{p_1^2+q_1^2}-p_1\right].$$

On the other hand the nine-point circle has radius-length equal to the distance from Z_0 to Z_4. From the formula

$$
\begin{aligned}
z_0-z_4 &= \tfrac{1}{2}(p_1+q_1\imath)\left[1-\frac{\imath}{2q_1}(p_1-1-q_1\imath)\right](z_3-z_2)\\
&= \frac{1}{4q_1}(p_1+q_1\imath)[q_1+(1-p_1)\imath](z_3-z_2)
\end{aligned}
$$

we note that

$$|z_0-z_4|=\frac{a}{4|q_1|}|p_1+q_1\imath||q_1+(1-p_1)\imath|,$$

and so *this is the radius-length of the nine-point circle.*

We also require *the distance between the centres of the inscribed and nine-point circles.* For this we note that

$$z_{13}-z_0$$
$$
\begin{aligned}
=&\left\{\frac{1}{2q_1}\left(1+\sqrt{p_1^2+q_1^2}-\sqrt{(p_1-1)^2+q_1^2}\right)\left(q_1+\left(\sqrt{p_1^2+q_1^2}-p_1\right)\imath\right)\right.\\
&\left.-\left(\frac{1}{2}(p_1+1+q_1\imath)-\frac{\imath}{4q_1}(p_1+q_1\imath)(p_1-1-q_1\imath)\right)\right\}(z_3-z_2)\\
=&\frac{1}{4q_1}\left\{\left(1+\sqrt{p_1^2+q_1^2}-\sqrt{(p_1-1)^2+q_1^2}\right)2q_1-(1+2p_1)q_1\right.\\
&\left.+\left[\left(1+\sqrt{p_1^2+q_1^2}-\sqrt{(p_1-1)^2+q_1^2}\right)2\left(\sqrt{p_1^2+q_1^2}-p_1\right)+p_1^2-p_1-q_1^2\right]\imath\right\}.
\end{aligned}
$$

It follows that

$$|z_0-z_{13}|^2=$$
$$
\begin{aligned}
\frac{a^2}{16q_1^2}&\left\{\left[\left(1+\sqrt{p_1^2+q_1^2}-\sqrt{(p_1-1)^2+q_1^2}\right)2q_1-(1+2p_1)q_1\right]^2\right.\\
&\left.+\left[\left(1+\sqrt{p_1^2+q_1^2}-\sqrt{(p_1-1)^2+q_1^2}\right)2\left(\sqrt{p_1^2+q_1^2}-p_1\right)+p_1^2-p_1-q_1^2\right]^2\right\}.
\end{aligned}
$$

The next feature which we wish to note is that if we denote by r_1 and r_2, respectively, the lengths of the radii of the nine-point circle and the incircle, then

$$|Z_0,Z_{13}|^2=(r_1-r_2)^2.$$

This can be verified on a computer; it can also be written out at length by writing each term in the form

$$u+v\sqrt{p_1^2+q_1^2}+w\sqrt{(p_1-1)^2+q_1^2}+x\sqrt{p_1^2+q_1^2}\sqrt{(p_1-1)^2+q_1^2},$$

where u, v, w and x are polynomials in p_1, q_1 and a. This is unsatisfactory as a method of proof but in the absence of some insight which will lead to a reasonable calculation it must suffice. It follows that

$$|Z_0, Z_{13}| = \pm(r_1 - r_2).$$

With these preparatory results, we can now show that *the nine-point circle and the incircle meet at just one point and they have a common tangent there.*

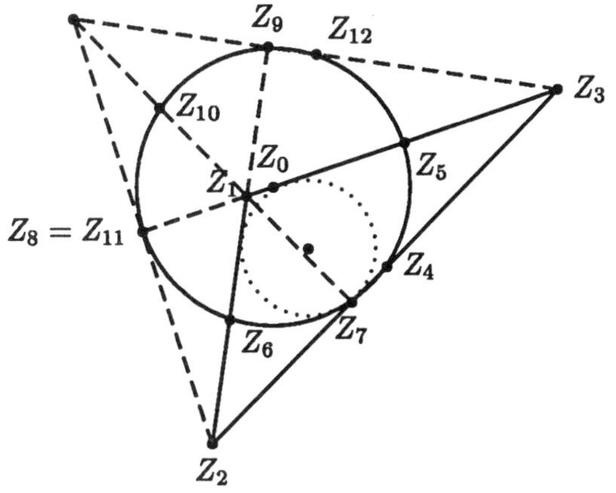

Figure 11.17. Feuerbach's theorem.

Suppose first that $r_2 \geq r_1$ so that $|Z_0, Z_{13}| = r_2 - r_1$. Let Z be the point on $[Z_{13}, Z_0$ such that $|Z_{13}, Z| = r_2$; then Z is a point on the incircle. As $r_2 - r_1 < r_2$ we have that $Z_0 \in [Z_{13}, Z]$ and so $|Z_0, Z| = r_2 - (r_2 - r_1) = r_1$. It follows that Z is also on the nine-point circle. It then follows that every other point of the nine-point circle is inside or on the incircle. But the incircle is contained in the triangle $[Z_1, Z_2, Z_3]$ and the nine-point circle is not (as it passes through the mid-points of the sides). Thus this gives a contradiction and we must have $r_1 > r_2$ and so $|Z_0, Z_{13}| = r_1 - r_2$. Now let Z be the point on $[Z_0, Z_{13}$ such that $|Z_0, Z| = r_1$; then Z is a point on the nine-point circle. As $r_1 - r_2 < r_1$ we have that $Z_{13} \in [Z_0, Z]$ and so $|Z_{13}, Z| = r_1 - (r_1 - r_2) = r_2$. It follows that Z is also on the incircle. Then every other point of the incircle is inside the nine-point circle and the line through Z perpendicular to $Z_0 Z_{13}$ is a tangent to both circles. This shows that the incircle and the nine-point circle of a triangle meet at just one point and they have a common tangent there.

If we modify our treatment of the incentre of the triangle $[Z_1, Z_2, Z_3]$, using the terminology of 5.5.1 points on the external bisector of $|Z_1 Z_2 Z_3|$ will have complex coordinates

$$z_2 + \frac{r}{2a}\left(1 - \frac{p_1 + q_1 \imath}{\sqrt{p_1^2 + q_1^2}}\right)(z_3 - z_2),$$

and points on the external bisector of $|Z_2 Z_3 Z_1|$ will have complex coordinates

$$z_2 + \left\{1 - \frac{s}{2a}\left(1 + \frac{p_1 - 1 + q_1 \imath}{\sqrt{(p_1 - 1)^2 + q_1^2}}\right)\right\}(z_3 - z_2).$$

It follows that the point of intersection of these lines has complex coordinate

$$z_2 + p_1 - \frac{\sqrt{p_1^2 + q_1^2} + q_1 \imath}{1 - \sqrt{p_1^2 + q_1^2} - \sqrt{(p_1 - 1)^2 + q_1^2}}(z_3 - z_2).$$

This point is equally distant from the side-lines of the triangle and is called an **ex-centre**. The circle with it as centre and which touches the side-lines of the triangle is called an **escribed circle** for the triangle.

Now

$$\frac{1}{2}\left[1 - \sqrt{p_1^2 + q_1^2} - \sqrt{(p_1 - 1)^2 + q_1^2}\right]\left[1 - \sqrt{p_1^2 + q_1^2} + \sqrt{(p_1 - 1)^2 + q_1^2}\right]$$

$$= \frac{1}{2}\left[\left(1 - \sqrt{p_1^2 + q_1^2}\right)^2 - ((p_1 - 1)^2 + q_1^2)\right]$$

$$= p_1 - \sqrt{p_1^2 + q_1^2}.$$

Then by a straightforward modification of our argument for the incircle, it follows that *the nine-point circle and this escribed circle meet at one point, where they have a common tangent.* As this argument is valid for the other two sides of the triangle as well, it follows that the two other escribed circles have this property also.

This combined result is known as FEUERBACH'S THEOREM.

11.7.2   The Wallace-Simson line, 1797

We take a triangle $[Z_1, Z_2, Z_3]$ and for a point Z let W_1, W_2, W_3 be the feet of the perpendiculars from Z to the side-lines $Z_2 Z_3, Z_3 Z_1, Z_1 Z_2$, respectively.

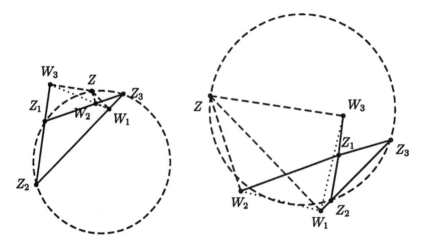

Figure 11.18(a). A Simson-Wallace line.   Figure 11.18(b). A right sensed duo-angle.

Using notation like that in 11.6.3, we suppose that

$$z_1 - z_2 = (p_1 + q_1 i)(z_3 - z_2), \quad z - z_2 = (p + q i)(z_3 - z_2).$$

Then $z = z_2 + p(z_3 - z_2) + q i(z_3 - z_2)$, and so $w_1 = z_2 + p(z_3 - z_2)$. Next, $w_2 = z_3 + s(z_1 - z_3)$, for some $s \in \mathbf{R}$. Hence $z - w_2 = (p - 1 + q i)(z_3 - z_2) - s(z_1 - z_3)$. But $z_1 - z_3 = (p_1 - 1 + q_1 i)(z_3 - z_2)$, so

$$z_3 - z_2 = \frac{1}{p_1 - 1 + q_1 i}(z_1 - z_3).$$

On inserting this, we have that

$$z - w_2 = \left[\frac{p - 1 + qi}{p_1 - 1 + q_1 i} - s \right] (z_1 - z_3).$$

We wish the coefficient of $z_1 - z_3$ to be purely imaginary, and so take

$$s = \frac{(p-1)(p_1-1) + qq_1}{(p_1-1)^2 + q_1^2}.$$

Hence

$$w_2 = z_3 + \frac{(p-1)(p_1-1) + qq_1}{(p_1-1)^2 + q_1^2}(z_1 - z_3).$$

Thirdly, $w_3 = z_2 + t(z_1 - z_2)$, for some $t \in \mathbf{R}$. Hence $z - w_3 = (p + qi)(z_3 - z_2) - t(z_1 - z_2)$. But

$$z_3 - z_2 = \frac{1}{p_1 + q_1 i}(z_1 - z_2),$$

and so

$$z - w_3 = \left[\frac{p + qi}{p_1 + q_1 i} - t \right] (z_1 - z_2).$$

We choose t so that the coefficient of $z_1 - z_2$ is purely imaginary. Thus

$$t = \frac{pp_1 + qq_1}{p_1^2 + q_1^2},$$

which yields

$$w_3 = z_2 + \frac{pp_1 + qq_1}{p_1^2 + q_1^2}(z_1 - z_2).$$

From these expressions for w_1, w_2, w_3 we note that

$$\frac{w_2 - w_1}{w_3 - w_1} = \frac{z_3 - z_2 + \frac{(p-1)(p_1-1)+qq_1}{(p_1-1)^2+q_1^2}(z_1 - z_3) - p(z_3 - z_2)}{\frac{pp_1+qq_1}{p_1^2+q_1^2}(z_1 - z_2) - p(z_3 - z_2)}$$

$$= \frac{1 - p + \frac{(p-1)(p_1-1)+qq_1}{(p_1-1)^2+q_1^2}(p_1 - 1 + q_1 i)}{\frac{pp_1+qq_1}{p_1^2+q_1^2}(p_1 + q_1 i) - p}.$$

The real part of this has numerator $(p_1^2+q_1^2-p_1)(p^2+q^2)-(p_1^2+q_1^2-p_1)p+q_1 q$, and the imaginary part has numerator $q_1(p^2+q^2) - q_1 p - (p_1^2+q_1^2-p_1)q$. If $\theta = \sphericalangle_{\mathcal{F}'} W_3 W_1 W_2$ then *for θ to have a constant magnitude* it is necessary and sufficient that

$$q_1(p^2+q^2) - q_1 p - (p_1^2+q_1^2-p_1)q$$
$$= k[(p_1^2+q_1^2-p_1)(p^2+q^2) - (p_1^2+q_1^2-p_1)p + q_1 q].$$

This can be re-written as

$$a^2 \left[\left(p - \frac{1}{2} \right)^2 + \left(q - \frac{p_1^2+q_1^2-p_1}{2q_1} \right)^2 \right] - \frac{a^2}{4} \left[1 + \frac{(p_1^2+q_1^2-p_1)^2}{q_1^2} \right]$$

$$= \frac{k(p_1^2+q_1^2-p_1)}{q_1} \left\{ a^2 \left[\left(p - \frac{1}{2} \right)^2 + \left(q + \frac{q_1}{2(p_1^2+q_1^2-p_1)} \right)^2 \right] \right.$$

$$\left. - \frac{a^2}{4} \left[1 + \frac{q_1^2}{(p_1^2+q_1^2-p_1)^2} \right] \right\}.$$

On using 11.6.1 we infer that as k varies *this gives the family of coaxal circles which pass through Z_2 and Z_3.*

For W_1, W_2, W_3 *to be collinear,* it is necessary and sufficient that the expression be real. On equating its imaginary part to 0 we obtain $q_1(p^2 + q^2) - q_1 p - (p_1^2 + q_1^2 - p_1)q = 0$. On writing this as

$$\frac{p^2 + q^2 - p}{q} = \frac{p_1^2 + q_1^2 - p_1}{q_1},$$

we note from the formula for a circumcentre in 11.6.4 that it holds when Z *lies on the circumcircle of the triangle* $[Z_1, Z_2, Z_3]$.

This latter result is due to WALLACE, but SIMSON'S name has for a long time been associated with it.

11.7.3   The incentre on the Euler line of a triangle

We suppose that we have the mobile coordinates $z_1 - z_2 = (p_1 + q_1\imath)(z_3 - z_2)$, where p_1 and q_1 are real numbers and $q_1 \neq 0$. Then $z_1 - z_3 = (p_1 - 1 + q_1\imath)(z_3 - z_2)$, and as in 11.6.3 we have

$$\frac{z_1 - z_2}{z_3 - z_2} = p_1 + q_1\imath = \frac{c}{a}\text{cis}\,\beta,$$

$$\frac{z_2 - z_3}{z_1 - z_3} = \frac{1 - p_1 + q_1\imath}{(1 - p_1)^2 + q_1^2} = \frac{a}{b}\text{cis}\,\gamma,$$

$$\frac{z_3 - z_1}{z_2 - z_1} = \frac{p_1 - 1 + q_1\imath}{p_1 + q_1\imath} = \frac{b}{c}\text{cis}\,\alpha,$$

where we are using our sandard notation. Then

$$p_1^2 + q_1^2 = \frac{c^2}{a^2}, \ (1 - p_1)^2 + q_1^2 = \frac{b^2}{a^2}.$$

We recall that the orthocentre, centroid and incentre have mobile coordinates

$$p_1 + \frac{p_1(1 - p_1)}{q_1}\imath, \ \frac{p_1 + 1}{3} + \frac{q_1}{3}\imath,$$

$$\frac{p_1 + \sqrt{p_1^2 + q_1^2}}{1 + \sqrt{p_1^2 + q_1^2}\sqrt{(1 - p_1)^2 + q_1^2}} + \frac{q_1}{1 + \sqrt{p_1^2 + q_1^2}\sqrt{(1 - p_1)^2 + q_1^2}}\imath,$$

respectively.

Now

$$1 - 2p_1 = \frac{b^2 - c^2}{a^2},$$

and so

$$p_1 = \frac{c^2 + a^2 - b^2}{2a^2}, \ 1 - p_1 = \frac{a^2 + b^2 - c^2}{2a^2}, \ p_1 + 1 = \frac{c^2 + a^2 - b^2 + 2a^2}{2a^2}.$$

Moreover

$$q_1^2 = \frac{c^2}{a^2} - p_1^2 = \frac{c^2}{a^2} - \left(\frac{c^2 + a^2 - b^2}{2a^2}\right)^2 = -\frac{(c^2 + a^2 - b^2)^2 - 4c^2a^2}{4a^4}$$

$$= -\frac{[(c + a)^2 - b^2][(c - a)^2 - b^2]}{4a^4},$$

while

$$p_1 + \sqrt{p_1^2 + q_1^2} = \frac{c^2 + a^2 - b^2}{2a^2} + \frac{c}{a} = \frac{(c+a)^2 - b^2}{2a^2},$$

$$1 + \sqrt{p_1^2 + q_1^2} + \sqrt{(1-p_1)^2 + q_1^2} = 1 + \frac{c}{a} + \frac{b}{a} = \frac{a+b+c}{a},$$

so that

$$\frac{p_1 + \sqrt{p_1^2 + q_1^2}}{1 + \sqrt{p_1^2 + q_1^2} + \sqrt{(1-p_1)^2 + q_1^2}} = \frac{(c+a)^2 - b^2}{2a(a+b+c)} = \frac{c+a-b}{2a},$$

$$\frac{q_1^2}{1 + \sqrt{p_1^2 + q_1^2} + \sqrt{(1-p_1)^2 + q_1^2}} = -\frac{1}{4a^3}\frac{[(c+a)^2 - b^2][(c-a)^2 - b^2]}{a+b+c}$$

$$= -\frac{1}{4a^3}(c+a-b)[(c-a)^2 - b^2].$$

The determinant for collinearity, on multiplying the middle column by q_1, is

$$\begin{vmatrix} \frac{c^2+a^2-b^2}{2a^2} & \frac{(c^2+a^2-b^2)(a^2+b^2-c^2)}{4a^4} & 1 \\ \frac{c^2+a^2-b^2+2a^2}{6a^2} & -\frac{[(c+a)^2-b^2][(c-a)^2-b^2]}{12a^4} & 1 \\ \frac{c+a-b}{2a} & -\frac{(c+a-b)[(c-a)^2-b^2]}{4a^3} & 1 \end{vmatrix},$$

and this is a non-zero multiple of

$$\begin{vmatrix} c^2 + a^2 - b^2 & (c^2 + a^2 - b^2)(a^2 + b^2 - c^2) & 1 \\ c^2 + a^2 - b^2 + 2a^2 & -[(c+a)^2 - b^2][(c-a)^2 - b^2] & 3 \\ a(c+a-b) & -a(c+a-b)[(c-a)^2 - b^2] & 1 \end{vmatrix},$$

the value of which is $4(ca^5 - c^3a^3 - ba^5 + a^3b^3 + c^3ba^2 - b^3ca^2)$. This factorizes as $4a^2(b-c)(c-a)(a-b)(a+b+c)$ and so *the incentre lies on the Euler line if and only if the triangle is isosceles.*

11.7.4   Miquel's theorem, 1838

Let Z_1, Z_2, Z_3 be non-collinear points, and $Z_4 \in Z_2Z_3$, $Z_5 \in Z_3Z_1$, $Z_6 \in Z_1Z_2$ be distinct from Z_1, Z_2 and Z_3. Let C_1, C_2, C_3 be the circumcircles of $[Z_1, Z_5, Z_6]$, $[Z_2, Z_6, Z_4]$, $[Z_3, Z_4, Z_5]$, respectively. Then C_1, C_2, C_3 have a point in common.

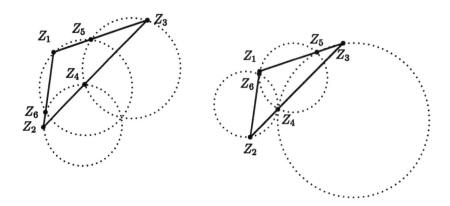

Figure 11.19(a). Miquel's theorem.   Figure 11.19(b). Miquel's theorem.

Proof. Suppose that these circles have centres the points W_1, W_2, W_3, respectively.

We first assume that C_2 and C_3 meet at a second point $Z_7 \neq Z_4$. If $Z_7 = Z_6$ the result is trivially true, so we may exclude that case. As Z_1, Z_2, Z_6 are collinear, we have $z_1 - z_6 = -\nu(z_2 - z_6)$, for some non-zero ν in $\mathbf{R}$. As Z_2, Z_4, Z_6, Z_7 are concyclic, by 10.9.3 we have

$$\frac{z_2 - z_6}{z_7 - z_6} = \rho \frac{z_2 - z_4}{z_7 - z_4},$$

for some non-zero ρ in $\mathbf{R}$. As Z_2, Z_3, Z_4 are collinear, we have $z_2 - z_4 = -\lambda(z_3 - z_4)$, for some non-zero λ in $\mathbf{R}$. As Z_3, Z_4, Z_5, Z_7 are concyclic, we have

$$\frac{z_3 - z_4}{z_7 - z_4} = \sigma \frac{z_3 - z_5}{z_7 - z_5},$$

for some non-zero σ in $\mathbf{R}$. On combining these we have

$$\frac{z_1 - z_6}{z_7 - z_6} = \nu\rho\lambda\sigma \frac{z_3 - z_5}{z_7 - z_5}.$$

It follows by 10.9.3 that Z_1, Z_5, Z_6, Z_7 are concyclic.

We suppose secondly that C_2 and C_3 have a common tangent at Z_2. It is convenient to suppose that $z_1 = z_2 + (p_1 + \imath q_1)(z_3 - z_2)$ and $w_2 = z_2 + (p + \imath q)(z_3 - z_2)$. Then the foot Z_7 of the perpendicular from W_2 to $Z_2 Z_3$ has complex coordinate $z_7 = z_2 + p(z_3 - z_2)$, and hence $z_4 = z_2 + 2p(z_3 - z_2)$. Then the mid-point Z_8 of Z_3 and Z_4 has complex coordinate $z_8 = z_2 + (p + \frac{1}{2})(z_3 - z_2)$. It follows that for the centre W_3 of C_3 we have $w_3 = z_2 + (p + \frac{1}{2} + \imath q')(z_3 - z_2)$, for some real number q'. But Z_4, W_2, W_3 are collinear, so that

$$\begin{vmatrix} 2p & 0 & 1 \\ p & q & 1 \\ p + \frac{1}{2} & q' & 1 \end{vmatrix} = 0,$$

and from this

$$q' = \frac{p - \frac{1}{2}}{p} q \imath(z_3 - z_2).$$

Hence

$$w_3 = z_2 + \left(p + \frac{1}{2} + \frac{p - \frac{1}{2}}{p}qi\right)(z_3 - z_2).$$

From this

$$w_3 - z_3 = \frac{p - \frac{1}{2}}{p}(p + qi)(z_3 - z_2),$$

and since

$$z_3 - z_2 = \frac{1}{p_1 - 1 + iq_1}(z_1 - z_3),$$

we have

$$w_3 = z_3 + \frac{p - \frac{1}{2}}{p}\frac{(p + iq)(p_1 - 1 - iq_1)}{(p_1 - 1)^2 + q_1^2}(z_1 - z_3).$$

Thus if Z_9 is the foot of the perpendicular from W_3 to $Z_3 Z_1$, we have

$$z_9 = z_3 + \frac{p - \frac{1}{2}}{p}\frac{p(p_1 - 1) + qq_1}{(p_1 - 1)^2 + q_1^2}(z_1 - z_3),$$

and so

$$z_5 = z_3 + \frac{2p - 1}{p}\frac{p(p_1 - 1) + qq_1}{(p_1 - 1)^2 + q_1^2}(z_1 - z_3).$$

Similarly

$$w_2 = z_2 + (p + iq)(z_3 - z_2), \quad z_3 - z_2 = \frac{1}{p_1 + iq_1}(z_1 - z_2),$$

and so

$$w_2 = z_2 + \frac{(p + iq)(p_1 - iq_1)}{p_1^2 + q_1^2}(z_1 - z_2).$$

It follows that for the foot Z_{10} of the perpendicular from W_2 to $Z_1 Z_2$ we have

$$z_{10} = z_2 + \frac{pp_1 + qq_1}{p_1^2 + q_1^2}(z_1 - z_2),$$

and so

$$z_6 = z_2 + 2\frac{pp_1 + qq_1}{p_1^2 + q_1^2}(z_1 - z_2).$$

We have

$$
\begin{aligned}
z_1 - z_5 &= \left[1 - \frac{2p - 1}{p}\frac{p(p_1 - 1) + qq_1}{(p_1 - 1)^2 + q_1^2}\right](z_1 - z_3), \\
z_1 - z_6 &= \left[1 - 2\frac{pp_1 + qq_1}{p_1^2 + q_1^2}\right](z_1 - z_2), \\
z_4 - z_6 &= 2p(z_3 - z_2) - 2\frac{pp_1 + qq_1}{p_1^2 + q_1^2}(z_1 - z_2) \\
&= 2\left[p - \frac{pp_1 + qq_1}{p_1^2 + q_1^2}(p_1 + iq_1)\right](z_3 - z_2), \\
z_4 - z_5 &= \left[(2p - 1)(z_3 - z_2) - \frac{2p - 1}{p}\frac{p(p_1 - 1) + qq_1}{(p_1 - 1)^2 + q_1^2}(z_1 - z_3)\right] \\
&= \frac{2p - 1}{p}\left[p - \frac{p(p_1 - 1) + qq_1}{(p_1 - 1)^2 + q_1^2}(p_1 - 1 + q_1)\right](z_3 - z_2).
\end{aligned}
$$

As $z_1 - z_2 = (p_1 + \imath q_1)(z_3 - z_2)$, from these combined we have that

$$\frac{(z_1 - z_5)(z_4 - z_6)}{(z_1 - z_6)(z_4 - z_5)}$$

is a real multiple of

$$\frac{\left[p - \frac{pp_1 + qq_1}{p_1^2 + q_1^2}(p_1 + \imath q_1)\right](p_1 - 1 + \imath q_1)}{\left[p - \frac{p(p_1 - 1) + qq_1}{(p_1 - 1)^2 + q_1^2}(p_1 - 1 + \imath q_1)\right](p_1 + \imath q_1)} = \frac{\left[p - \frac{pp_1 + qq_1}{p_1 - \imath q_1}\right](p_1 - 1 + \imath q_1)}{\left[p - \frac{p(p_1 - 1) + qq_1}{p_1 - 1 - \imath q_1}\right](p_1 + \imath q_1)}$$

$$= \frac{-q_1(\imath p + q)[(p_1 - 1)^2 + q_1^2]}{-q_1(\imath p + q)[p_1^2 + q_1^2]} = \frac{(p_1 - 1)^2 + q_1^2}{p_1^2 + q_1^2},$$

and this is real. It follows that Z_1, Z_4, Z_5, Z_6 are concyclic.
   This is known as MIQUEL'S THEOREM.

11.8

11.8.1   Isogonal conjugates

Definition. Given non-collinear points Z_1, Z_2, Z_3, we say that half-lines $[Z_1, Z_4$, $[Z_1, Z_5$ are **isogonal conjugates** with respect to the angle-support $|Z_2 Z_1 Z_3$ if the sensed angles $\measuredangle_F Z_2 Z_1 Z_4$, $\measuredangle_F Z_5 Z_1 Z_3$, have equal magnitudes.

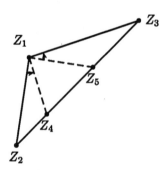

Figure 11.20. Isogonal conjugates.

To see how this operates, we first suppose that Z_4 and Z_5 are both on the line $Z_2 Z_3$ and that

$$Z_4 = \frac{1}{1 + \lambda_1} Z_2 + \frac{\lambda_1}{1 + \lambda_1} Z_3, \quad Z_5 = \frac{1}{1 + \lambda_2} Z_2 + \frac{\lambda_2}{1 + \lambda_2} Z_3,$$

for real numbers λ_1 and λ_2. We recall that then

$$\frac{\overline{Z_2 Z_4}}{\overline{Z_4 Z_3}} = \lambda_1, \quad \frac{\overline{Z_2 Z_5}}{\overline{Z_5 Z_3}} = \lambda_2.$$

Then

$$\frac{z_4 - z_1}{z_2 - z_1} \Big/ \frac{z_3 - z_1}{z_5 - z_1} = \frac{\frac{1}{1+\lambda_1} z_2 + \frac{\lambda_1}{1+\lambda_1} z_3 - z_1}{z_2 - z_1} \cdot \frac{\frac{1}{1+\lambda_2} z_2 + \frac{\lambda_2}{1+\lambda_2} z_3 - z_1}{z_3 - z_1}$$

is positive and so on multiplying across by $(1 + \lambda_1)(1 + \lambda_2)$

$$\frac{z_2 - z_1 + \lambda_1(z_3 - z_1)}{z_2 - z_1} \frac{z_2 - z_1 + \lambda_2(z_3 - z_1)}{z_3 - z_1}$$

$$= \frac{z_2 - z_1}{z_3 - z_1} + \lambda_1 \lambda_2 \frac{z_3 - z_1}{z_2 - z_1} + \lambda_1 + \lambda_2$$

is real. On subtracting $\lambda_1 + \lambda_2$ it follows that

$$\frac{z_2 - z_1}{z_3 - z_1} + \lambda_1 \lambda_2 \frac{1}{(z_2 - z_1)/(z_3 - z_1)}$$

is real. We write

$$\frac{z_2 - z_1}{z_3 - z_1} = u + v\imath, \quad (u, v \in \mathbf{R}),$$

where $v \neq 0$ as $Z_1 \notin Z_2 Z_3$. With this notation we have that

$$u + v\imath + \lambda_1 \lambda_2 \frac{u - v\imath}{u^2 + v^2}$$

is real so that

$$v - \lambda_1 \lambda_2 \frac{v}{u^2 + v^2} = 0,$$

and hence $\lambda_1 \lambda_2 = u^2 + v^2$. We deduce that

$$\lambda_1 \lambda_2 = \frac{|Z_1, Z_2|^2}{|Z_3, Z_1|^2}.$$

As $\lambda_1 \lambda_2 >)$, the points Z_4 and Z_5 must be both in $[Z_2, Z_3]$ or both in $Z_2 Z_3 \setminus [Z_2, Z_3]$.

11.8.2   Concurrency

Given non-collinear points Z_1, Z_2, Z_3, suppose now that for $Z_4, Z_5 \in Z_2 Z_3$ the half-lines $[Z_1, Z_4 \, , [Z_1, Z_5$ are isogonal conjugates with respect to the angle-support $|Z_2 Z_1 Z_3$, for $Z_6, Z_7 \in Z_3 Z_1$ the half-lines $[Z_2, Z_6 \, , [Z_2, Z_7$ are isogonal conjugates with respect to the angle-support $|Z_3 Z_2 Z_1$, and for $Z_8, Z_9 \in Z_1 Z_2$ the half-lines $[Z_3, Z_8 \, , [Z_3, Z_9$ are isogonal conjugates with respect to the angle-support $|Z_1 Z_3 Z_2$. Then by 11.8.1, with an obvious notation we have

$$\lambda_4 \lambda_5 \mu_6 \mu_7 \nu_8 \nu_9 = \frac{|Z_1, Z_2|^2 |Z_2, Z_3|^2 |Z_3, Z_1|^2}{|Z_3, Z_1|^2 |Z_1, Z_2|^2 |Z_2, Z_3|^2} = 1.$$

Now if $Z_1 Z_4$, $Z_2 Z_6$, $Z_3 Z_8$ are concurrent, by Ceva's theorem we have $\lambda_4 \mu_6 \nu_8 = 1$. It follows that $\lambda_5 \mu_7 \nu_9 = 1$, which is the condition (11.5.1) for $Z_1 Z_5$, $Z_2 Z_7$, $Z_3 Z_9$. Then if two of these are concurrent, the third must pass through their point of intersection.

11.8.3 Symmedians

If in 11.8.2 $[Z_1, Z_4$, $[Z_2, Z_6$, $[Z_3, Z_8$ are the median half-lines of $[Z_1, Z_2, Z_3]$, then their isogonal conjugates $[Z_1, Z_5$, $[Z_2, Z_7$, $[Z_3, Z_9$ are called the **symmedians** of the triangle. Now the mid-points Z_4, Z_6, Z_8 lie in the segments $[Z_2, Z_3]$, $[Z_3, Z_1]$, $[Z_1, Z_2]$, respectively. Hence the points Z_5, Z_7, Z_9 lie in the segments $[Z_2, Z_3]$, $[Z_3, Z_1]$, $[Z_1, Z_2]$, respectively. By the cross-bar theorem the symmedian lines $Z_1 Z_5$, $Z_2 Z_7$ must meet and then $Z_3 Z_9$ must pass through their point of intersection. This point of concurrency is called the **symmedian point** of the triangle.

To identify the symmedians further, suppose that O be the centre of the circumcircle of the triangle $[Z_1, Z_2, Z_3]$, and W_1 the mid-point of $\{Z_2, Z_3\}$. Let the line OW_1 meet the circumcircle, on the opposite side of $Z_2 Z_3$ from Z_1, at the point U_1, $[Z_1, W_1$ meet the circle again at Z_4, and the line through Z_4 parallel to $Z_2 Z_3$ meet the circle again at Z_5. Then $|Z_2 Z_1 Z_3$ and $|Z_4 Z_1 Z_5$ have the same mid-line $Z_1 U_1$ as each other. Thus $[Z_1, Z_5$, is isogonal conjugate of the median half-line $[Z_1, Z_4$, and so is the corresponding symmedian.

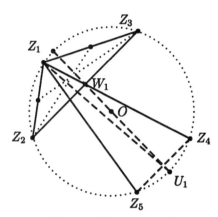

Figure 11.21. Symmedians.

Exercises

11.1 Show that if $Z_1 \neq Z_2$, the points Z on the perpendicular bisector of $[Z_1, Z_2]$ are those for which $\overrightarrow{OZ} = \frac{1}{2}(\overrightarrow{OZ_1} + \overrightarrow{OZ_2}) + t(\overrightarrow{OZ_2} - \overrightarrow{OZ_1})^\perp$, for some $t \in \mathbf{R}$.

11.2 If $l = Z_1 Z_2$, then $Z_4 = \pi_l(Z_3)$ if and only if $Z_4 = Z_1 + s(Z_2 - Z_1)$ where

$$s = \frac{(\overrightarrow{OZ_3} - \overrightarrow{OZ_1}).(\overrightarrow{OZ_2} - \overrightarrow{OZ_1})}{\|\overrightarrow{OZ_2} - \overrightarrow{OZ_1}\|^2},$$

and also if and only if $\overrightarrow{OZ_4} = \overrightarrow{OZ_3} + t(\overrightarrow{OZ_2} - \overrightarrow{OZ_1})^\perp$, where

$$t = -\frac{(\overrightarrow{OZ_3} - \overrightarrow{OZ_1}).(\overrightarrow{OZ_2} - \overrightarrow{OZ_1})^\perp}{\|\overrightarrow{OZ_2} - \overrightarrow{OZ_1}\|^2}.$$

11.3 If $l = Z_1 Z_2$, then $Z_4 = s_l(Z_3)$ if and only if $\overrightarrow{OZ_4} = \overrightarrow{OZ_3} + t(\overrightarrow{OZ_2} - \overrightarrow{OZ_1})^\perp$, where

$$t = -2\frac{(\overrightarrow{OZ_3} - \overrightarrow{OZ_1}).(\overrightarrow{OZ_2} - \overrightarrow{OZ_1})^\perp}{\|\overrightarrow{OZ_2} - \overrightarrow{OZ_1}\|^2}.$$

11.4 If $Z_2 \notin [Z_0, Z_1$ and l is the mid-line of the angle support $|Z_1 Z_0 Z_2$, then $Z \in l$ if and only if

$$\overrightarrow{OZ} = \overrightarrow{OZ_0} + t\left[\frac{1}{\|\overrightarrow{OZ_2} - \overrightarrow{OZ_0}\|}(\overrightarrow{OZ_2} - \overrightarrow{OZ_0}) - \frac{1}{\|\overrightarrow{OZ_1} - \overrightarrow{OZ_0}\|}(\overrightarrow{OZ_1} - \overrightarrow{OZ_0})\right]^\perp,$$

for some $t \in \mathbf{R}$.

11.5 Prove that $\overrightarrow{OZ'} - \overrightarrow{OZ_1} = \cos\alpha(\overrightarrow{OZ} - \overrightarrow{OZ_1}) + \sin\alpha(\overrightarrow{OZ} - \overrightarrow{OZ_1})^\perp$, represents a rotation about the point Z_1, through the angle α.

11.6 For any vector $\overrightarrow{OZ}$, with $Z \equiv_\mathcal{F} (x, y)$, define

$$\overrightarrow{OZ}^\vdash = \overrightarrow{OU}, \quad \overrightarrow{OZ}^\dashv = \overrightarrow{OV},$$

where $U \equiv_\mathcal{F} (x, -y)$, $V \equiv_\mathcal{F} (y, x)$, so that $U = s_{OI}(Z)$, $V = s_l(Z)$, l being the mid-line of $|IOJ$. Prove that

(i)
$$(\overrightarrow{OZ_1} + \overrightarrow{OZ_2})^\vdash = \overrightarrow{OZ_1}^\vdash + \overrightarrow{OZ_2}^\vdash,$$

(ii)
$$(k\overrightarrow{OZ})^\vdash = k(\overrightarrow{OZ})^\vdash,$$

(iii)
$$(\overrightarrow{OZ}^\vdash)^\vdash = \overrightarrow{OZ},$$

(iv)
$$(\overrightarrow{OZ_1} + \overrightarrow{OZ_2})^\dashv = \overrightarrow{OZ_1}^\dashv + \overrightarrow{OZ_2}^\dashv,$$

(v)
$$(k\overrightarrow{OZ})^\dashv = k(\overrightarrow{OZ})^\dashv,$$

(vi)
$$(\overrightarrow{OZ}^\dashv)^\dashv = \overrightarrow{OZ},$$

(vii)
$$(\overrightarrow{OZ}^\vdash)^\dashv = \overrightarrow{OZ}^\perp,$$

(viii)
$$(\overrightarrow{OZ}^\dashv)^\vdash = -\overrightarrow{OZ}^\perp.$$

Prove that axial symmmetry in the line through the origin, with angle of incli-nation α, is given by $\overrightarrow{OZ'} = \cos 2\alpha \overrightarrow{OZ}^\vdash + \sin 2\alpha \overrightarrow{OZ}^\dashv$.

11.7 For non-collinear points Z_1, Z_2, Z_3, suppose that

$$\overrightarrow{OZ_4} = \overrightarrow{OZ_2} + (\overrightarrow{OZ_1} - \overrightarrow{OZ_2})^{\perp}, \quad \overrightarrow{OZ_5} = \overrightarrow{OZ_2} - (\overrightarrow{OZ_3} - \overrightarrow{OZ_2})^{\perp}.$$

Show that then $Z_1 Z_5 \perp Z_3 Z_4$, and in fact

$$\overrightarrow{OZ_5} - \overrightarrow{OZ_1} = \left(\overrightarrow{OZ_4} - \overrightarrow{OZ_3}\right)^{\perp}.$$

11.8 Prove that $\delta_{\mathcal{F}}(Z_1, Z_2, Z_3) + \delta_{\mathcal{F}}(Z_1, Z_4, Z_3) = \delta_{\mathcal{F}}(Z_1, Z_5, Z_3)$, where $(Z_1, Z_2) \uparrow (Z_4, Z_5)$.

11.9 If a point Z be taken on the circumcircle of an equilateral triangle $[Z_1, Z_2, Z_3]$, prove that when Z is on the opposite side of $Z_2 Z_3$ from Z_1, $|Z, Z_1| = |Z, Z_2| + |Z, Z_3|$.

11.10 For a triangle $[Z_1, Z_2, Z_3]$, if Z_4 is the foot of the perpendicular from Z_1 to $Z_2 Z_3$, show that Z_4 is the mid-point of the orthocentre and the point where the line $Z_1 Z_4$ meets the circumcircle again.

11.11 Deduce the Pasch property of a triangle from Menelaus' theorem.

12

Trigonometric functions in calculus; mensuration of discs and circles

COMMENT. In 3.7.1 and 9.3.1 we extended degree-measure to reflex angles, but the only use we have made of this hitherto has been for the purposes of notation in Chapter 9. In this chapter we relate it to the length of an arc of a circle, defining π in the process. In doing this we make a start on calculus for the cosine and sine functins, introduce radian measure of angles and can go on to derive Maclaurin series. We also derive the area of a disk. In doing this we assume from the calculus definition of the length of a rectifiable curve, the Riemann integral and a formula for area in terms of a line integral; this is done so as to save labour, although it would be possible to handle arc-length of a circle and the area of a disk more elementarily.

12.1 REPEATED BISECTION OF AN ANGLE

12.1.1

Addition $\oplus$ is associative on $\mathcal{A}^*(\mathcal{F})$.
  Proof. On using 9.3.2 instead of 9.3.3, as in the proof of 9.3.3(iii) we see that

$$\cos[(\alpha \oplus \beta) \oplus \gamma] = \cos[\alpha \oplus (\beta \oplus \gamma)],$$
$$\sin[(\alpha \oplus \beta) \oplus \gamma] = \sin[\alpha \oplus (\beta \oplus \gamma)].$$

Thus $(\alpha \oplus \beta) \oplus \gamma$ and $\alpha \oplus (\beta \oplus \gamma)$ are the angles in $\mathcal{A}^*(\mathcal{F})$ with the same cosine and the same sine, so they are the same angle except perhaps when $P_3 = Q$ and either angle could be $0_{\mathcal{F}}$ or $360_{\mathcal{F}}$. Thus $(\alpha \oplus \beta) \oplus \gamma = \alpha \oplus (\beta \oplus \gamma)$, except perhaps when either side is a null or full angle. But when each of α, β, γ is $0_{\mathcal{F}}$, then each of $(\alpha \oplus \beta) \oplus \gamma$, $\alpha \oplus (\beta \oplus \gamma)$ is $0_{\mathcal{F}}$ and so they are equal. If any of α, β, γ is non-null, and either of $(\alpha \oplus \beta) \oplus \gamma$, $\alpha \oplus (\beta \oplus \gamma)$ is $0_{\mathcal{F}}$ or $360_{\mathcal{F}}$ then both of them must be $360_{\mathcal{F}}$, and again they are equal.

Definition. Given any angle $\alpha \in \mathcal{A}^*(\mathcal{F})$, in 9.4.1 we defined $\frac{1}{2}\alpha$. Now for any integer $n \geq 1$, $\frac{1}{2^n}\alpha$ is defined inductively by

$$\frac{1}{2^1}\alpha = \frac{1}{2}\alpha, \quad \text{and for all} \quad n \geq 1, \quad \frac{1}{2^{n+1}}\alpha = \frac{1}{2}[\frac{1}{2^n}\alpha].$$

For all $\alpha \in \mathcal{A}^*(\mathcal{F})$, $|\frac{1}{2}\alpha|^\circ = \frac{1}{2}|\alpha|^\circ$.

Proof. We use the notation of 9.4.1. From the definition of $\text{ml}(|QOP)$ this is immediate when $P \in \mathcal{H}_1$, so we suppose that $P \in \mathcal{H}_2$, $P \neq S$, $P \neq Q$. Then

$$\begin{aligned}|\alpha|^\circ &= 180 + |\angle SOP|^\circ = |\angle QOP'|^\circ + |\angle P'OS|^\circ + |\angle SOP|^\circ \\ &= |\angle QOP'|^\circ + |\angle P'OP|^\circ = 2|\angle QOP'|^\circ\end{aligned}$$

by 3.7.1.

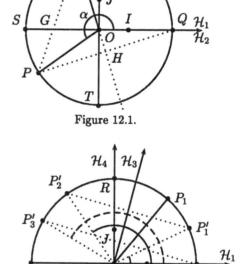

For this, note that $[P, P']$ meets OQ in a point G. If $H = \text{mp}(Q, P)$, then $P' \in OH$ and if $\mathcal{H}_5, \mathcal{H}_6$ are the closed half-planes with common edge OH, with $Q \in \mathcal{H}_5$, then as $H \in [Q, P]$ we have $P \in \mathcal{H}_6$. Then $[P', P] \subset \mathcal{H}_6$ so $G \in \mathcal{H}_6$. But $[0, Q \subset \mathcal{H}_5$, $[0, S \subset \mathcal{H}_6$ so $G \in [0, S$. Thus $[0, S \subset \mathcal{IR}(|P'OP)$.

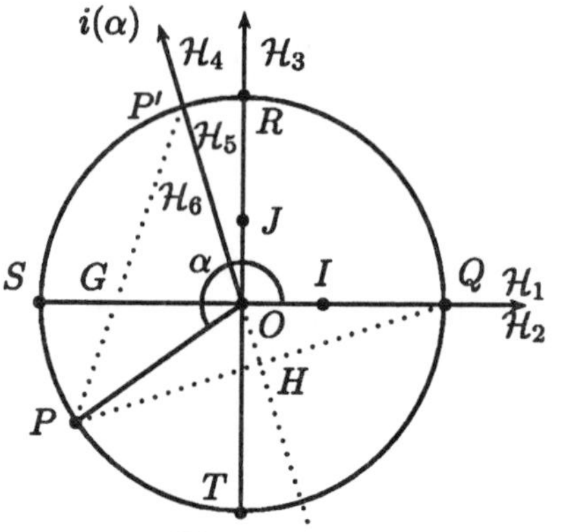

Figure 12.1.

If $\alpha, \beta \in \mathcal{A}^*(\mathcal{F})$ are such that $|\alpha|^\circ + |\beta|^\circ \leq 360$, then

$$\tfrac{1}{2}(\alpha \oplus \beta) = \tfrac{1}{2}\alpha \oplus \tfrac{1}{2}\beta.$$

Proof. By the commutativity and associativity of $\oplus$,

$$\begin{aligned}(\tfrac{1}{2}\alpha \oplus \tfrac{1}{2}\beta) &\oplus (\tfrac{1}{2}\alpha \oplus \tfrac{1}{2}\beta) \\ = (\tfrac{1}{2}\alpha \oplus \tfrac{1}{2}\alpha) &\oplus (\tfrac{1}{2}\beta \oplus \tfrac{1}{2}\beta) \\ = \alpha \oplus \beta,&\end{aligned}$$

so we show that under the condition $|\alpha|^\circ + |\beta|^\circ \leq 360$, if $\frac{1}{2}\alpha \oplus \frac{1}{2}\beta$ has support $|QOP'_3|$ then $P'_3 \in \mathcal{H}_1$.

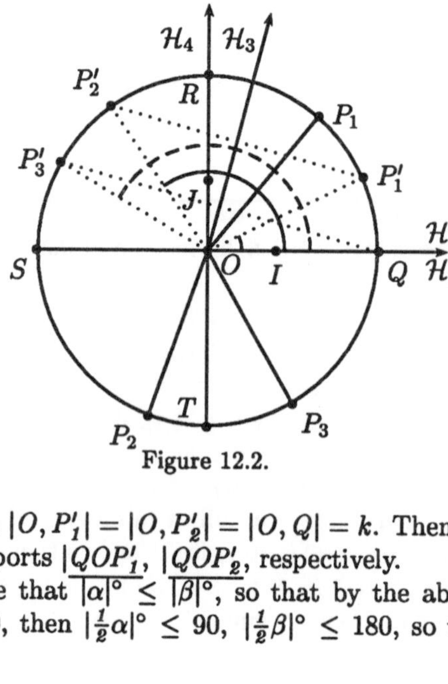

Figure 12.2.

Let $[0, P'_1 = i(\alpha)$, $[0, P'_2 = i(\beta)$ with $|0, P'_1| = |0, P'_2| = |0, Q| = k$. Then $\frac{1}{2}\alpha$ and $\frac{1}{2}\beta$ are the angles in $\mathcal{A}^*(\mathcal{F})$ with supports $|QOP'_1|$, $|QOP'_2|$, respectively.

Without loss of generality we suppose that $\overline{|\alpha|^\circ} \leq \overline{|\beta|^\circ}$, so that by the above, $|\frac{1}{2}\alpha|^\circ \leq |\frac{1}{2}\beta|^\circ$. As $|\frac{1}{2}\alpha|^\circ + |\frac{1}{2}\beta|^\circ \leq 180$, then $|\frac{1}{2}\alpha|^\circ \leq 90$, $|\frac{1}{2}\beta|^\circ \leq 180$, so that

$P_1' \in \mathcal{H}_1$, $P_2' \in \mathcal{H}_1$ and $[O, P_1' \subset \mathcal{IR}(|QOP_2')$. Let the angle in $\mathcal{A}^*(\mathcal{F})$ with degree measure $\frac{1}{2}(|\alpha|° + |\beta|°)$ have support $|\overline{QOP_3'}$. Then as $\frac{1}{2}(|\alpha|° + |\beta|°) \leq 180$ we have $P_3' \in \mathcal{H}_1$.

Let the angle in $\mathcal{A}^*(\mathcal{F})$ with degree measure $\frac{1}{4}(|\alpha|° + |\beta|°)$ have support $|\overline{QOW}$, where $|O, W| = k$. Then $W \in \mathcal{H}_1$ and $[O, P_1' \subset \mathcal{IR}(|QOW)$ as $\frac{1}{2}|\alpha|° \leq \frac{1}{4}(|\alpha|° + |\beta|°)$. Thus

$$|\angle P_1' OW|° = |\angle QOW|° - |\angle QOP_1'|° = \tfrac{1}{4}(|\alpha|° + |\beta|°) - \tfrac{1}{2}|\alpha|° = \tfrac{1}{4}(|\beta|° - |\alpha|°).$$

As $\frac{1}{4}(|\alpha|° + |\beta|° \leq \frac{1}{2}|\beta|°$ we have $[O, W \subset \mathcal{IR}(|QOP_2')$. Then

$$|\angle WOP_2'|° = |\angle QOP_2'|° - |\angle QOW|° = \tfrac{1}{2}|\beta|° - \tfrac{1}{4}(|\alpha|° + |\beta|°) = \tfrac{1}{4}(|\beta|° - |\alpha|°).$$

As $|\angle P_1' OW|° = |\angle WOP_2'|°$ we have $OW = \text{ml}(|P_1' OP_2')$.

By a similar argument

$$|\angle QOW|° = \tfrac{1}{4}(|\alpha|° + |\beta|°) = |\angle WOP_3'|°,$$

so $OW = \text{ml}(|QOP_3')$. Then $P_3' = s_{OW}(Q)$ and so $\frac{1}{2}\alpha \oplus \frac{1}{2}\beta$ has support $|QOP_3'$.

When $\frac{1}{2}(|\alpha|° + |\beta|°) < 180$ this gives that $\frac{1}{2}\alpha \oplus \frac{1}{2}\beta = \frac{1}{2}(\alpha \oplus \beta)$. The remaining case is when $|\alpha|° + |\beta|° = 360$. Then $\frac{1}{2}(|\alpha|° + |\beta|°) = 180$ so $\frac{1}{2}\alpha \oplus \frac{1}{2}\beta = 180_\mathcal{F}$, while $\alpha \oplus \beta = 360_\mathcal{F}$ so $\frac{1}{2}(\alpha \oplus \beta) = 180_\mathcal{F}$.

COROLLARY 1. *If $\alpha, \beta \in \mathcal{A}^*(\mathcal{F})$ are such that $|\alpha|° + |\beta|° \leq 360$, then for all integers $n \geq 1$,*

$$\frac{1}{2^n}(\alpha \oplus \beta) = \frac{1}{2^n}\alpha \oplus \frac{1}{2^n}\beta.$$

COROLLARY 2. *If $\alpha, \beta \in \mathcal{A}^*(\mathcal{F})$ are such that $|\alpha|° + |\beta|° \leq 360$, then*

$$|\alpha \oplus \beta|° = |\alpha|° + |\beta|°.$$

Proof. In the course of the proof we saw that, under the given conditions,

$$|\tfrac{1}{2}\alpha \oplus \tfrac{1}{2}\beta|° = \tfrac{1}{2}(|\alpha|° + |\beta|°), \quad \tfrac{1}{2}\alpha \oplus \tfrac{1}{2}\beta = \tfrac{1}{2}(\alpha + \beta).$$

From these $|\frac{1}{2}(\alpha \oplus \beta)|° = \frac{1}{2}(|\alpha|° + |\beta|°)$, so that $|\alpha \oplus \beta|° = |\alpha|° + |\beta|°$.

12.2   CIRCULAR FUNCTIONS

12.2.1

Definition. Let $x \in \mathbf{R}$ satisfy $0 \leq x \leq 360$. Then if for $\alpha \in \mathcal{A}^*(\mathcal{F})$ we have $|\alpha|° = x$, we write $c(x) = \cos\alpha$, $s(x) = \sin\alpha$. By 9.6, c and s do not depend on $\mathcal{F}$.

The functions c and s have the following properties:-

(i) *For $0 \leq x \leq 360, 0 \leq y \leq 360, x + y \leq 360$,*

$$c(x + y) = c(x)c(y) - s(x)s(y), \quad s(x + y) = s(x)c(y) + c(x)s(y).$$

(ii) *For $0 \leq x \leq 360$, $c(x)^2 + s(x)^2 = 1$.*

(iii) *For* $0 \leq x \leq 360$, $c(x) = 1 - 2s(\frac{1}{2}x)^2$, $s(x) = 2c(\frac{1}{2}x)s(\frac{1}{2}x)$.

Proof. These follow immediately from the definition of c and s, and the formulae for $\cos(\alpha \oplus \beta)$ and $\sin(\alpha \oplus \beta)$.

Definition. For $0 \leq x \leq 360$ and integers $n \geq 3$, let

$$u_n(x) = 2^n s\left(\frac{x}{2^{n+1}}\right), \quad v_n(x) = 2^n \frac{s\left(\frac{x}{2^{n+1}}\right)}{c\left(\frac{x}{2^{n+1}}\right)}.$$

COMMENT. When $n \geq 1$, u_n is the area of the union of 2^{n+1} non-overlapping congruent triangles, each having vertex at the centre of the circle/arc, and each having the end-points of its base on the circle/arc.

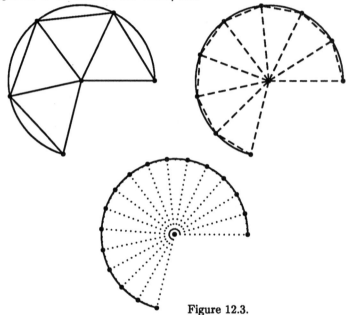

Figure 12.3.

When $n \geq 1$, v_n is the area of the union of 2^n non-overlapping congruent triangles, each having vertex at the centre of the circle/arc, and each having its base-line a tangent to the circle/arc. Our reason for taking $n \geq 3$ is that then

$$\frac{x}{2^{n+1}} \leq \frac{360}{16} = 25 < 45.$$

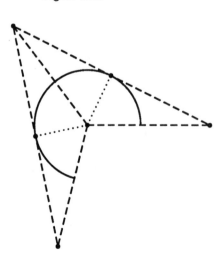

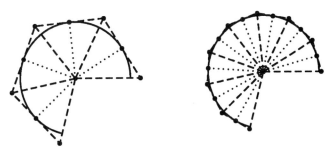

Figure 12.4

The sequences u_n, v_n have the following properties:-

(i) (u_n) *is non-decreasing.*

(ii) (v_n) *is non increasing.*

(iii) *For $0 < x \le 360$ and $n \ge 2$, we have $0 < u_n(x) \le v_n(x)$.*

(iv) *As $n \to \infty, u_n(x)$ and $v_n(x)$ tend to a common limit $\mu(x)$ and*

$$u_n(x) \le \mu(x) \le v_n(x).$$

Proof.
(i) For

$$u_n(x) = 2^{n+1} s\left(\frac{x}{2^{n+2}}\right) c\left(\frac{x}{2^{n+2}}\right) \le u_{n+1}(x).$$

(ii) For

$$v_n(x) = \frac{2^{n+1} s\left(\frac{x}{2^{n+2}}\right) c\left(\frac{x}{2^{n+2}}\right)}{2c\left(\frac{x}{2^{n+2}}\right)^2 - 1} = v_{n+1}(x) \frac{c\left(\frac{x}{2^{n+2}}\right)^2}{2c\left(\frac{x}{2^{n+2}}\right)^2 - 1} > v_{n+1}(x)$$

as $c(\frac{x}{2^{n+2}})^2 < 1$ and so $2c(\frac{x}{2^{n+2}})^2 - 1 < c(\frac{x}{2^{n+2}})^2$, while $c(\frac{x}{2^{n+1}}) > 0$ and thus $2c(\frac{x}{2^{n+2}})^2 - 1 > 0$.
(iii) For

$$u_n(x) = v_n(x)c\left(\frac{x}{2^{n+1}}\right) \le v_n(x).$$

(iv) By (i), (ii) and (iii), $u_n(x) \le v_3(x)$ so we have a non-decreasing sequence which is bounded above. Hence there is some $\mu(x)$ such that $\lim_{n\to\infty} u_n(x) = \mu(x)$. It follows that $s(\frac{x}{2^{n+1}}) \to 0 \ (n \to \infty)$ and hence

$$c\left(\frac{x}{2^{n+1}}\right) = 1 - 2s\left(\frac{x}{2^{n+2}}\right)^2 \to 1 \ (n \to \infty).$$

But then

$$\frac{v_n(x)}{u_n(x)} = c\left(\frac{x}{2^{n+1}}\right)^{-1} \to 1 \ (n \to \infty),$$

and so as well $\lim_{n\to\infty} v_n(x) = \mu(x)$. As $u_n(x)$ is non-decreasing and $v_n(x)$ is non-increasing we must have $u_n(x) \le \mu(x) \le v_n(x)$.

For $x \geq 0$, $y \geq 0$, $x + y \leq 360$,

$$\mu(x + y) = \mu(x) + \mu(y).$$

Proof. For

$$\begin{aligned}
u_n(x + y) &= 2^n s\left(\frac{x+y}{2^{n+1}}\right) = 2^n s\left(\frac{x}{2^{n+1}} + \frac{y}{2^{n+1}}\right) \\
&= 2^n s\left(\frac{x}{2^{n+1}}\right) c\left(\frac{y}{2^{n+1}}\right) + c\left(\frac{x}{2^{n+1}}\right) 2^n s\left(\frac{y}{2^{n+1}}\right) \\
&\to \mu(x).1 + 1.\mu(y) \ (n \to \infty).
\end{aligned}$$

COMMENT. This equation for μ is referred to as DARBOUX'S FUNCTIONAL EQUA-
TION.

There is a number $p > 0$ such that for $0 \leq x \leq 360$, $\mu(x) = px$.
Proof. We start by writing out a well known proof that any function μ which
satisfies Darboux's functional equation on $[0, 360]$ satisfies $\mu(r) = r\mu(1)$ for all rational
numbers r in $[0, 360]$. On first taking $y = x$ we have $\mu(2x) = 2\mu(x)$; then taking
$y = 2x$, we get $\mu(3x) = 3\mu(x)$; next taking $y = 3x$, we get $\mu(4x) = 4\mu(x)$; and
proceeding this way we evidently get

$$\mu(nx) = n\mu(x) \text{ for all } n \in \mathbf{N} \text{ and } 0 \leq x \leq \frac{360}{n}.$$

This can be confirmed by induction on n.
On applying this with $n = m$, $x = 1/m$ for any $m \in \mathbf{N}$ we deduce that

$$\mu(1) = m\mu\left(\frac{1}{m}\right)$$

so that

$$\mu\left(\frac{1}{m}\right) = \frac{1}{m}\mu(1).$$

Now on taking n and $x = 1/m$ we see that

$$\mu\left(\frac{n}{m}\right) = \mu\left(n.\frac{1}{m}\right) = n\mu\left(\frac{1}{m}\right) = n.\frac{1}{m}\mu(1) = \frac{n}{m}\mu(1).$$

Thus $\mu(r) = r\mu(1)$ for every rational number r with $0 < r \leq 360$, and this extends
to the case $r = 0$ also as putting $x = y = 0$ in Darboux's functional equation gives
$\mu(0) = 0$.
It remains to show that $\mu(y) = y\mu(1)$ for each irrational number y with $0 < y <
360$. Choose a decreasing sequence of rational numbers $t_n(y)$ which converge to y.
By the case for rational numbers, we have $\mu(t_n) = t_n\mu(1)$. Now $t_n = (t_n - y) + y$ so
by Darboux's functional equation

$$\mu(t_n) = \mu(t_n - y) + \mu(y),$$

and so

$$\mu(y) = t_n\mu(1) - \mu(t_n - y).$$

We wish to make $n \to \infty$ in this and take limits.

Now as defined $\mu(z) > 0$ for $0 < z < 360$. Then if $0 < x < y < 360$ as $\mu(x) + \mu(y - x) = \mu(y)$ and $\mu(y - x) > 0$, we must have $\mu(x) < \mu(y)$. Thus μ is an increasing function. Hence as z decreases to 0, $\mu(z)$ decreases and is positive and so the one-sided limit

$$l = \lim_{z \to 0+} \mu(z) = \lim_{z \to 0, z > 0} \mu(z)$$

must exist. But the formula $\mu(\frac{1}{n}) = \frac{\mu(1)}{n}$ shows that $\mu(\frac{1}{n}) \to 0$ $(n \to \infty)$ and so we must have $l = 0$. As $t_n - y \to 0$ it follows that $\mu(t_n - y) \to 0$ $(n \to \infty)$ and then

$$\lim_{n \to \infty} \mu(y) = \lim_{n \to \infty} [t_n \mu(1)] + \lim_{n \to \infty} \mu(t_n - y)$$

yields $\mu(y) = y\mu(1) + 0$.

To conclude we take $p = \mu(1)$ and note that this is positive as shown above.

12.2.2   Definition of π

Definition. We denote $\mu(360)$ by π. Then $360p = \pi$ so that $p = \frac{\pi}{360}$ and so

$$\mu(x) = \frac{\pi}{360} x.$$

12.3   DERIVATIVES OF COSINE AND SINE FUNCTIONS

12.3.1

With the notation of 12.1,

$$\lim_{t \to 0+} \frac{s(t)}{t} = \frac{\pi}{180}.$$

Proof. We have seen above that for $0 < x \leq 360$, taking $n = 2$ in 12.2.1(iv),

$$0 < s\left(\frac{x}{8}\right) \leq \frac{\pi}{180} \frac{x}{8} \leq \frac{s\left(\frac{x}{8}\right)}{c\left(\frac{x}{8}\right)},$$

and so for $0 < t < 45$,

$$0 < s(t) \leq \frac{\pi}{180} t \leq \frac{s(t)}{c(t)}.$$

From the first two inequalities here we infer that $s(t) \to 0$ $(t \to 0+)$ and hence

$$c(t) = 1 - 2s\left(\frac{t}{2}\right)^2 \to 1 \ (t \to 0+).$$

As

$$\frac{\pi}{180} c(t) \leq \frac{s(t)}{t} \leq \frac{\pi}{180},$$

the result follows.

For $0 < x < 360$ the derivatives $c'(x)$ and $s'(x)$ exist and are given by

$$c'(x) = -\frac{\pi}{180} s(x), \ s'(x) = \frac{\pi}{180} c(x).$$

Proof. Suppose that $0 < x < 360$, and $h > 0$ is so small that $x+h < 360$, $x-h > 0$. Then

$$\frac{c(x+h) - c(x)}{h} = \frac{c(x)c(h) - s(x)s(h) - c(x)}{h} = c(x)\frac{c(h) - 1}{h} - s(x)\frac{s(h)}{h}.$$

But $\frac{c(h)-1}{h} = -\frac{2s(\frac{h}{2})^2}{h} \to 0$ $(h \to 0+)$ and so

$$\frac{c(x+h) - c(x)}{h} \to -\frac{\pi}{180}s(x) \ (h \to 0+).$$

Also

$$
\begin{aligned}
c(x) &= c(x - h + h) = c(x - h)c(h) + s(x - h)s(h), \\
s(x) &= s(x - h + h) = c(x - h)s(h) + s(x - h)c(h).
\end{aligned}
$$

As $c(h)^2 + s(h)^2 = 1$ we can solve these equations by Cramer's rule to obtain

$$c(x - h) = c(x)c(h) + s(x)s(h), \ s(x - h) = s(x)c(h) - c(x)s(h).$$

From the first of these, by a similar argument to that above, it follows that

$$\frac{c(x - h) - c(x)}{-h} \to -\frac{\pi}{180}s(x) \ (h \to 0+).$$

These results combined show that $c'(x) = -\frac{\pi}{180}s(x)$.
The other result is proved similarly.

12.4 PARAMETRIC EQUATIONS FOR A CIRCLE

12.4.1 Area of a disk

The disk $\mathcal{D}(O; k) = \{Z : |O, Z| \le k\}$ has area πk^2.
   Proof. We use a well-known expression for area as the value of a line integral; see e.g. Apostol [1, Volume II, page 383].
   If γ is the path with equation $x = kc(t),\ y = ks(t)$ $(0 \le t \le 360)$, then γ traverses the circle $C(O, k)$ once. Moreover if $O \equiv (0,0)$, $P \equiv (kc(t), ks(t))$, $T \equiv (kc'(t), ks'(t)) = \left(-\frac{\pi ks(t)}{180}, \frac{\pi kc(t)}{180}\right)$, we have that

$$\delta_{\mathcal{F}}(O, P, T) = \frac{\pi}{360}k^2[c^2(t) + s^2(t)] = \frac{\pi}{360}k^2 > 0.$$

Now $C(O; k)$ is the boundary of $\mathcal{D}(O; k)$ and so the area of $\mathcal{D}(O; k)$ is given by

$$
\begin{aligned}
\frac{1}{2}\int_\gamma x \, dy - y \, dx &= \frac{1}{2}\int_0^{360} k^2[c(t)s'(t) - s(t)c'(t)] \, dt \\
&= \frac{1}{2}\int_0^{360} \frac{\pi}{180}k^2 \, dt = \pi k^2.
\end{aligned}
$$

12.4.2   Length of an arc of a circle

For $P \in C(O; k)$ let α be an angle in $\mathcal{A}^*(\mathcal{F})$ with support $|QOP$. Let $|\alpha|^\circ = \theta$ and γ be the curve with parametric equations $x = kc(t)$, $y = ks(t)$ $(0 \leq t \leq \theta)$. Then γ has the end-points Q and P and has length $\frac{\pi}{180}\theta k$.

Proof. The length of γ is given by

$$\int_0^\theta \sqrt{[kc'(t)]^2 + [ks'(t)]^2}\, dt = \frac{\pi}{180}\theta k.$$

NOTE. In the last result, when $\theta = 360$ γ is the circle $C(O; k)$ and has length 2π. When $\theta = 180$, γ is a semi-circle and has length π. For other cases, we note that the line QP has equation

$$s(\theta)(x - k) - [c(\theta) - 1]y = 0.$$

The left-hand side in this has the value $-ks(\theta)$ when $x = y = 0$, and when $x = kc(t)$, $y = ks(t)$ the value

$$\begin{aligned}
k\{s(\theta)[c(t) - 1] - [c(\theta) - 1]s(t)\} &= k[-2s(\theta)s^2(\tfrac{1}{2}t) + 2s^2(\tfrac{1}{2}\theta)s(t)] \\
&= 4ks(\tfrac{1}{2}\theta)s(\tfrac{1}{2}t)\left[s(\tfrac{1}{2}\theta)c(\tfrac{1}{2}t) - s(\tfrac{1}{2}t)c(\tfrac{1}{2}\theta)\right] \\
&= 4ks(\tfrac{1}{2}\theta)s(\tfrac{1}{2}t)s(\tfrac{1}{2}\theta - \tfrac{1}{2}t) \\
&> 0,\ \text{for } 0 < t < \theta.
\end{aligned}$$

When $\theta < 180$, we have $-s(\theta) < 0$ so that $Z \equiv (kc(t), ks(t))$ is in the closed half-plane $\mathcal{H}_6$ with edge QP which does not contain O. Thus, recalling 7.5, γ is the minor arc with end-points Q and P.

When $180 < \theta < 360$, $-s(\theta) > 0$ so Z is in the closed half-plane $\mathcal{H}_5$ with edge QP which contains O. Thus γ is the major arc with end-points Q and P.

12.4.3   Radian measure

Definition. If, for any angle α, $x = |\alpha|^\circ$ then $\frac{\pi}{180}x$ is called the **radian measure** of α, and denoted by $|\alpha|^r$. We also define, for $0 \leq x \leq 2\pi$,

$$C(x) = c\left(\frac{180}{\pi}x\right),\ \ S(x) = s\left(\frac{180}{\pi}x\right),$$

and then by the chain rule have

$$C'(x) = -S'(x),\ S'(x) = C(x),$$

for $0 < x < 2\pi$.

12.5   EXTENSION OF DOMAINS OF COSINE AND SINE

12.5.1

NOTE. If we define $E(x)$ by $E(x) = C(x) + \imath S(x)$ $(0 \leq x \leq 2\pi)$, where $\imath^2 = -1$, then by 12.2.1(i)

$$E(x + y) = E(x)E(y)\ \ (0 \leq x \leq 2\pi), \tag{12.5.1}$$

and by 12.4.3
$$E'(x) = \imath E(x) \ (0 < x < 2\pi).$$

It is natural to enquire if the definition of E can be extended from $[0, 2\pi]$ in such a way as to preserve its basic properties.

From (12.5.1) we note that in particular

$$E(x + \pi) = -E(x) \ (0 \le x \le \pi),$$

as $E(\pi) = -1$, and if we specify that this is to hold for all $x \in \mathbf{R}$ then the domain of definition of E becomes extended to $\mathbf{R}$.

It follows that $E(x + 2\pi) = E(x)$ for all $x \in \mathbf{R}$ and so for all $n \in \mathbf{Z}$, $E(x + 2n\pi) = E(x)$. By the chain rule, for $0 < x < 2\pi$,

$$E'(x + 2n\pi) = E'(x) = \imath E(x) = \imath E(x + 2n\pi).$$

Thus for each $n \in \mathbf{Z}$, $E'(x)$ exists and satisfies $E'(x) = \imath E(x)$ for $2n\pi < x < (2n+2)\pi$. Moreover for $0 < h < 2\pi$,

$$\frac{E(2n\pi + h) - E(2n\pi)}{h} = \frac{E(h) - 1}{h} = \frac{C(h) - 1}{h} + \imath \frac{S(h)}{h} \to \imath \ (h \to 0),$$

while for $-2\pi < h < 0$,

$$\frac{E(2n\pi + h) - E(2n\pi)}{h} = \frac{E(2\pi + h) - 1}{h} = \frac{-E(\pi + h) - 1}{h}$$

$$= -\frac{E(\pi + h) - E(\pi)}{h} \to -E'(\pi) \ (h \to 0)$$

$$= \imath.$$

Thus $E'(2n\pi)$ exists for each $n \in \mathbf{Z}$ and as its value there is $\imath = \imath E(2n\pi)$, we see that $E'(x) = \imath E(x)$ for all $x \in \mathbf{R}$.

We note that $|E(x)| = \sqrt{C(x)^2 + S(x)^2} = 1$ for $0 \le x \le 2\pi$ and then as E has period 2π, $|E(x)| = 1$ for all x. Hence $E(x) \ne 0$ for all $x \in \mathbf{R}$.

Then for any fixed $y \in \mathbf{R}$,

$$\frac{d}{dx} \frac{E(x + y)}{E(x)} = \frac{E(x)\imath E(x + y) - E(x + y)\imath E(x)}{E(x)^2} = 0,$$

so for some complex constant c,

$$E(x + y) = cE(x), \text{ for all } x \in \mathbf{R}.$$

On putting $x = 0$ we see that $c = E(y)$ and so $E(x + y) = E(x)E(y)$ for all $x, y \in \mathbf{R}$.

From $E'(x) = \imath E(x) \ \forall \ x \in \mathbf{R}$, we deduce that $C'(x) = -S(x)$, $S'(x) = C(x)$, $\forall \ x \in \mathbf{R}$. From these the usual Maclaurin series for $C(x)$ and $S(x)$ (or $E(x)$) can be derived and tables of approximations to them prepared; see e.g. Apostol [1, Volume I, pages 435–436]. By introducing $\arctan x$, π could be expressed as the sum of infinite series and these used to prepare approximations efficiently; see e.g. Apostol [1, Volume I, pages 253–255, 284–285].

12.5.2

Given a frame of reference $\mathcal{F}$, let $Q \equiv (1,0)$ and consider the points Z of the unit circle $\mathcal{C}(O;1)$ and the angles $\theta = \angle_{\mathcal{F}} QOZ$ in $\mathcal{A}(\mathcal{F})$. If $s = |\theta|^r$ then $Z \sim \operatorname{cis} \theta = C(s) + \imath S(s)$. Any number $t \in \mathbf{R}$ such that

$$C(t) + \imath S(t) = C(s) + \imath S(s)$$

is called a radian measure of θ; collectively we shall call them the **multiple radian measures** of θ. They have the form $t = s + 2n\pi\imath$ where $n \in \mathbf{Z}$. We write $t \sim \angle_{\mathcal{F}} QOZ$.

Suppose now that $t_1 \sim \angle_{\mathcal{F}} QOZ_1 = \theta_1$ and $t_2 \sim \angle_{\mathcal{F}} QOZ_2 = \theta_2$. Then

$$
\begin{aligned}
E(t_1) = C(t_1) + \imath S(t_1) &= \cos\theta_1 + \imath \sin\theta_1, \\
E(t_2) = C(t_2) + \imath S(t_2) &= \cos\theta_2 + \imath \sin\theta_2,
\end{aligned}
$$

and from this by multiplication we have that

$$E(t_1 + t_2) = E(t_1)E(t_2) = \cos(\theta_1 + \theta_2) + \imath \sin(\theta_1 + \theta_2).$$

Thus $t_1 + t_2 \sim \angle_{\mathcal{F}} QOZ_1 + \angle_{\mathcal{F}} QOZ_2$ and so addition of multiple measures corresponds to addition of angles.

Given any $t \in \mathbf{R}$, then $\gamma(u) = E(ut)$ $(0 \le u \le 1)$ gives a curve on the unit circle, with initial point Q and terminal point Z. If $t \notin [-2\pi, 2\pi]$, γ winds around the origin and the curves generated in this way give us some pictorial feel for these multiple radian measures of angles.

Appendix A

List of Axioms

AXIOM A_1. *Each line is a proper non-empty subset of Π. For each set $\{A, B\}$ of two distinct points in Π, there is a unique line in Λ to which A and B both belong.*

AXIOM A_2. *Each natural order $\leq_l$ has the properties:-*

(i) *$A \leq_l A$ for all points $A \in l$;*

(ii) *if $A \leq_l B$ and $B \leq_l C$ then $A \leq_l C$;*

(iii) *if $A \leq_l B$ and $B \leq_l A$, then $A = B$;*

(iv) *for any points $A, B \in l$, either $A \leq_l B$ or $B \leq_l A$.*

AXIOM A_3. *Open half-planes $\mathcal{G}_1, \mathcal{G}_2$ with common edge l have the properties:-*

(i) *$\Pi \setminus l = \mathcal{G}_1 \cup \mathcal{G}_2$;*

(ii) *$\mathcal{G}_1$ and $\mathcal{G}_2$ are both convex sets;*

(iii) *if $P \in \mathcal{G}_1$ and $Q \in \mathcal{G}_2$, then $[P, Q] \cap l \neq \emptyset$.*

AXIOM A_4. *Distance has the following properties:-*

(i) *$|A, B| \geq 0$ for all $A, B \in \Pi$;*

(ii) *$|A, B| = |B, A|$ for all $A, B \in \Pi$;*

(iii) *if $Q \in [P, R]$, then $|P, Q| + |Q, R| = |P, R|$;*

(iv) *given any $k \geq 0$ in $\mathbf{R}$, any line $l \in \Lambda$, any point $A \in l$ and either natural order $\leq_l$ on l, there is a unique point $B \in l$ such that $A \leq_l B$ and $|A, B| = k$, and a unique point $C \in l$ such that $C \leq_l A$ and $|A, C| = k$.*

AXIOM A_5. *Degree-measure $|\quad|^\circ$ of angles has the following properties:-*

(i) *In all cases $|\alpha|^\circ \geq 0$;*

(ii) *if α is a straight-angle, then $|\alpha|° = 180$;*

(iii) *if $\angle BAC$ is a wedge-angle and the point $D \neq A$ lies in the interior region $\mathcal{IR}(|\underline{BAC})$, then*

$$|\angle BAD|° + |\angle DAC|° = |\angle BAC|°,$$

while if $|\underline{BAC}$ is a straight angle-support and $D \notin AB$, then

$$|\angle BAD|° + |\angle DAC|° = 180;$$

(iv) *if $B \neq A$, if $\mathcal{H}_1$ is a closed half-plane with edge AB, and if the half-lines $[A,C$ and $[A,D$ in $\mathcal{H}_1$ are such that $|\angle BAC|° = |\angle BAD|°$, then $[A,D = [A,C$;*

(v) *if $B \neq A$, if $\mathcal{H}_1$ is a closed half-plane with edge AB, and if $0 < k < 180$, then there is a half-line $[A,C$ in $\mathcal{H}_1$ such that $|\angle BAC|° = k$.*

AXIOM A_6. *If triangles T and T', with vertices $\{A,B,C\}$ and $\{A',B',C'\}$, respectively, are such that*

$$|C,A| = |C',A'|, \ |A,B| = |A',B'|, \ |\angle BAC|° = |\angle B'A'C'|°,$$

then $T \underset{(A,B,C)\overset{\equiv}{\Rightarrow}(A',B'C')}{} T'.$

AXIOM A_7. *Given any line $l \in \Lambda$ and any point $P \notin l$, there is at most one line m such that $P \in m$ and $l \parallel m$.*

References

[1] T.M. Apostol. *Calculus*, volume I and II. Blaisdell, second edition, 1967.

[2] J.K. Baumgart et al., editors. *Historical Topics for the Mathematics Classroom*. National Council of Teachers of Mathematics, second edition, 1989.

[3] F. Cajori. *A History of Elementary Mathematics*. Macmillan, 1896.

[4] H.S.M. Coxeter. *Introduction to Geometry*. Wiley, third edition, 1969.

[5] H.S.M. Coxeter and S.L. Greitzer. *Geometry Revisited*. Mathematical Association of America, 1967.

[6] G.H. Forder. *The Foundations of Euclidean Geometry*. Cambridge, Dover, 1931, 1958.

[7] G.H. Forder. *Higher Course Geometry*. Cambridge, second edition, 1949.

[8] L. Hahn. *Complex Numbers and Geometry*. Mathematical Association of America, 1994.

[9] R.A. Johnson. *Advanced Euclidean Geometry (Modern Geometry)*. Houghton Mifflin, Dover, 1929, 1960.

[10] W. Ledermann and S. Vajda. *Handbook of Applicable Mathematics: Combinatorics and Geometry*, volume V, part A. Wiley, 1985.

[11] C.M. McGregor, J.J.C. Nimmo, and W.W. Stothers. *Fundamentals of University Mathematics*. Horwood Publishing (begun as Albion Publishing, renamed), Chichester, West Sussex, 1994.

[12] G.C. Smith. *Introductory Mathematics: Algebra and Analysis*. Springer, 1998.

[13] R.E. Wheeler and E.R. Wheeler. *Modern Mathematics for Elementary School Teachers*. Brooks/Cole, ninth edition, 1995.

Index

Horwood Series: Mathematics and Applications

GEOMETRY OF NAVIGATION
ROY WILLIAMS, Master Mariner, BSc, PhD, FRIN, AFIMA
ISBN: 1-898563-46-2   *ca*. 144 pages   Hardback   1999

Contents: Geometrical representation of the earth: Mathematics of chart projections: Navigating along Rhumb lines: Shortest paths on the surface of a sphere: Shortest paths on the surface of an ellipsoid: Paths between nearly antipodean points: Great ellipse on the surface of an ellipsoid: Navigating along the arc of a small ellipse: Surface position from astronomical observation: Surface position from satellite data: Appendices: Table of latitude parts (meridian distance): Transformation between equations: Direct cubic spline approximation.

Journal of Navigation (Professor J. Kemp)
"Sets out [his] results in the form of 'computational procedures' which can be followed by non-specialists. Roy Williams has undertaken a difficult, but much-needed, task in applying modern mathematical methods to both old and new navigational problems. He has met this challenge with considerable success in this well-produced and clearly presented book."

MATHEMATICAL ANALYSIS AND PROOF
DAVID S. G. STIRLING, Senior Lecturer in Mathematics, University of Reading
ISBN 1-898563-36-5   256 pages   Paperback   1997

Contents: Setting the scene; Logic and deduction; Mathematical induction; Sets and numbers; Order and inequalities; Decimals; Limits; Infinite series; Structure of real number system; Continuity; Differentiation; Functions defined by power series; Integration; Functions of several variables.

The Mathematical Gazette (K.L. McAvaney, Deakin University, Geelong, Australia)
"A fine line between accuracy and exactitude. David Stirling treads it carefully ... a thorough a comprehensive introduction ... very much in the classical mould but written in a chatty style with the common student misunderstandings in mind. It should be in your undergraduate reference library."

Choice: American College Library Association, USA (D. Robbins, Trinity College, CT)
"Get down to serious analysis at the level of Walter Rudin ... standard analytic topics are treated ... the level is roughly that of Bartle, and Stirling makes an effort to explain the plans of attack for certain proofs as well as presenting them."

FUNDAMENTALS OF UNIVERSITY MATHEMATICS, Second Edition
C. McGREGOR, J. NIMMO & W.W. STOTHERS, Mahematics Department, Glasgow University
ISBN: 1-898563-10-1   540 pages   Hardback   2000

The Mathematical Gazette
"This book sits firmly on the university side of the difficult and tricky borderline between school and university mathematics. Definitions, theorems and corollaries, figures and notes are all carefully numbered for reference. General layout is exemplary and the book is beautifully printed, with theorems provided clearly in a modern format with helpful and plentiful figures. If you are looking for a first year university text you should carefully look at this one. It is a unifier of mathematical ideas at this level that I found it most valuable. I felt after working my way through it that I saw mathematics at this level more clearly as a single subject and less as a disparate collection of topics."

ENGINEERING MATHEMATICS

H. GRETTON and N. CHALLIS, Department of Engineering, Sheffield Hallam University

ISBN: 1-898563-65-9   250 pages   Hardback   1999

This book aims to help students to take a modern approach to practical engineering mathematical techniques, fully embracing modern technology – hand-held machines, spreadsheets, symbol manipulators. With so much technological power available, the emphasis broadens rom mechanics of solution to include specifying the problem, asking of the answer is appropriate and convincing oneself and others of this. The book encourages a range of solution approaches using SONG (anacronym of Symbolic, Oral, Numerical and Graphical), reflecting the Harvard Reform Calculus movement. This addresses the richness of the ideas, deepening understanding, and allowing confirmation of solutions. Development of key skills is integrated: communication both written and oral, IT use, problem-solving and modelling. The structure should help students become more conscious of and responsible for their own learning, for example using self-assessments.

Contents: Numbers, algebra and functions; Differential and integral calculus; Linear simultaneous equations and matrices; Ordinary differential equations; Sequences and series; Basic statistics and probability.

MODELLING AND MATHEMATICS EDUCATION: ICTMA 9:
Applications in Science and Technology

J.F. MATOS, University of Lisbon; S.H. HOUSTON, University of Ulster; W. BLUM, University of Kassel; and S.P. CARREIRA, University of Lisbon

ISBN: 1-898563-66-7   300 pages   Hardback   2000

This book records the 1999 Lisbon Conference of ICTMA (the International Conference on Mathematical Modelling & Applications). It contains the selected and edited content of the conference and makes a significant contribution to mathematical modelling which is the significant investigative preliminary to all scientific and technological applications, from machinery to satellites and docking of spaceships.

Contents: Weak derivatives; Problems of research on teaching and learning of applications and modelling; Enacting possible worlds making sense of (human) nature; There is more to mathematical modelling than just producing a model; The metaphorical nature of mathematical models; The secondary school curriculum for mathematical modelling and applications.

MATHEMATICAL MODELLING:
Teaching and Assessment in a Technology-Rich World

P. GALBRAITH, University of Queensland, Brisbane, Australia; W. BLUM, The University of Kassel, Germany; G. BOOKER, Griffith University, Brisbane, Australia; IAN D .HUNTLEY, University of Bristol

ISBN 1-898563-42-X   368 pages   Hardback   1997

This book contributes to the teaching, learning and assessing of mathematical modelling in this era of rapidly expanding technology. It addresses all levels of education, from secondary schools through teacher training colleges, colleges of technology, universities, and state and national departments of mathematical education and research groups.

Contents: Theme A; Issues and alternatives in assessing modelling; Theme B: Technologically enriched mathematical modelling; Theme C: Real world: Models and applications; Theme D: Applications and modelling in teaching and learning; Theme E: Applications and modelling in a system or national context.

GAME THEORY: Mathematical Models of Conflict
A.J. JONES

ISBN: 1-898563-14-4   300 pages   Hardback   2000

A modern, up-to-date text for senior undergraduate and graduate students (and teachers and professionals) of mathematics, economics, sociology; and operational research, psychology, defence and strategic studies, and war games. Engagingly written with agreeable humour, this account of game theory can be understood by non-mathematicians. It shows basic ideas of extensive form, pure and mixed strategies, the minimax theorem, non-cooperative and co-operative games, and a "first class" account of linear programming, theory and practice. The book is self-contained with comprehensive references from source material.

TEACHING AND LEARNING MATHEMATICAL MODELLING:
Innovation, Investigation and Application

Editors: S.K. HOUSTON, University of Ulster, Northern Ireland; W.BLUM, University of Kassel, Germany; IAN HUNTLEY, University of Bristol, England; N.T.NEILL, University of Ulster, Northern Ireland

ISBN- 1-898563-29-2   416 pages   Hardback   1997

Sponsored by the organising committee of International Conferences on the Teaching of Mathematical Modelling and Applications (ICTMA), this book contributes to teaching and learning mathematical modelling in universities throughout degree study, colleges of technology, teachers' training colleges, high schools and sixth form colleges.

Contents: Reflections and investigations; Assessment at undergraduate level; Secondary courses and case studies; Tertiary case studies; Tertiary courses.

The Mathematics Teacher (Barry E. Shealy, Buffalo, USA)
"The greatest value would be for people in postsecondary contexts conceptualising courses centred around mathematical applications and modelling."

Choice: American College Library Association (D.V. Chopra, Wichita State University, USA)
"Of great value to those teaching mathematical modelling in high schools, colleges, and universities."

Zentralblatt für Mathematik und ihre Grenzgebiete/Mathematics Abstracts, Germany
"Deals with assessment, particularly at undergraduate level, secondary education and case studies of good practice with examples of courses and how they are taught in a variety of countries including Russia, the Netherlands, the USA and the UK. Concludes with descriptions of ideas relating to undergraduate modelling courses."

EXPERIMENTAL DESIGN TECHNIQUES IN STATISTICAL PRACTICE
W.P. GARDINER, Department of Mathematics, Glasgow Caledonian University, *and* G. GETTINBY, Department of Statistics and Modelling Science, University of Strathclyde, Glasgow

ISBN 1-898563-35-7   256 pages   Hardback   1997

This book describes classical and modern design structures together with new developments which now play a significant role for quality improvement and product development, principally for interpretation of data from industrial and scientific research. The material is reinforced with software output for analysis purposes, and covers much detail of value to users, such as model building and derivation of expected mean squares.

Contents: Introduction; Inferential data analysis for simple experiments; One-factor designs; One-factor blocking designs; Factorial experimental designs; Hierarchical designs; Two-level fractional factorial designs; Two-level orthogonal arrays; Taguchi methods; Response-surface methods; Appendices: Statistical tables; Glossary; Problems and answers.

DECISION AND DISCRETE MATHEMATICS

prepared by THE SPODE GROUP with Ian Hardwick, Truro School, Cornwall

ISBN 1-898563-27-6   240 pages   Paperback   1996

A complete coverage in the decision mathematics (or Discrete Mathematics) module of A-level examination syllabuses. Also suitable for first year undergraduate courses in qualitative studies or operational research, or for access courses. Reflects the combined teaching skills and experience of authors within The Spode Group. The text is modular, explaining concepts used in decision mathematics and related operational research, and electronics. Emphasises techniques and algorithms in real life situations and working problems. Clear diagrams; plentiful worked examples; Exam-standard questions; Many exercises.

Contents: Introduction to networks; Recursion; Shortest route; Dynamic programming; Network flows; Critical path analysis; Linear programming: Graphical; Linear programming: Simplex method; The transportation problem; Matching and assignment problems; Game theory; Recurrence relations; Simulation; Iterative processes; Sorting; Algorithms; Appendices; Answers; Glossary.

Choice: American College Library Association (J. Johnson)
"Topics include networks, recursive and iterative processes, critical path analysis, linear programming, the transporation problem, recurrence, game theory simulation, and sorting/packing algorighms ... explores both key concepts and fundamental algorithme, but also relates the area of decision mathematics to real-life situations ... lower-division undergraduates, two-year technical program students."

A MATHEMATICAL KALEIDOSCOPE:
Applications in Industry, Business and Science
BRIAN CONOLLY, Emeritus Professor of Mathematics, (Operational Research), University of London *and* STEVEN VAJDA, Visiting Professor at Sussex University, *formerly* Professor of Operational Research, Department of Engineering Production, University of Birmingham

ISBN: 1-898563-21-7   276 pages   Paperback   1995

Contents: Miscellaneous fantasies; Finance; Games; Mathematical programming; Search, pursuit, rational outguessing; Organisation and management; Mathematical teasers; Triangular geometry.

Financial Risk in Insurance (D.R. Marshall)
"A wide variety of topics, all related to applied mathematics ... readership is advanced undergraduates and postgraduates in applied mathematics, statistics and operational research; researchers and applied mathematicians in professional practice; and careers officers."

London Mathematical Society Newsletter (Dr F. Oliveira Pinto)
"Precise, sometimes amusing, reflecting the unusual background of the topics ... original in essence or presentation ... since the authors are skilful in manipulating mathematical expressions, the reader cannot escape from vigorous mental exercise!"

Journal of the Royal Statistical Society (Dr John Bather, University of Sussex)
"Here we have two very experienced players describing aspects they hav enjoyed. It is light-hearted and helps to transmit the enthusiasm of the authors."

SURFACE TOPOLOGY, Third Edition
PETER A. FIRBY and CYRIL F. GARDINER, Department of Mathematics, University of Exeter

ISBN: 1-898563-77-2   *ca.* 350 pages   Hardback   2001

This updated and revised edition of a widely acclaimed and successful text for undergraduates examines topology of recent compact surfaces through the development of simple ideas in plane geometry. The approach allows for a straightforward treatment of an area particularly attractive for its wealth of applications and variety of interactions with branches of mathematics, linked with surface topology, graph theory, group theory, vector field theory, and plane Euclidean and non-Euclidean geometry.

The authors' obvious enthusiasm for their subject makes delightful reading with their gently presented introductions to graph theory, group theory, algebraic topology and hyperbolic geometry. The unique blend of readability of topics discussed, and their multi-faceted range of applications, appeal to a wide readership including undergraduate and postgraduate students in pure and applied mathematics, researchers, and even non-specialists.

Contents: Intuitive ideas; Plane models of surfaces; Surfaces as plane diagrams; Distinguishing surfaces; Patterns on surfaces; Maps and graphs; Vector fields on surfaces; Plane tessellation representations of compact surfaces; Some applications of tessellation representations; Introducing the fundamental group. Appendices: Graphs and groups; Tutorial solutions to exercises; Further reading; References.

ORDINARY DIFFERENTIAL EQUATIONS & APPLICATIONS:
Mathematical methods for applied mathematicians, physicists, engineers and bioscientists
W.S. WEIGLHOFER and K.A. LINDSAY, Department of Mathematics, University of Glasgow

ISBN: 1-898563-57-8   224 pages   Paperback   1999

This advanced undergraduate text provides the basis for a semester/module of 20-25 lectures for students of applied mathematics and the applied sciences of physics, engineering, chemistry and biology. It approaches the study of ordinary differential equations: the more traditional setting, first-order differential equations and their solutions; followed by a practical and modern approach to mathematical modelling. Thereby it emphasises how various types of differential equations arise in simple modelling scenarios in applied mathematics and physics, the engineering and biological sciences.

Contents: Introduction and revision; Modelling applications; Linear differential equations of second order; Oscillatory motion; Miscellaneous solution techniques; Lapalace transform; High order initial value problems; System of first order linear equations; Boundary value problems; Optimization; Calculus variation; Appendix A: Self study projects; Appendix B: Extended tutorial solutions.